PREFACE

Welcome to the realm of "Engineering Design and Graphics," a field that plays a pivotal role in various engineering disciplines. This book has been meticulously crafted to guide you through the essential principles, techniques, and practices of creating accurate and comprehensive engineering drawings and graphics.

"Engineering Design and Graphics" is more than just lines and shapes on paper; it's a language that allows engineers, designers, and professionals to communicate complex ideas, designs, and specifications with precision and clarity. Whether you're a student stepping into the dynamic world of engineering or a seasoned professional aiming to refine your skills, this book aims to equip you with the tools and knowledge needed to excel in the realm of technical drawing and graphics.

Within this book, you'll explore orthographic projections, graphics systems, various perspectives, symbols, and notations used in engineering and technical drawings. You'll learn how to create detailed visual representations, convey dimensions and tolerances accurately, and understand the significance of different projection methods. A balance between theoretical concepts and practical applications is maintained, offering step-by-step instructions, illustrative examples, and exercises to help you develop a strong foundation in drawing and graphics techniques.

It's important to acknowledge that "Engineering Design and Graphics" is a dynamic field shaped by technological advancements and evolving industry standards. As such, this book reflects knowledge up to its publication date, and we encourage you to stay curious, explore emerging trends, and adapt your skills to the changing landscape.

We extend our gratitude to the engineering community, educators, and students who have contributed to the development of this book. We hope that this resource will serve as a valuable companion on your journey to mastering engineering drawings and graphics.

We wish you an insightful and rewarding learning experience!

Chapter 1: Basic Elements of Drawing

1.1 INTRODUCTION

Drawings can be created quickly and precisely using drawing tools. The quality of the tools has a significant impact on how accurate the drawings are. It is simple to achieve desired accuracy using high-quality devices. Therefore, it is crucial to get instruments that are as high-quality as possible.

Below is the list of minimum drawing instruments and other drawing materials which every student must possess:

1. Drawing board
2. T-square
3. Set-squares - 45° and 30°- 60°
4. Scales
5. Protractor
6. French curves
7. Drawing papers
8. Drawing pencils
9. Eraser (Rubber)
10. Drawing pins, clips or adhesive tapes
11. Drafting machine
12. Roll-n-draw

We shall now describe each of the above in details with their uses:

1). DRAWING BOARDS

Drawing boards are scale plastic boards designed specifically for technical drawing. They are used to draw parallel lines easily and precisely, e.g., for three-dimensional projections and much more. They feature a mechanical device with rail-mounted rulers that run horizontally and vertically at an exact 90° angle.

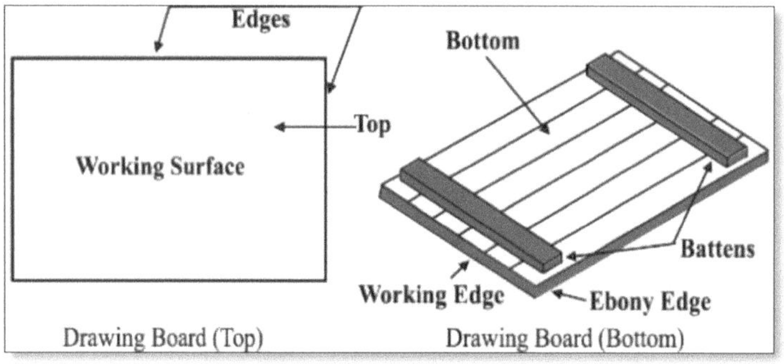

Figure 1.1

Drawing board is made in various sizes. Its selection depends upon the size of the drawing paper to be used. The sizes of drawing boards recommended by the Bureau of Indian Standards (IS:1444-1989) are tabulated in table 1-1. For use in schools and colleges, the last two sizes of the drawing boards are more convenient. Large-size boards are used in drawing offices of engineers and engineering firms.

The drawing board is placed on the table in front of the student, with its working edge on his left side. It is more convenient if the table-top is sloping downwards towards the student. If such a table is not available, the necessary slope can be obtained by placing a suitable block of wood under the distant longer edge of the board.

2). T-SQUARE

T-Square is basically a scale attached to a vertical head and as per definition, it is a tool used to draw horizontal lines on the Drawing Sheet, apart from this it is also used to support Set Square which can be used to produce lines in multiple of 15°.

The measuring portion is called Blade and is used to draw horizontal lines on Drawing Sheet. The edge of the scale is called Working Edge**.** The remaining black portion is called Head/Stock which is used to support the T-Square against the slotted edge of the Drawing Board so that it can facilitate drawing parallel horizonal lines as shown in figure 1.2

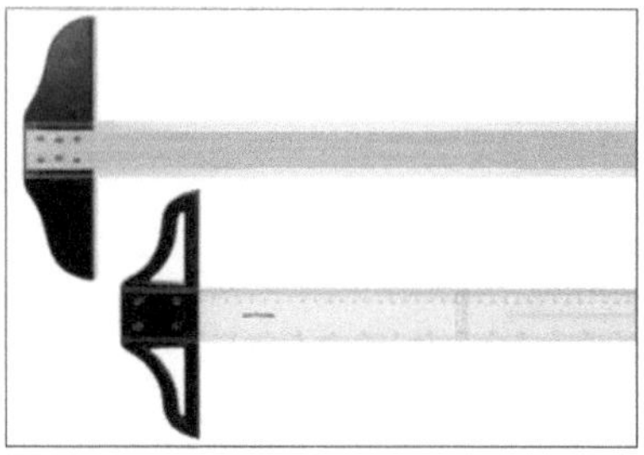

Figure 1.2

T-SQUARE is basically available in four different size for different size of sheet namely T_0, T_1, T_2, T_3.

Type	Scale Length (mm)
T_0	1500
T_1	1000
T_2	700
T_3	500

Figure 1.3

3). SET-SQUARES

A triangle-shaped tool that is often made of plastic or metal and is used to draw angles and straight lines. Set squares can also be used to draw perpendicular lines, angles, and complete shapes, such as squares, rectangles, and triangles. They're particularly helpful for checking to see if a shape has a right angle - a set square will always contain a 90-degree angle, so you can use them to identify these angles in shapes.

Two different types of set squares are present in a geometry box, one with 90-45-45 degree angles and the other with 30-60-90 degree angles.

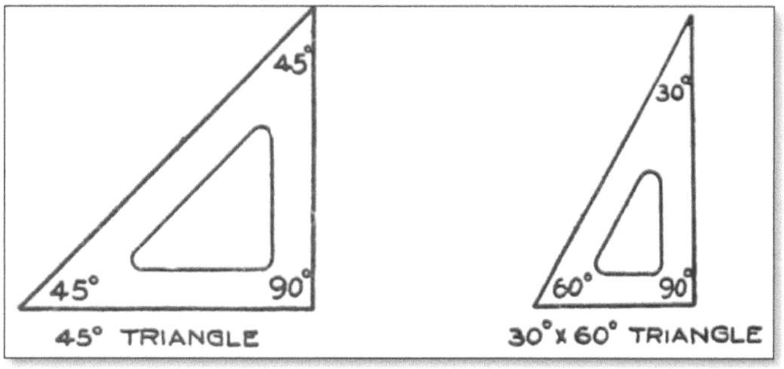

Figure 1.4

The set-squares are made of wood, tin, celluloid or plastic. Those made of transparent celluloid or plastic are commonly used as they retain their shape and accuracy for a longer time. Two forms of set-squares are in general use. A set-square is triangular in shape with one of the angle as right angle. The 30°-60° set-square of 250 mm length and 45° set-square of 200 mm length are convenient sizes for use in schools and colleges.

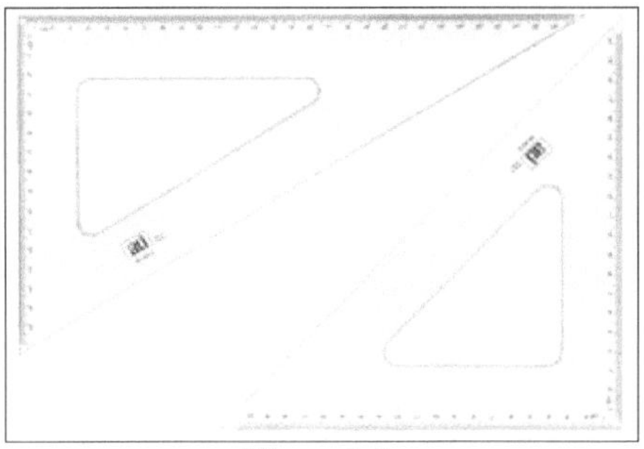

Figure1.5

4). SCALES

The scales are the ratio of the linear dimension of an element of an object as represented in the original drawing to the actual linear dimension of the same element of the object itself.

The scales in engineering drawing are a set of levels or numbers which are used in a particular system as a measuring or comparing parameter. The engineering scales are made in a variety of graduation to meet the requirements of many different kinds of engineering works. The engineering scale adopted for drawing should have a bearing on the degree of accuracy required of scaled measurements on prepared drawings.

Full scale, Enlarging scale, and Reducing scales in Engineering drawing

Drawing an object at its true size is not always possible. A depiction of a particularly large object, for instance, cannot be plotted at full scale since it would require adjusting the size of the drawing on the drawing page. Objects that are very small cannot be drawn at full size because they would be difficult to read and draw. The following scale types are used in engineering drawings so that things can be accommodated and conveniently plotted and read:

Full scale: When an engineering drawing is prepared to the actual size of the object, the scale used is termed as full-size scale and the drawing is known as full-size drawing.

A scale with a ratio of 1:1 is known as a full-size scale in engineering drawing.

Example: A 30 mm radius plain disc is represented on the drawing by a circle of 30 mm radius.

Enlarging scale: When a very small object such as components of a wristwatch, is enlarged in some regular proportion to accommodate its drawing. So, when the drawing is prepared larger than the actual size, the scale is said to be an enlarged scale, and the drawing is said to be an enlarged-sized drawing. A scale in which the ratio is larger than 1:1

Reducing scale: When the object is of large size, the actual dimensions of the object have to be reduced to accommodate the drawn object in the drawing sheet. So, when a drawing is prepared smaller than the actual size of the object, the scale used is termed as reducing scale and the drawing is known as a reduced-sized drawing.

A scale in which the ratio is smaller than 1:1

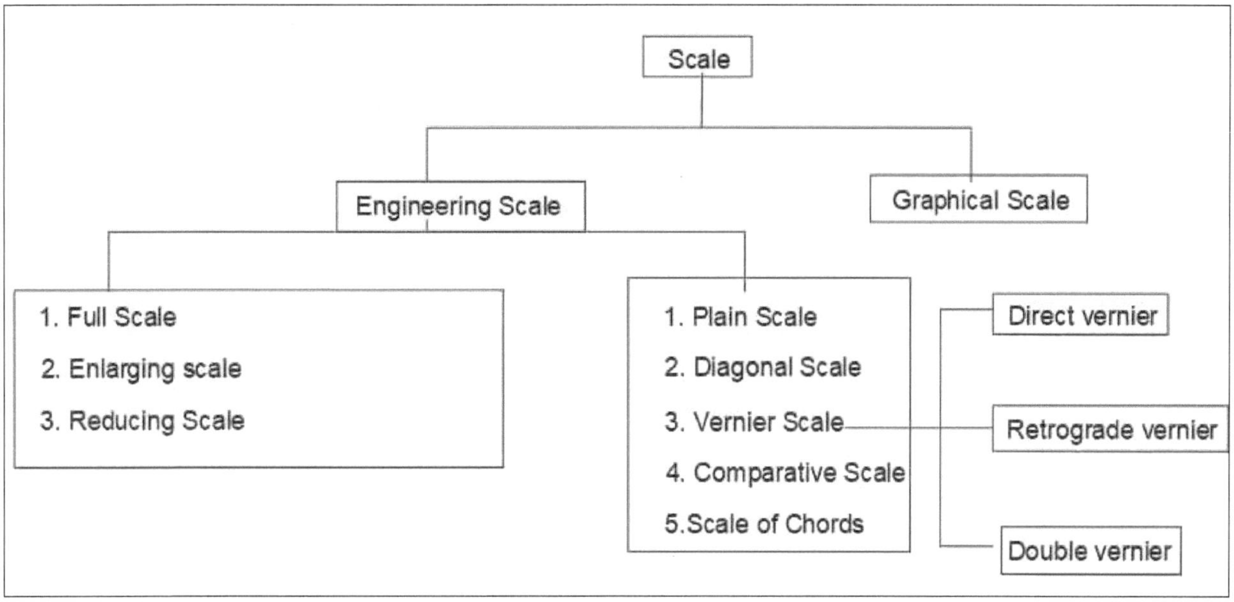

Figure 1.6

Designation of Scales in Engineering Drawing

The complete designation of an engineering scale shall consist of the word "SCALE' followed by the mention of its ratio, as follows:
SCALE 1:1 for full size
SCALE X:1 for enlargement scales
SCALE 1:X for reduction scales

Representative Fraction (RF)

The representative fraction is defined as the ratio of the dimension of an element of an object in the drawing to its actual linear dimension of the same element of the object itself.

$$R.F. = \frac{\text{Length of an object on the drawing}}{\text{Actual Length of the object}}$$

The value of R.F. is always expressed in the form of x:y, in which both x and y are lengths converted into the same units. Hence, RF has no decimal and it is unitless.
RF for full-sized scale = 1
RF for reducing scale < 1
RF for enlarging scale > 1
It is also highlighted that RF is the ratio of linear dimensions and not the ratio of areas or volumes. In case, the quantities are given in units of areas, then take the square root to convert them into lengths units. Similarly, if the quantities are given in units of volume, then take the cube root to convert them into lengths unit.

BIS recommended Scales in engineering drawing

The standard values of RF for scales in engineering drawing as recommended by BIS are as follows:
Full Scales – 1:1
Reducing Scales – 1:2, 1:5, 1:10, 1:50, 1:100, 1:200, 1:1000, 1:2000, 1:5000
Enlarging Scales – 50:1, 20:1, 10:1, 5:1, 2:1
The scales in engineering drawing have to be chosen for depends upon various factors like the size of the object, functionality of the drawing.

The requirement for the construction of a scales in engineering drawing

The information required for the construction of plain, diagonal, or a vernier scales are as follows:
RF of the Scale
The maximum length to be measured by the scale
The minimum length to be measured by the scale (i.e., Least count)
The length of the scales in engineering drawing is determined by the formula
Length of the scale = representative fraction x Maximum length to be measured.

Types of scale in engineering drawing

The scales in engineering drawing are classified as the following:

1. Plain scale
2. Diagonal scale
3. Vernier scale
4. Comparative scale
5. Scale off chords

Plain scale

In a plain scale a line divided into a suitable number of equal parts or units, the first of which is subdivided into smaller parts.

The use of plain scale in engineering drawing is to measure up to two consecutive units, i.e., a unit and its immediate subdivision. For example – meter and decimetre, feet and inches, etc.

Diagonal scale

A diagonal scale in engineering drawing is used when the measurements are required in three consecutive units; i.e., a unit and its immediate two subdivisions. A diagonal scale's measurement is more accurate than a plain scale's measurement.

For example, meter, decimetre and centimetre.

Vernier scale

Three successive units on a metric scale are measured using a vernier scale. As a result, the vernier scale's accuracy in measurement is similar to a diagonal scale. In engineering drawings, the vernier scales are used to show the distances between a unit and its two adjacent subdivisions.

The main scale and the vernier scale are the two components of the vernier scales. The vernier is an additional scale that is connected to the main scale at one end. The minor divisions of the main scale are divided with the aid of the vernier because doing so naturally is difficult.

The vernier scale is typically employed on length measuring tools like micrometres, screw gauges, and vernier callipers.

Comparative scale

A pair of scales that are graduated to read measurements in various unit systems but share a common representative fraction are called comparative scales. An engineering scale that reads in kilometres and hectometres, for instance, can be directly converted to miles and furlongs with the aid of a comparative scale built using the same Representative fraction, or RF.

Scale of chords:

1. In the absence of a protractor, a scale of chords may be used to measure the angle or to set the required angle.
2. The constructions based on the lengths of chords of angles measured on the same arc.

5). PROTRACTOR

Wood, tin, or celluloid are used to make protractor. Transparent celluloid protractors are frequently used. They have a flat, semi-circular, or circular shape. The most popular protractor is semi-circular and has a diameter of about 100 mm. Its circumferential edge is numbered every ten degrees, graduated into one-degree divisions, and legible from both ends. The base of the protractor is the diameter of the semicircle (i.e., the straight line 0-180°), and the centre O is indicated by a line perpendicular to that line.

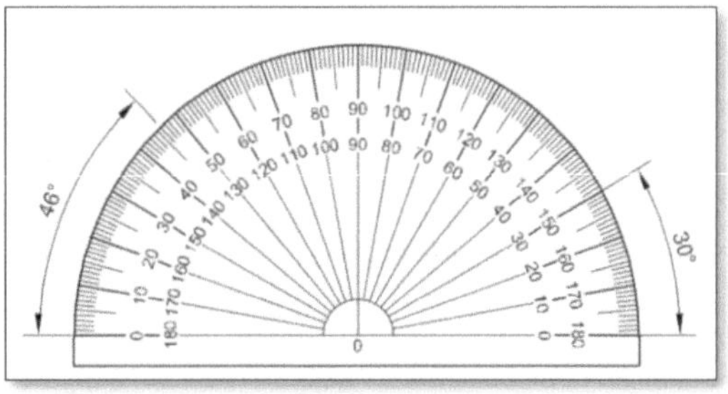

Figure 1.7

6). FRENCH CURVES

These are made in many different shapes, normally come in sets of 6, 12, 16, etc. French curves are best suited to draw smooth curves/arcs (which cannot be drawn by a compass) with ease. To draw a smooth curve using French curve first set it by trial against a part of the line to be drawn, then shift it to the next portions.

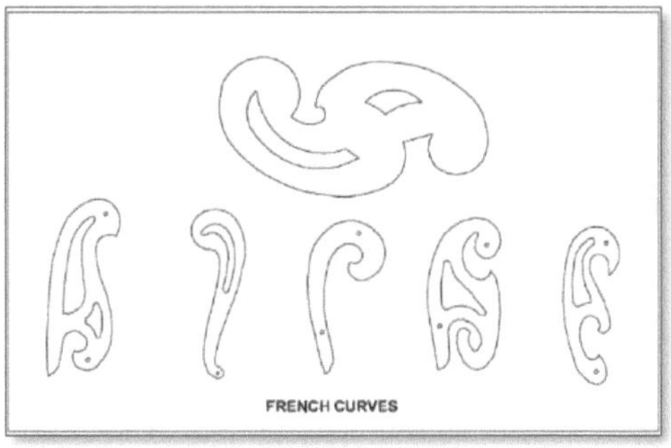

Figure 1.8

7). DRAWING PAPERS

Drawing sheet is a white paper on which an object is drawn which is available in various sizes. The sheet used for engineering should be of good quality. It should be white in colour with uniform thickness with must resist the easy torn of paper. The surface of sheet must be smooth.

These are of two types:

- Hand-made paper
- Mill-made paper

Hand-made papers have rough surfaces, pale in colour and not used for regular work, but meant for charts.

8). DRAWING PENCILS

Pencil is used to draw on the paper. Any type of pencil is not suitable for drawing. There are some limitations, the drawing appearance should be very neat and understandable.

Every line of the drawing should indicate its importance. Pencil is used to draw on the paper.

Any type of pencil is not suitable for drawing. There are some limitations, the drawing appearance should be very neat and understandable. Every line of the drawing should indicate its importance.

It depends upon the hardness of pencil. Based on the hardness quality pencils are classified into 18 grades and they are following:

GRADE OF PENCILS	HARDNESS OF PENCIL
9H	Hardest
6H, 5H, 4H	Extremely Hard
3H	Very Hard
2H	Hard
H	Moderately Hard
F	Firm
HB	Medium Hard
B	Moderately Soft and Black
2B	Soft and Black
3B	Very Soft and Black
4B, 5B, 6B	Very Soft and very Black
7B	Softest

Figure 1.9

GRADE OF PENCILS	HARDNESS OF PENCIL
9H	Hardest
6H, 5H, 4H	Extremely Hard
3H	Very Hard
2H	Hard
H	Moderately Hard
F	Firm
HB	Medium Hard
B	Moderately Soft and Black
2B	Soft and Black
3B	Very Soft and Black
4B, 5B, 6B	Very Soft and very Black
7B	Softest

Figure 1.9

Out of the above 18 grades of pencils, following grades are used in engineering drawings.

Grade of Pencil	Used to Draw
3H	Construction Lines

2H	Dimension lines, centre lines, sectional lines, hidden lines
H	Object lines, Lettering
HB	Dimensioning, boundary Lines

Figure 1.10

9). ERASER (RUBBER)

Soft India-rubber is the most suitable kind of eraser for pencil drawings. It should be such as not to spoil the surface of the paper. Frequent use of rubber should be avoided by careful planning.

10). DRAWING PINS, CLIPS OR ADHESIVE TAPES

Drawing clips or pins are used to fix the drawing sheet on the drawing board at the required place. Frequent use of pins cause formation of impressions of pin pricks on the board, thus spoiling the surface of the board.
The present trend is to go in for steel clips, if the size of the drawing paper is the same as that of the drawing board. Clips are used at all the four corners of the drawing board to clamp the paper. Adhesive tapes are also used for fixing the drawing sheet.

11). DRAFTING MACHINE

A drafting machine is a tool used in technical drawing, consisting of a pair of scales mounted to form a right angle on an articulated protractor head that allows an angular rotation. The protractor head (two scales and protractor mechanism) is able to move freely across the surface of the drawing board, sliding on two guides directly or indirectly anchored to the drawing board. These guides, which act separately, ensure the movement of the set in the horizontal or vertical direction of the drawing board, and can be locked independently of each other.

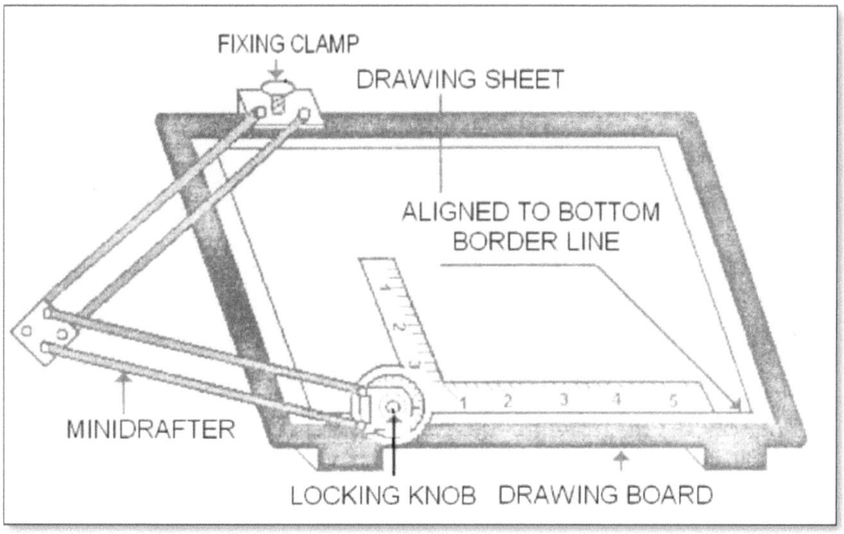

Figure 1.11

12). ROLL-N-DRAW

It consists of graduated roller, scale of 16 centimetre and protractor. It is ideal for drawing vertical lines, horizontal lines, parallel lines, angles and circles.

1.2 DRAWING INSTRUMENT BOX, CONTAINING

The drawing instrument box contains the following as mentioned earlier:
(1) Large-size compass with interchangeable pencil and pen legs
(2) Lengthening bar
(3) Small bow divider
(4) Small bow compass

(5) Small bow ink-pen
(6) Large-size divider
(7) Inking pen

1.2.1 LARGE-SIZE COMPASS WITH INTERCHANGEABLE PENCIL AND PEN LEGS

Compass usage is required to draw arcs and circles. It has two legs, which are joined at the top by a hinge. One leg's lower end is fitted with a sharp needle, and the other leg's lower end is fitted with a pencil lead.

To draw a circle, adjust the opening of the legs of the compass to the required radius. Hold the compass with the thumb and the first two fingers of the right hand and place the needle point lightly on the centre, with the help of the left hand.

Bring the pencil point down on the paper and swing the compass about the needle-leg with a twist of the thumb and the two fingers, in clockwise direction, until the circle is completed.

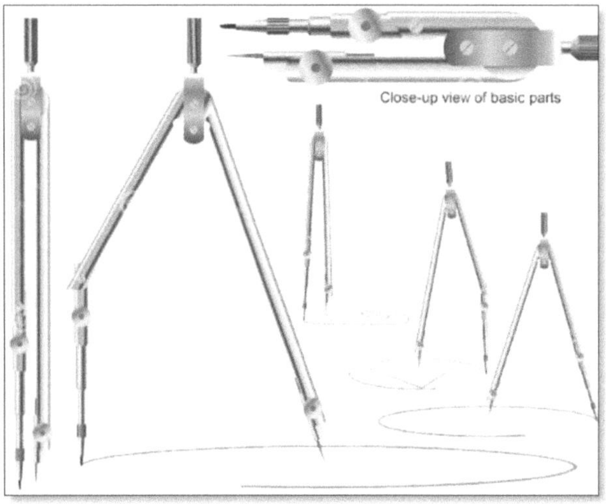

Figure 1.12

1.2.2 LENGTHENING BAR

To draw larger circles, it is fitted to the compass. The pencil point or pen point is inserted to its end.

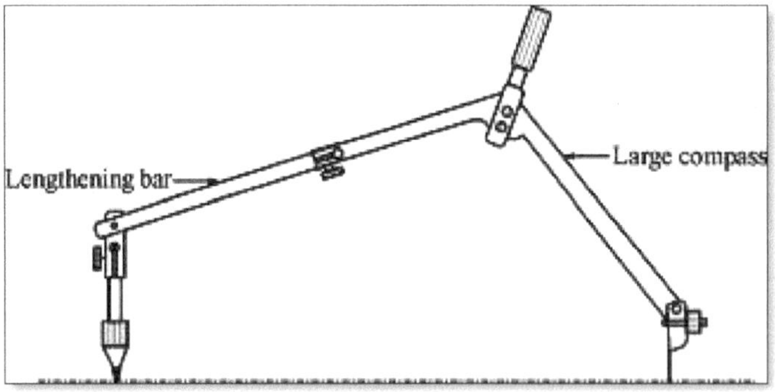

Figure 1.13

1.2.3 SMALL BOW DIVIDER

The small bow divider is adjusted by a nut and is very convenient for marking minute divisions and large number of short equal distances.

1.2.4 SMALL BOW COMPASS

Small bow compass is used to draw small circles and arcs with a radius of less than 25 mm, and is especially useful for drawing numerous small-circle with the same diameter Art.

In Drawing Tools and applications the compass's curves should be drawn with the same degree of blackness as its straight lines. It is challenging to apply the same amount of pressure to the compass lead as to a pencil.
To maintain consistent blackness throughout all of the lines, it is therefore preferable to use a slightly softer kind of lead (approximately one grade lower, HB or HJ in the compass than the pencil used for drawing straight lines).

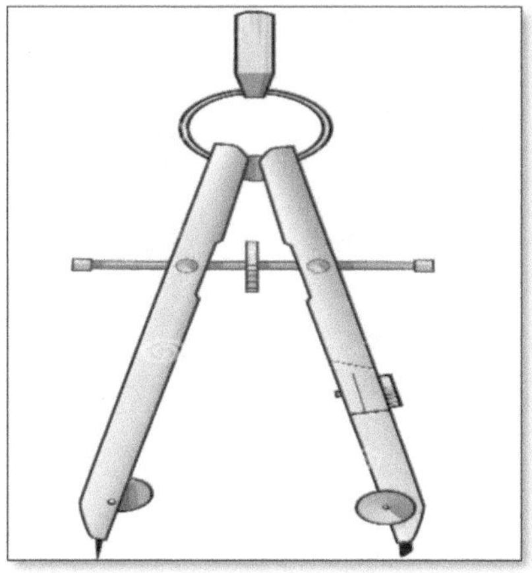

Figure 1.14

1.2.5 SMALL BOW INK-PEN

It is used for drawing small circles and arcs in ink.

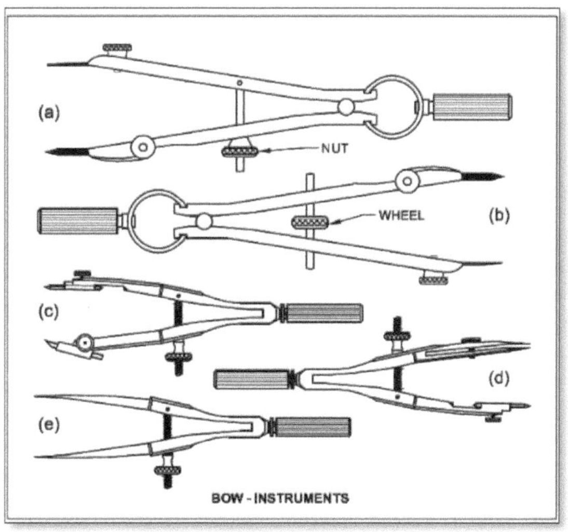

Figure 1.15

1.2.6 LARGE-SIZE DIVIDER

It is a drawing tool used to divide measurements and dividing lines into a number of equal parts. Instead of basic legs, a divider with adjustable joints is preferred.

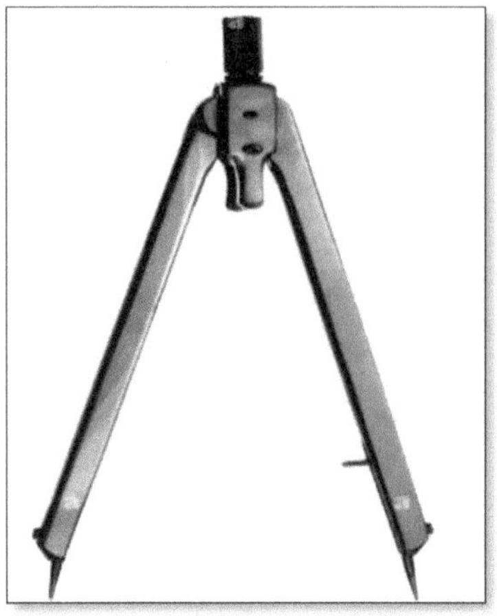

Figure 1.16

1.2.7 INKING PEN

It is one of the drawing instrument(s) which is used to ink the straight lines drawn with the instruments but never for free hand lines or lettering.

It consists of a pair of steel nibs fitted to a holder made of metal or ivory. Ink is filled between the two nibs to about 6 mm length by means of a quill which is usually fitted to the cork of the ink bottle. The gap between the nibs through which the ink flows and upon which the thickness of the line depends is adjusted by means of the screw 5.

The pen should be kept sloping at about 60° with the paper in the direction of drawing the line and the ends of the nibs should be slightly away from the edge of the T-square or set-square. The screw should be on the side, farther from the T-square.

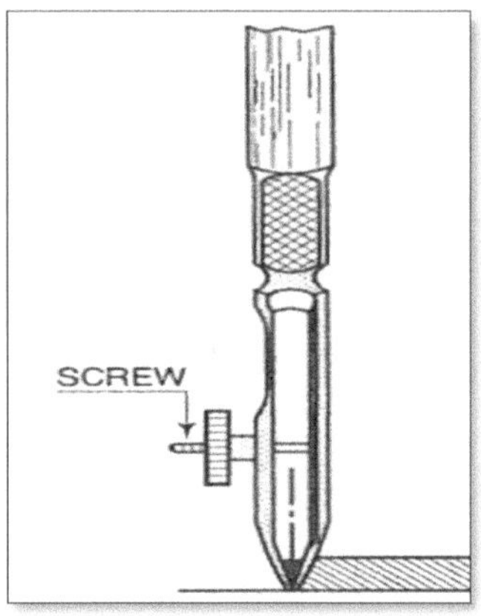

Figure 1.17

1.3 CONVENTION OF LINES

Engineering drawing lines are one type of line that makes up a drawing. Similar to how a language uses grammar and alphabets. Since a line serves as the foundation for engineering drawings, an alphabet of lines could be thought of as a set of conventional symbols that represent all the lines required for various purposes. In a well-done drawing, each line serves a specific purpose.

Lines of different thickness and features are used for specific use.

Technical drawings are drawn with different types of lines. By proper choice and application of lines product features can be correctly defined in a drawing.

Line	*Description*	*General Applications*
A	Continuous thick	A1 Visible outlines
B	Continuous thin (straight or curved)	B1 Imaginary lines of intersection B2 Dimension lines B3 Projection lines B4 Leader lines B5 Hatching lines B6 Outlines of revolved sections in place B7 Short centre lines
C	Continuous thin, free-hand	C1 Limits of partial or interrupted views and sections, if the limit is not a chain thin
D	Continuous thin (straight) with zigzags	D1 Line (see Fig. 2.5)
E	Dashed thick	E1 Hidden outlines
G	Chain thin	G1 Centre lines G2 Lines of symmetry G3 Trajectories
H	Chain thin, thick at ends and changes of direction	H1 Cutting planes
J	Chain thick	J1 Indication of lines or surfaces to which a special requirement applies
K	Chain thin, double-dashed	K1 Outlines of adjacent parts K2 Alternative and extreme positions of movable parts K3 Centroidal lines

Figure 1.18

These lines differ from each other in two respects:

1. Their thickness or weight, e.g., thick, medium and thin.
2. Their shape or construction, depending upon their conventional use.

Lines specified thick should be about 3 times as thick and lines specified medium should be 1.5 times as thick as lines specified thin. The line group used on a drawing is identified from the thickest lines, i.e., visible outlines. The line group used depends upon the size of the drawing and should be chosen from one of the line groups.

VISIBLE OUTLINE OR OBJECT LINE:

Any drawing's standout feature is its outline. These long, thick lines serve as a representation of an object's general shape or outline. These lines represent every edge or surface that is visible.

SECTION LINES:

These uniformly spaced, thin, continuous lines are used to depict a section of a part. Usually, these lines are drawn at a 45° angle to the horizontal (but they could also be drawn at a 60° or 30° angle). These lines cover the entire surface where they are drawn, from edge to edge.

CONSTRUCTION LINES:
These are very thin continuous lines that are used in drawing construction to project or transfer location from one view to another. With the exception of geometrical drawings, these lines are not visible in the final products. These are not ink drawings.

HIDDEN LINES:
It is possible to depict in hidden detail a shape that cannot be seen when viewed from outside an object. Not every hidden detail in views needs to be depicted in a drawing. It is only necessary to include a small enough amount of the hidden detail to clearly demonstrate the object's precise shape. When unnecessary, hidden lines can be removed to maintain the drawing's clarity. The following guidelines should be followed when including hidden detail lines in a drawing:

Hidden detail lines are **medium thick lines** consisting of short dashes, about 3 mm long with gaps of nearly half their length between them. The length of the dashes varies slightly in relationship to the size of the drawing.

1. Hidden detail lines forming angles should be drawn with dashes touching at the angle.
2. In a hidden detail, if a line meets a curved line, a dash should touch at the point of tangency.
3. To indicate that hidden detail lines cross visible outlines, either a dash should be drawn directly through the visible outline, or dashes should be drawn with a space between them. The hidden detail would reveal that the detail drawn begins at the visible outline if this rule were not followed. To represent these invisible or hidden details or parts on drawings, uniformly spaced short dashes are used. Compared to the object lines, these dashes are about 60% thinner.

DIMENSION LINES:
These are continuous thin lines.

EXTENSION LINES:
These are continuous thin lines.

OUTLINES OF REVOLVED SECTIONS AND ADJACENT PARTS:
These are continuous thin lines.

CUTTING PLANE LINES:
These lines, which are drawn as centre lines but are thicker at the ends and have arrows touching them, are made up of alternating long dashes and short dashes in the ratio of roughly 6:1 to 4:1. When an object is sectioned to reveal hidden details, these are used to show the edge of the cutting plane. The arrows at the ends of these lines point in the direction that the section should be viewed.

BREAK LINES:

- **SHORT BREAKS AND IRREGULAR BOUNDARY LINES:**
  These are **continuous thin lines** drawn freehand and are used to show a break of an object which extends over a short length of drawing or to indicate the irregular boundary of an object. These result in a saving in space and time used for drawing without loss of detail.

- **LONG BREAKS:**
  These are **continuous thin ruled lines** having freehand zigzags at suitable intervals and are used when the brake extend over a considerable length of the drawing.

LONG CHAIN THICK LINES:
These are composed of alternate long and short dashes and it is used to indicate surfaces which are to receive additional treatment.

1.4 REPRESENTATIVE FRACTIONS

The proportion of units on the map to those on the ground is shown by a representative fraction. The relationship between one "unit" on the map and however many "units" of the same type there are in reality is known as a representative fraction.

$$R.F. = \frac{\text{Length of an object on the drawing}}{\text{Actual Length of the object}}$$

When a 1 cm long line in a drawing represents 1 meter length of the object.

$$R.F = \frac{1cm}{1m} = \frac{1cm}{1x\,100cm} = \frac{1}{100}$$

1.5 DIMENSIONING TECHNIQUES AS PER SP-46:2003

Dimensions are numerical values that are graphically represented on engineering drawings with lines, symbols, and notes in the appropriate unit of measurement. To define the size characteristics, such as length, height, breadth, diameter, radius, angle, etc., these are shown on the engineering drawing. In its bulletin IS 11669:1986 (Reaffirmed 1999), the Bureau of Indian Standards (BIS) recommends the general dimensioning principle for engineering drawings.

Importance of dimensioning in engineering drawing:

To provide a clear and complete description of the features of an object, proper dimensioning in engineering drawings is needed.

Types of Dimensioning in engineering drawings:

Dimensions are classified into the following types:

1. Functional dimensions
2. Non-functional dimensions
3. Auxiliary dimensions

Functional dimensions (F)

This is the dimensioning in engineering drawing that is essential to the function of the piece or space.

Non-functional dimensions (NF)

This is the dimensioning in engineering drawing that is not essential to the function of the piece or space.

Auxiliary dimensions (AUX)

The dimensioning in engineering drawing which is given for information purpose only. It is derived from other values shown on the drawing or in related documents. An auxiliary dimension is given in parentless and no tolerance applied to it.

Dimensioning Terminology

The various terminologies related to the dimensioning in engineering drawing are as follows:

1. Dimension value

This is a numerical value that is assigned to the size, shape, or location of the feature being dimensioned. Dimensions are expressed in specific units on the drawing with relevant information.

2. Dimension Lines

The slender continuous lines known as dimension lines display the direction and extent of a dimension. These lines should be evenly spaced 6 to 8 millimetres apart and 8 to 10 millimetres away from the drawing's outline. It is preferable to place the dimension values close to the centre of the dimension lines.

3. Projection Lines

Projection lines are thin, continuous lines that are 2 to 3 mm beyond the dimension lines and stretched out from the outlines for dimensioning. These lines to be drawn perpendicular to the feature that needs to be dimensioned. If possible, avoid having the dimension and projection lines cross over other lines.

4. Leaders

Leaders are thin continuous lines and terminated by an arrowhead or dot referring to a feature and note. Figure and notes are written above the extended dimension line.

To avoid confusion leaders should not be –

1. inclined at an angle less than 30°
2. parallel to adjacent projection lines
3. parallel or adjacent dimensions

Leaders are never drawn horizontal, vertical, curved, or freehand. They are generally drawn at any convenient angle 30°, 45° and 60°. Long leaders are generally avoided.

5. Arrowheads

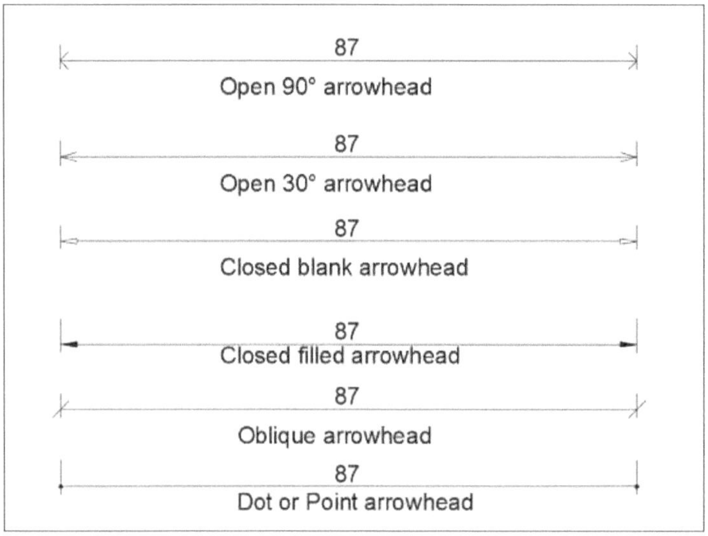

Figure 1.19

The termination of dimension lines is indicated by arrowheads. The arrowheads may be open at a convenient angle of 30° to 90°, closed filled, or close blank. The closed filled arrowheads have a length of about three times the width/depth and are generally preferred in the engineering drawing. Generally, the length of the arrowhead for small drawing is 3 mm and for large drawing the length of the arrowhead is 5 mm.

When the space to accommodate the termination of arrowheads is insufficient, oblique stroke and points may be used.

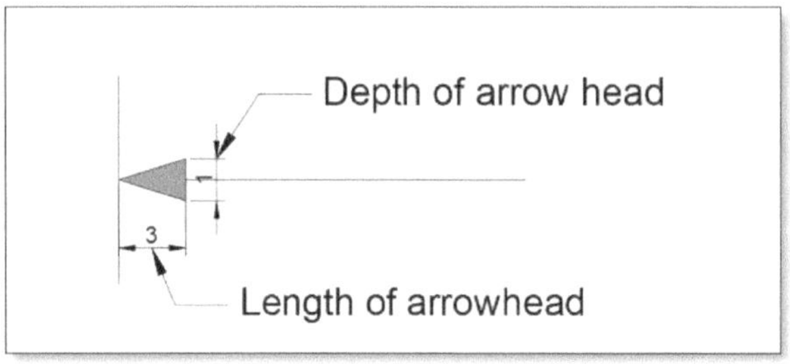

Figure 1.20

Methods of Dimensioning in Engineering drawing:

The dimensions are done in a drawing as per one of the following methods as recommended in SP 46-2003. SP 46 (2003): Engineering Drawing Practice for Schools and Colleges.

The two recommended system of placing the dimensions are as follows:

1. Aligned system of dimensioning in engineering drawing:

In this method of dimensioning, the dimension text is placed parallel to the dimension line and above it preferably in the middle.

The dimension texts are placed in such a way that they can be read either from the bottom or from the right-hand side of the drawing.

Linear dimensioning

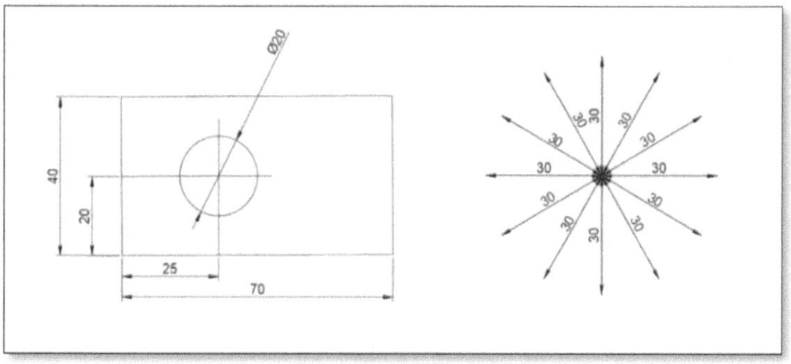

Figure 1.21

Angular dimensioning

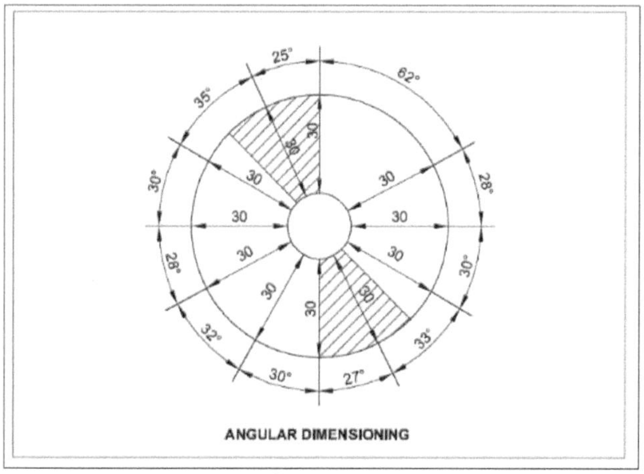

Figure 1.22

2. Unidirectional system of dimensioning in Engineering drawing:

In the unidirectional method of dimensioning, the dimensions are indicated so that they can be placed in the horizontal direction only.

Linear dimensioning

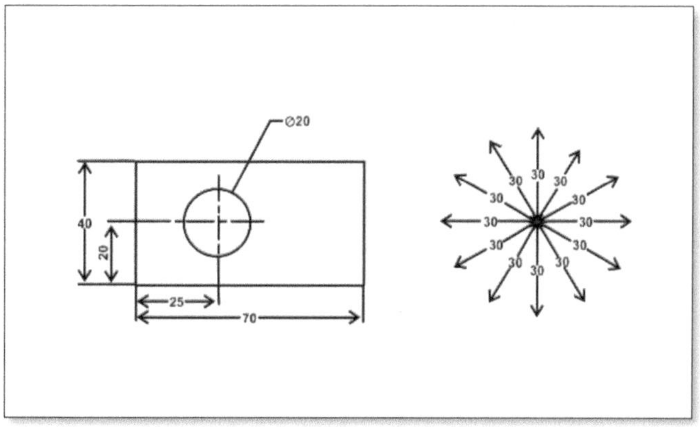

Figure 1.23

Angular dimensioning

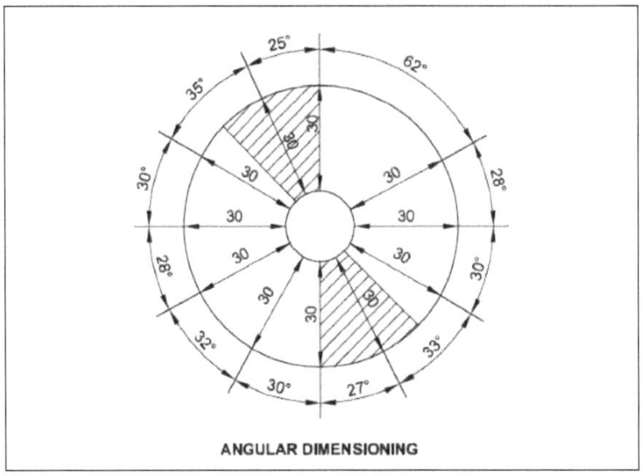

Figure 1.24

Arrangements of Dimensions

When several dimensions are to be placed in an engineering drawing, those dimensions are to be arranged in such a way that it provides a unique explanation.

The classification of dimensioning on the basis of arrangement is as follows:

1. Chain dimensioning or Continuous dimensioning

In this method, successive dimensions are arranged in a continuous straight line.

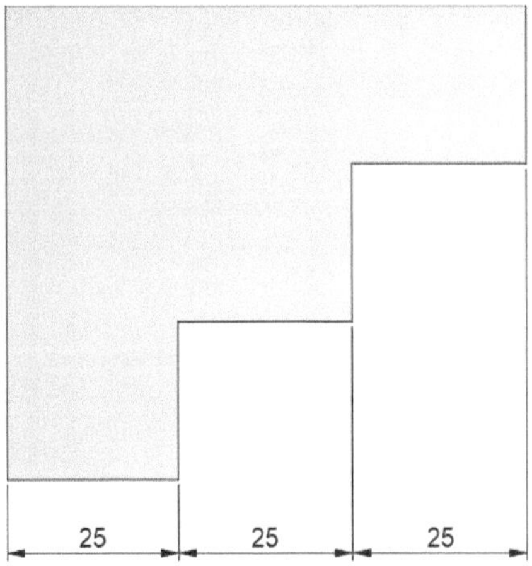

Figure 1.25

2. Parallel dimensioning or progressive dimensioning

This method involves placing a number of single dimensions parallel to one another from a single origin. When several dimensions share a common origin, this method is used. This approach can help you avoid the cumulative error.

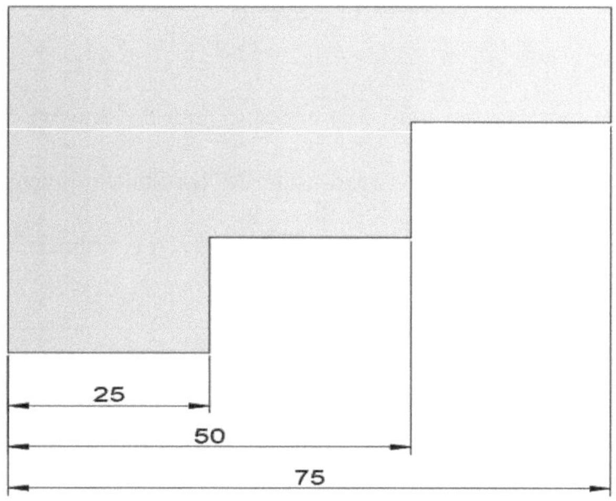

Figure 1.26

3. Running dimensioning or Superimposed dimensioning
In this method of dimensioning, all the dimensions begin from a common origin, which is indicated by a small circle of 3 mm diameter, and terminated with arrowheads where individual dimension ends.
The dimension texts are rotated through 90° and placed in-line with the expansion line or above the dimension line near the arrowhead.

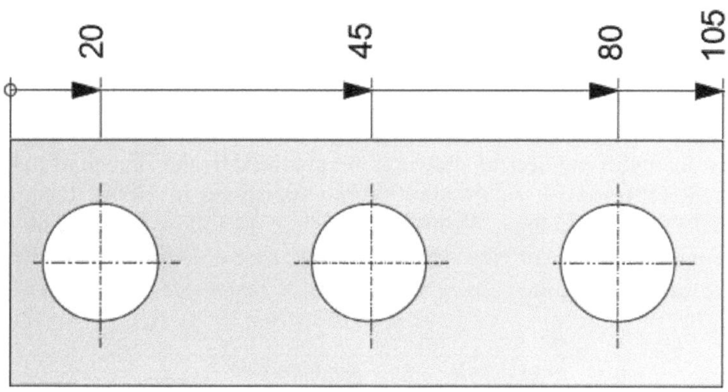

Figure 1.27

4. Combined dimensioning
In this arrangement of dimensions, there is the simultaneous use of parallel dimensioning, chain dimensioning, and running dimensioning in a single drawing.

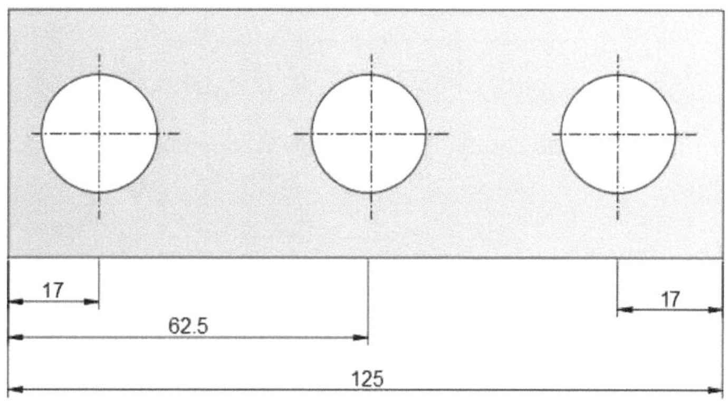

Figure 1.28

5. Coordinate dimensioning

Whenever a part is having too many dimensions, it is easier to read the drawing by the coordinate dimensioning method.

Example: A plate with many holes can be easily read using the coordinate dimensioning method. Other styles of dimensioning will provide an overly cluttered result in this case.

In this system, origin and other features are to be numerically levelled. A coordinate table contains all the dimensional details to be placed near the drawing.

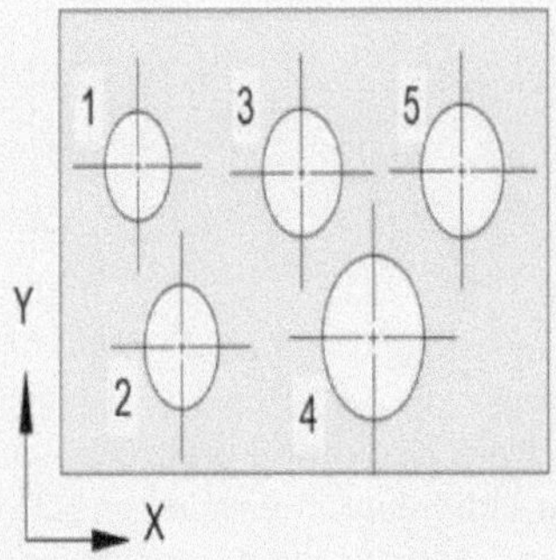

	X	Y	Ø
1	20	50	15
2	25	20	18
3	35	50	18
4	45	20	25
5	65	50	18

Figure 1.29

Rules of Dimensioning in Engineering Drawing

It is very important to adhere to a set of standard dimensioning rules because dimensions are used to define the size characteristics of a drawn object with lines, notes, and symbols, such as length, height, breadth, radius, diameter, angle, etc.

The guidelines for dimensioning in engineering drawings are provided by the Bureau of Indian Standards (BIS) in its bulletin IS 11669:1986 (Reaffirmed 1999). IS 11669: 1986 (reiterated in 1999): Basic Dimensioning Rules for Technical Drawings Rules for Dimensioning: When dimensioning, the following rules should be followed:

1. Mark the dimension outside the figure or view. However, in the case of Dimensioning the radius of an arc or circle, it may be shown inside.

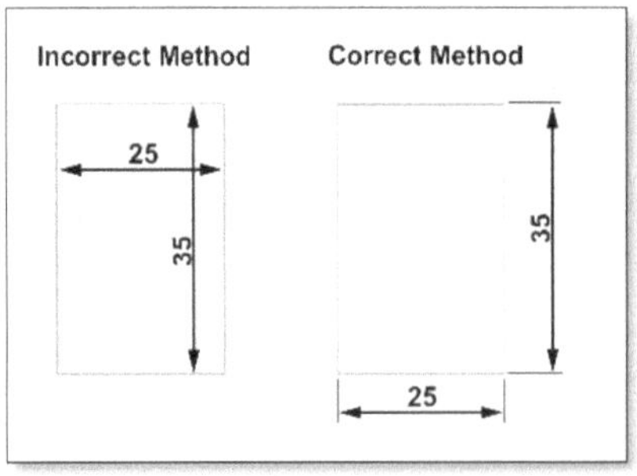

Figure 1.30

2. Dimensioning rules for Circle: A circle has to dimensioned by its diameter symbol Ø. The symbol Ø has to place before the dimension value.

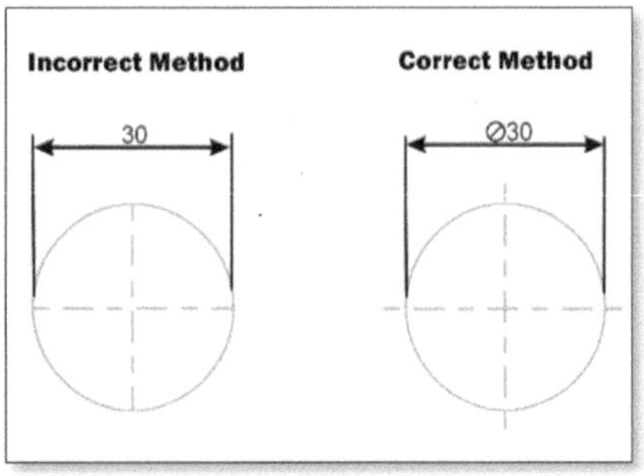

Figure 1.31

An arc has to dimensioned by its Radius symbol R.

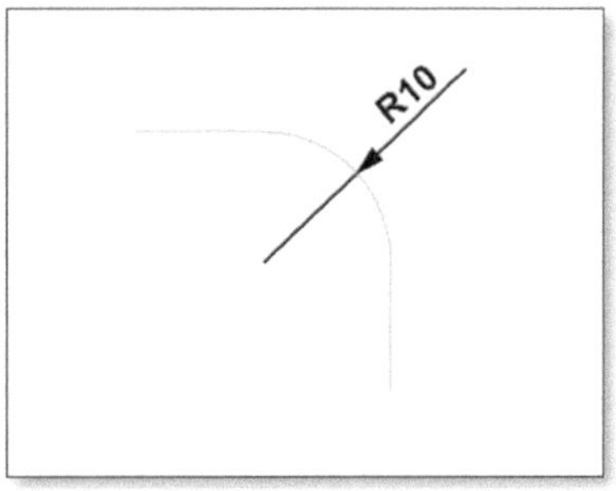

Figure 1.32

3. Dimensioning rules for Centreline:
i). The axis or centreline shall not be used as a dimension line with arrowheads at its ends.

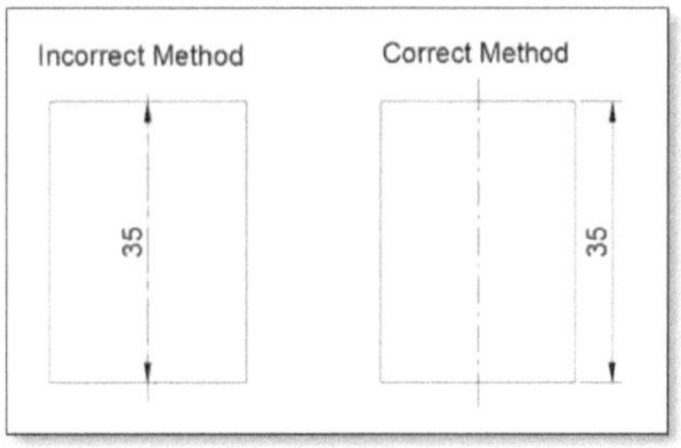

Figure 1.33

ii). The axes or centre lines must extend approximately 3 mm beyond the boundary of the part whose details they indicate.

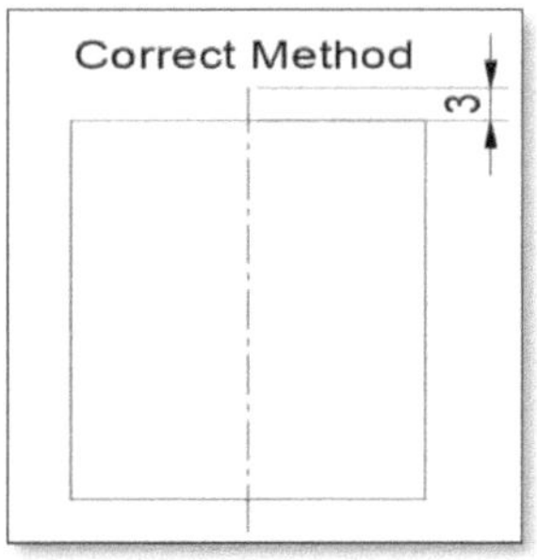

Figure 1.34

iii). Marking a dimension from a centreline is incorrect, except when the centreline passes through the centre of a hole. Also, an extended centreline may be served as an extension line.

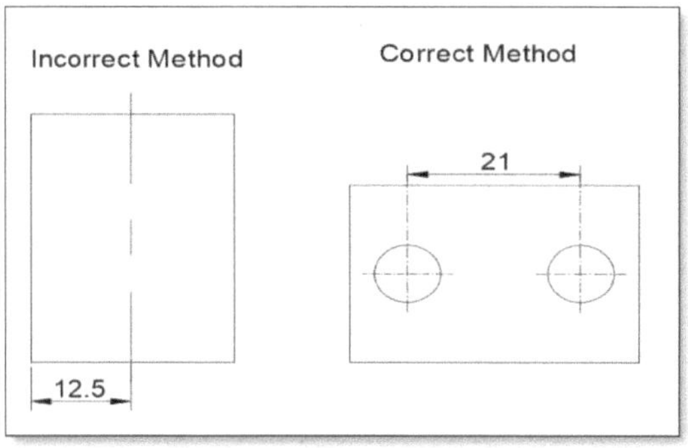

Figure 1.35

4. Dimensioning rules for Holes:
The centre-to-centre distance between the holes has to dimension in the view in which the holes are visible.

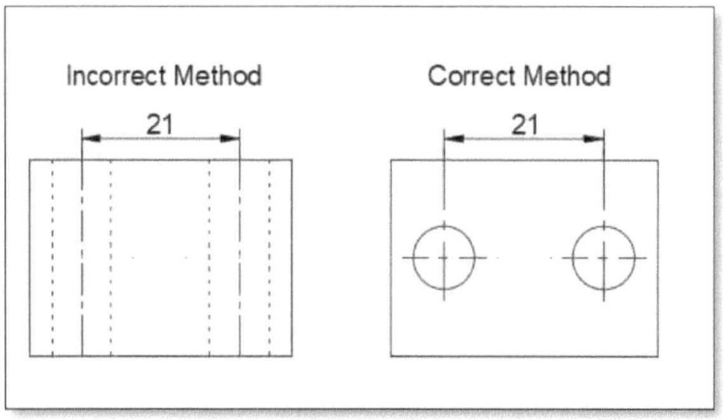

Figure 1.36

5. Dimensioning shall be done to the visible lines and not to the invisible or hidden lines.

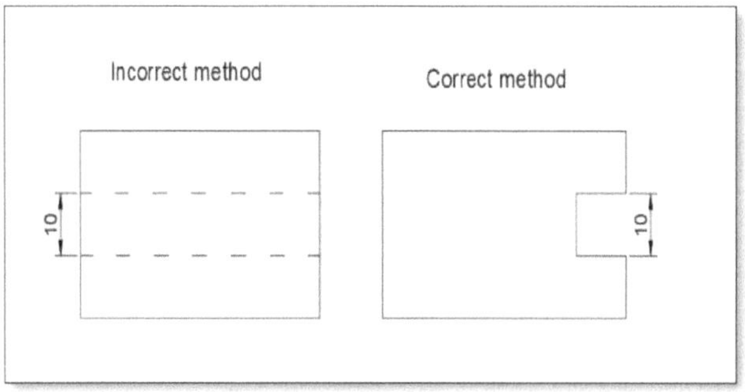

Figure 1.37

6. The dimension value must be placed approximately 2 mm above the dimension line.

7. The extension line must extend 2 mm beyond the dimension line.

8. Dimensioning rules for Conical taper section:
The conical taper on diameter must be dimensioned as shown.
Conical Taper = (D-d)/L= 1 in n

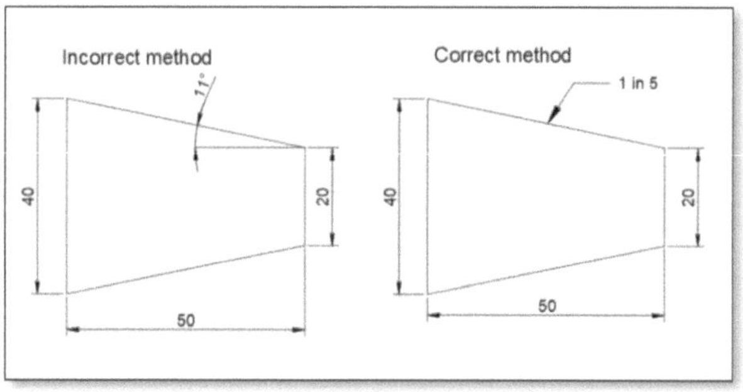

Figure 1.38

9. Dimensioning rules for Hatched portion:
When dimensioning within a hatched portion of a drawing, the hatching line should not cut the dimension text.

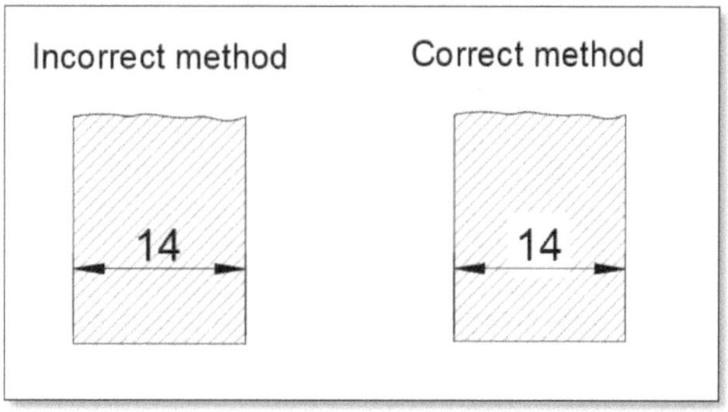

Figure 1.39

10. Dimensioning rules for Repeated features:
As far as possible dimensioning of repeated features of the same size is avoided. The notes have to be written horizontally. The leader lines have to be inclined at an angle of 30°, 45° and 60° to the horizontal.

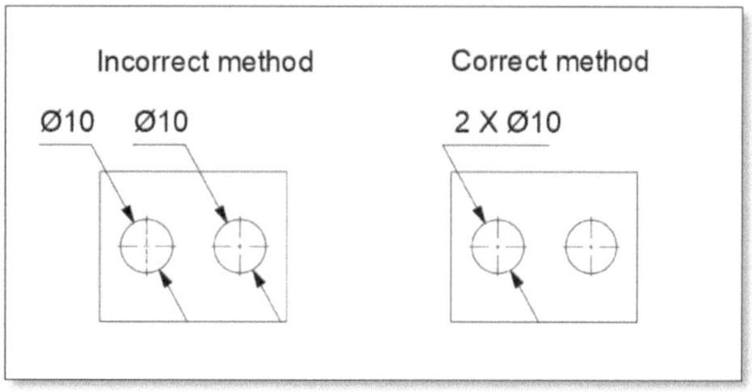

Figure 1.40

11. Overall dimensioning rules:
The overall dimensions have to be placed outside the intermediate dimensions.
The smaller dimensions must be placed nearer the view and the larger dimensions further away so that the extension lines or projection lines do not cross the dimension lines.

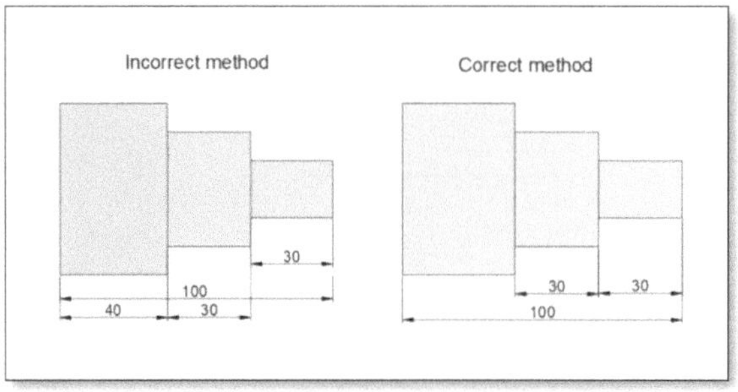

Figure 1.41

12. Dimension rules for a broken feature:
The dimension line should be shown unbroken where the figure to which it represents is shown broken.

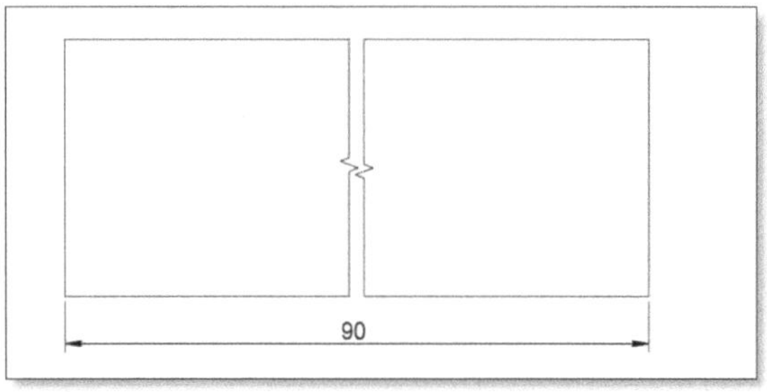

Figure 1.42

13. When a dimension line cannot be completely drawn to its normal termination point, the free end must be terminated in a double arrowhead.

Figure 1.43

14. As far as possible, all the dimensions in a particular drawing should be in one unit only.

15. Dimensions indicated in one view need not be repeated in another view, except for the case of clarity and identification.

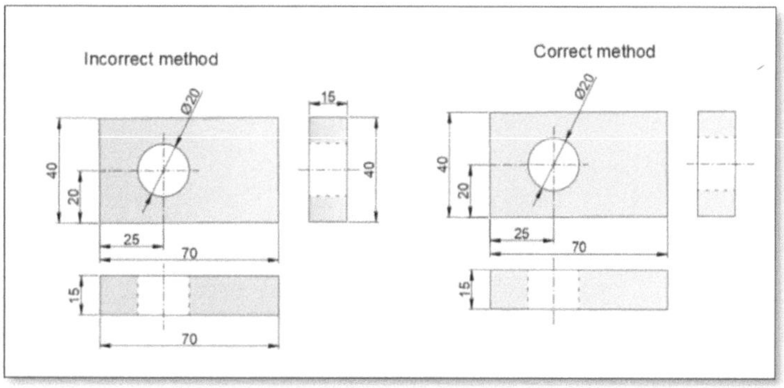

Figure 1.44

16. The dimensions must be placed on the view that most clearly shows the corresponding features.

17. The drawing line should never be used as a dimension line or it should not coincide with a dimension line.
The dimension line should be uniformly spaced throughout the drawing.
Dimension lines should be 8 mm to 10 mm from the outline of the drawing and 6mm to 8mm from each other.

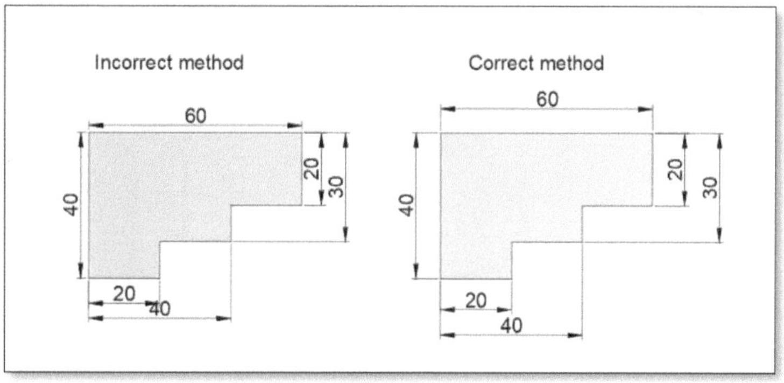

Figure 1.45

18. The dimension text should be placed preferably near the middle. If unavoidable due to lack of space, the dimension text may be placed above the extended portion of the dimension line beyond the arrowheads, preferably on the right-hand side.

1.6 GEOMETRY AND TANGENCY CONSTRUCTION

Dealing with problems involving two- and three-dimensional geometrical constructions can be challenging for students. The reader can gain confidence by copying some of these examples on a drawing board or using CAD software. They can use it to help them visualize and place the lines in space that make up each segment of a view or the perimeter of a three-dimensional object. Being able to correctly justify each line and dimension that appears on a drawing is a necessary component of draughtsmanship. Numerous computer programs will provide tools for a variety of constructions, including tangents, ellipses, and irregular curves.
In the examples that follow, use these features whenever you can. Assume all fundamental dimensions when appropriate.

To Bisect a Given Angle AOB -
1. With centre O, draw an arc to cut OA at C and OB at D.
2. With centres C and D, draw equal radii to intersect at E.
3. Line OE bisects angle AOB.

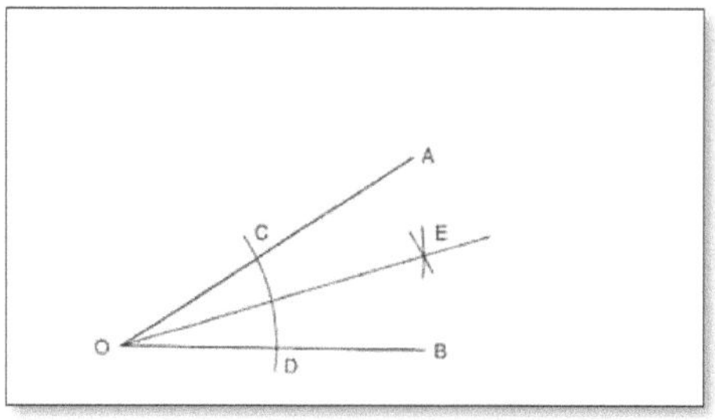

Figure 1.46

To Bisect a Given Straight Line AB -
1. With centre A and radius greater than half AB, describe an arc.
2. Repeat with the same radius from B, the arcs intersecting at C and D.
3. Join C to D and this line will be perpendicular to and bisect AB.

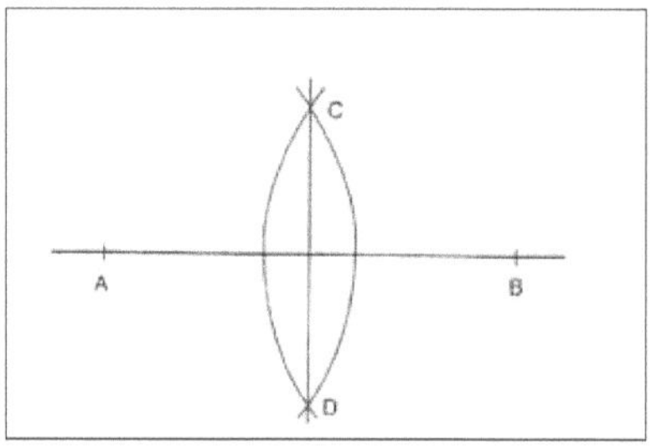

Figure 1.47

To Find the Centre of a Given Arc AB -
1. Draw two chords, AC and BD.
2. Bisect AC and BD as shown; the bisectors will intersect at E.
3. The centre of the arc is point E.

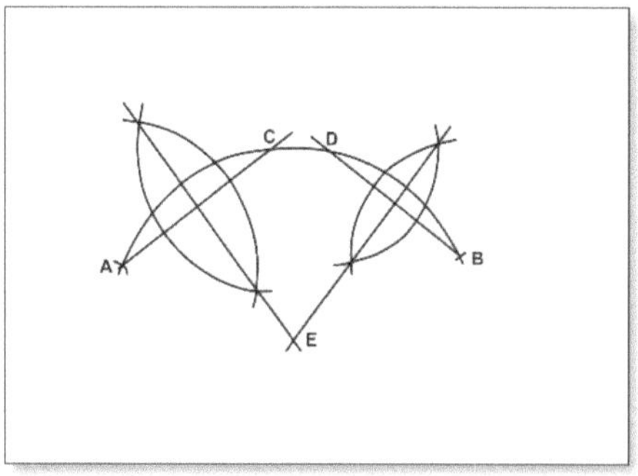

Figure 1.48

To Inscribe a Circle in a Given Triangle ABC -
1. Bisect any two of the angles as shown so that the bisectors intersect at D.
2. The centre of the inscribed circle is point D.

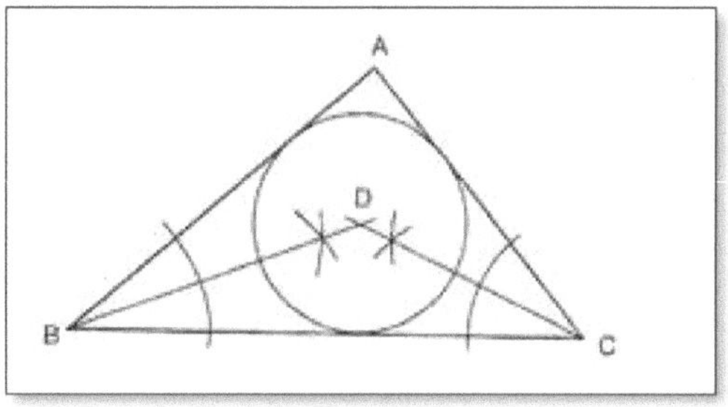

Figure 1.49

To Draw a Regular Polygon, Given the Length of the Sides -
Note that a regular polygon is defined as a plane figure which is bounded by straight lines of equal length and which contains angles of equal size. Assume the number of sides is seven in this example.

1. Draw the given length of one side AB, and with radius AB describe a semi-circle.
2. Divide the semi-circle into seven equal angles, using a protractor, and through the second division from the left join line A2.
3. Draw radial lines from A through points 3, 4, 5, and 6.
4. With radius AB and centre on point 2, describe an arc to meet the extension of line A3, shown here as point F.
5. Repeat with radius AB and centre F to meet the extension of line A4 at E.
6. Connect the points as shown, to complete the required polygon.

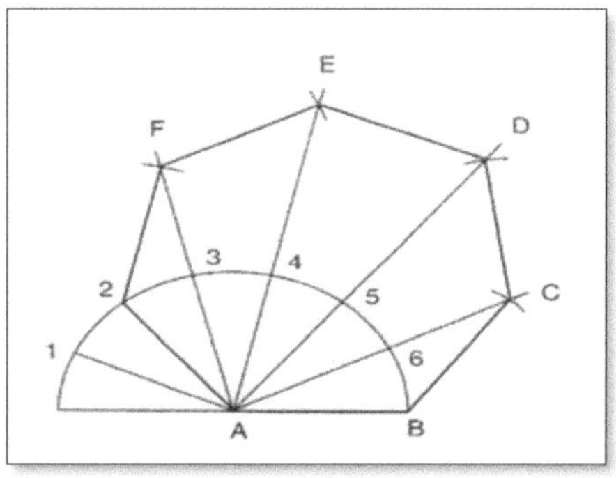

Figure 1.50

TANGENCY

A disc will touch the surface at one point if it is balanced on its edge and placed on a flat surface. As seen in Figure 1.51, this location is referred to as the point of tangency, and the straight line that symbolizes the flat plane is referred to as a tangent.

The tangent forms a 90° angle with the normal, which is a line drawn from the point of tangency to the disc's centre. The constructions that follow demonstrate various tangent drawing techniques.

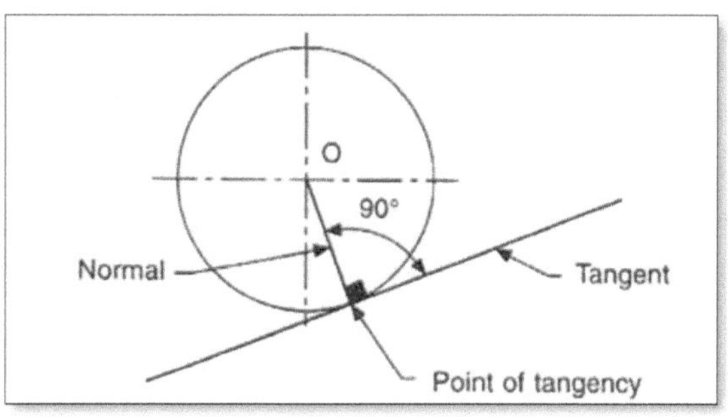

Figure 1.51

To Draw a Tangent to a Point A on the Circumference of a Circle, Centre O:
Join OA and extend the line for a short distance. Erect a perpendicular at point A by the method shown.

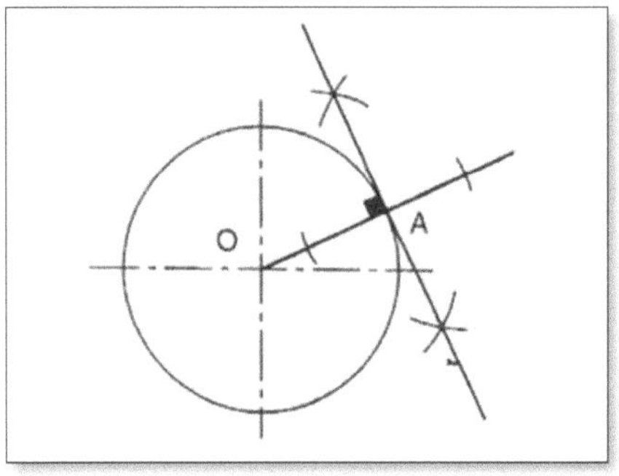

Figure 1.52

To Draw an External Tangent to Two Circles:
Join the centres of the circles by line AB, bisect AB, and draw a semi-circle. Position point E so that DE is equal to the radius of the smaller circle. Draw radius AE to cut the semi-circle at point G.
Draw line AGH so that H lies on the circumference of the larger circle. Note that angle AGB lies in a semi-circle and will be 90°. Draw line HJ parallel to BG.
Line HJ will be tangential to the two circles and lines BJ and AGH are the normal.

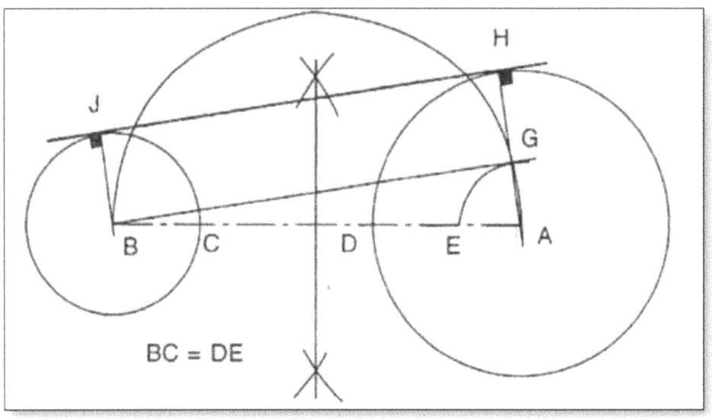

Figure 1.53

1.7 GENERAL SUGGESTIONS FOR DRAWING A SHEET

The arrangement of drawing sheets has a significant impact on the creation of a tidy, lovely, and appropriate drawing. The engineer must adhere to the normative guidelines and conventions for drawing sheet layout as recommended by the Bureau of Indian Standards through their specifications. In addition to that, the engineer should be familiar with the office standards for indicating special aspects.
There is a margin all the way around the drawing sheet in a layout for drawing sheets, and the title block space is set. On different sheet sizes, the margin space may change, but the title block space is typically fixed. Every drawing has a title block, which is repeatedly provided so that the information in the title block makes it simple and quick to read the drawing.

LAYOUT OF DRAWING SHEET:

After selecting the size of the drawing sheet and to divide the proper space for margin, title block, parts list, revision panel, folding marks and selecting a suitable scale on the drawing sheet is known as layout of drawing sheet.

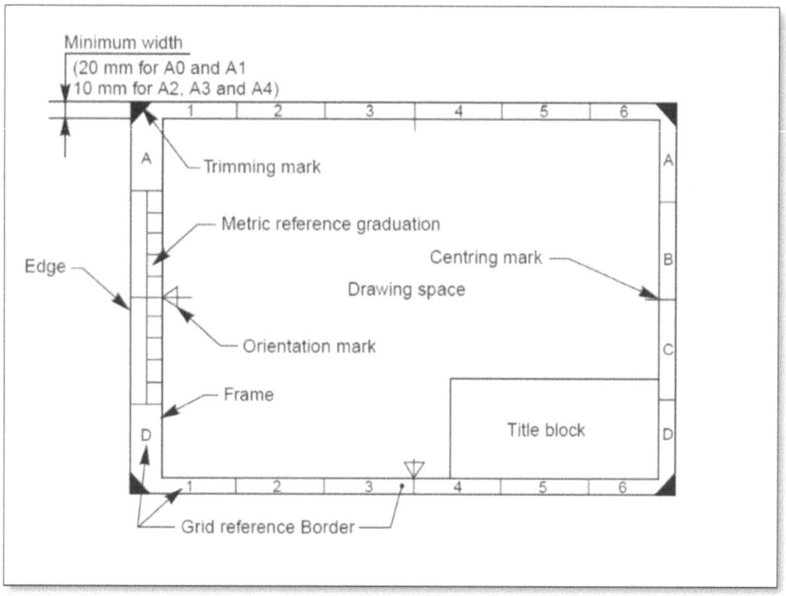

Figure 1.54

DIVIDING DRAWING SHEET INTO ZONES:
Marking a zone on a drawing sheet serves to highlight the precise spot in a drawing where a modification or alteration has since been incorporated. On commercial drawings, the zoning system is frequently used to highlight changes from the original drawing. The length of the longer side of the drawing side is divided into an even number of segments and marked 1, 2, 3, etc. for this purpose. The shorter side of the drawing sheet is also divided into equal spaces and lettered according to the alphabet (A, B, C, D, E, etc.) on both sides. By extending the number zone and alphabet zone to coincide at the altered site, it is possible to identify the area of the drawing that has been altered.

MARGIN:
The space between extreme outside boundaries and the working space is a margin provided around the sheet by drawing margin lines. The margin facilitates easy filing and binding. The margin ensures that if the sheet edge becomes damaged by handling, the drawing within the margin enables prints to be trimmed. Prints after trimming conforms to the sizes of trimmed sheets.

TITLE BLOCK:

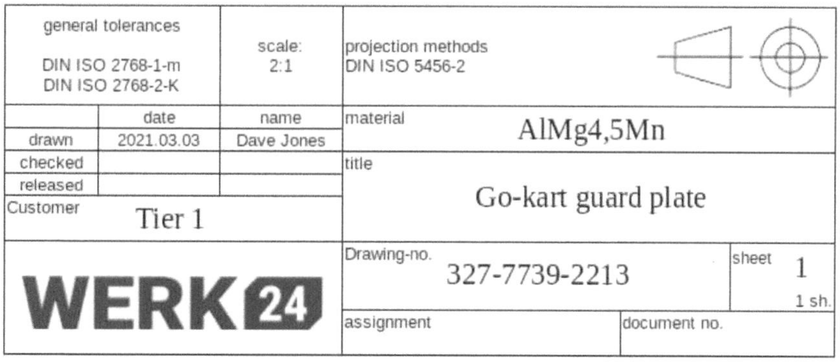

Figure 1.55

The title space should be within the drawing space. The title block is a recommended space of size 185 mm x 65 mm and should be placed at the right hand, bottom corner of the sheet. The title block is an important feature, because it contains the following information:
Name of title of drawing.
Drawing and sheet number.

Scale used.
Symbols denoting the method of projection.
Name of the firm.
Dated initials of staff designing, drawing, checking for standards and approving.
The material from which the part is made and it finish.

FOLDING MARKS:
Folding marks are to made on the drawing sheet to facilitate of prints in an easy manner.

BORDERS AND FRAMES:
Borders enclosed by the edges of the trimmed sheet and the frame limiting the drawing space should be provided with all sizes. It is recommended that these borders have the minimum width of 20 mm for A0 and A1 and a minimum width of 10 mm for size A2, A3 and A4.

CENTERING MARKS:
Four centering marks should be provided on all drawing in order to facilitate the positioning of the drawing when reproduced or microfilmed.

GRID REFERENCES:
The provision of grid reference system is recommended for all sizes in order to permit easy location on the drawing of details, additions, modifications, etc. The number of divisions should be divisible by two and be chosen in relation to the complexity of the drawing. It is recommended that the length of any side of the rectangle comprising the grid shall not be less than 25 mm and not more than 75 mm. The rectangles of the grid should be referred by means of capital letters along one edge and the numerals on the other edge.

REVISION PANEL:
The revision panel means that the drawing should record all additions and alterations which have been incorporated in the drawing from time to time. To record the revision, one revision panel is drawn at the top right-hand corner of the sheet. In the columns, the revision number or letter, date zone and signature of approving authority are given.

NUMBERING OF SHEETS:
If there is more than one drawing sheets to complete the assembly and detail to complete one drawing, the numbering on each sheet should show the numbers of the particular sheet. For instance, if there are four sheets to cover one drawing, then the first sheet shall be numbered as below:

"Sheet 1 of 4 or 2 of 4"

This sheet number should be recorded below the drawing in the space meant for sheet number.

ITEM REFERENCES:
The item references should be assigned in sequential order to each component part shown in assembly and/or each detailed item on the drawing. Further identical parts shown in the same item reference.
Item references should be composed of Hindu-Arabic numerals only. All items reference on the same drawing shall be clearly distinguishable from all other indications. This can be achieved by:
using characters of a larger height.
encircling the characters of each item reference. In that case all such circles shall have the same diameter and to be drawn with continuous thin line.
combining the above two methods.
The item references should be placed outside the general outlines of the items concerned and each item reference should be connected to the associated item by a leader line.
Leader line should not intersect, should be kept as small as practicable, and should be drawn at an angle to the item references, the leader line should be directed towards the centre of the circle.
Item reference of related items may be shown against the same leader line. While the item references identical items need only be shown once, provided there is no risk of ambiguity.

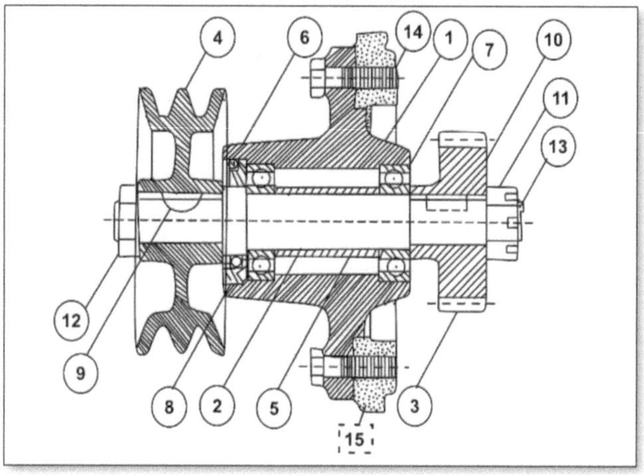

Figure 1.56

ADDITIONAL INFORMATION:

After completing the drawing sheet details, the additional information if specifically necessary to mention, must be recorded so that, the drawing is complete and self-contained as far as possible. The information should cover the following:

List of material for assembly and details.

Quantity of numbered parts for assembly and details.

Job number or order number.

Complete specification of material.

Key to machining and other symbols used in the drawing.

Treatment, finish, marks, tolerance on dimensions.

Other information regarding manufacturing procedures and relevant special requirement.

PART NO	NAME	MATERIAL	REQUIRED	NOTES
1	BASE	CL	1	-
2	SHAFT	103OSTL	1	-
3	GEAR	32OSTL	1	-
4	DOUBLE V-PULLEY	310ALLUM	1	-
5	SLEEVE	1020STL	1	-
6	BEARINGS	-	2	NEW DEPARTURE
7	COTTON PIN	-	1	.5 SAE
8	BEARING RETAINER	102STL	1	-

Figure 1.57

Assignment

1. In engineering drawing, which type of line indicates that there is a change of plane?

a) Continuous thin wavy

b) Long chain thin
c) Continuous thick
d) Medium thick short dashes
2. Which of the following lines are used to show that the object is cut and then viewed?
a) Hidden lines
b) Leader lines
c) Centre lines
d) Hatching Lines
3. What do hidden lines in orthographic projections denote?
a) Holes or slots
b) Change of Plane
c) Position of cut
d) Centre of a circle or cylinder
4. From the below figure, what is the name of the line X?

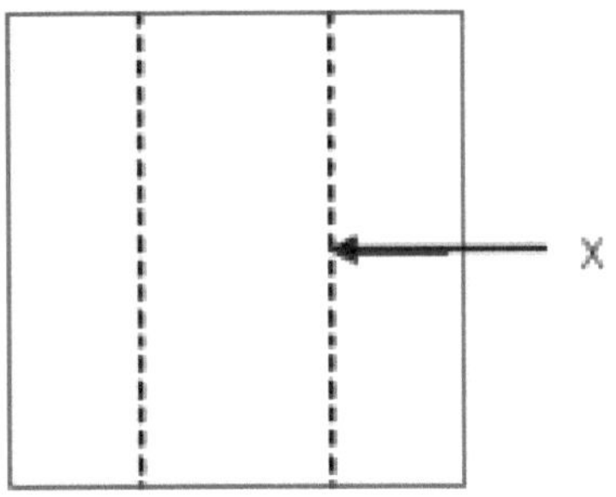

a) Outline
b) Section line
c) Hidden line
d) Hatching
5. The axis of the cylinder or sphere is denoted by which of the following line?
a) Section line
b) Centre line
c) Hidden line
d) Leader line
6. What is the length of the short dashes of the centre lines?
a) 5 mm
b) 2 mm
c) 1 mm
d) 3 mm
7. Among the following, which of the representation is wrong?

a)

b)

c)

d)

8. What is the difference between the section line and the centre line?
a) The length of the long dashes
b) The length of short dashes
c) The width of the gap
d) The two ends of the lines
9. The accuracy of the drawing depends on the quality of the instruments used.
a) True
b) False
10. Which of the following instrument is made of thin strips of wood arranged in a line to form a rectangle and on which, the drawing is made?
a) Mini – drafter
b) Drawing Board
c) Protractor
d) Scale
11. Finally the standards SP: 46- 2003 and SP: 46 – 1988 are revised and replaced with earlier __________ drawing standards.
a) IS: 698
b) IS: 697
c) IS: 696
d) IS: 695
12. The committee again reviewed the position and finally recommended revised ____________ for implementation of 1st angle projection.
a) SP: 46 – 1988
b) SP: 45 - 1988
c) SP: 46 – 1955
d) SP: 46 - 1960
13. The second revised version of the B.I.S. published in _______
a) 1960
b) 1955
c) 2003
d) 1973
14. In which year the B.I.S. standard got revised and recommended the use of third angle projection method?
a) 1960
b) 1955
c) 2003
d) 1973
15. The preferred size of the drawing sheets is recommended by the ______
a). B.I.S.
b) ASME
c) ASTM
d) NIST
16. The untrimmed size for _______ sheet is 240 mm x 330 mm.
a) A1
b) A2
c) A3
d) A4
17. SP: 46 (2003) recommends the borders of _______ mm width for the sheet sizes A0 and A1, and _______ mm for the sizes A2, A3, A4 and A5.
a) 10, 20
b) 15, 20
c) 20, 10
d) 15, 10
18. __________ is not an essential thing for free-hand sketching.
a) A soft-grade pencil
b) French curves
c) A soft rubber-eraser

d) A paper in the form of sketch-book or a pad
19. How many battens will be there for a Drawing board?
a) 1
b) 2
c) 3
d) 4
20. The angle which we can't make using a single Set-square is ________
a) 45 Degree
b) 30 Degree
c) 60 Degree
d) 75 Degree
21. The angle which we can't make using both the Set-squares is _____________
a) 15 Degree
b) 105 Degree
c) 165 Degree
d) 125 Degree
22. Which is not the use of divider?
a) To divide curved or straight lines into the desired number of equal parts
b) To draw circles
c) To transfer dimensions from one part of the following to another part
d) To set-off given distances from the scale to the drawing
23. _________ is used to draw curves which are not circular.
a) Compass
b) Protractor
c) French curves
d) Pro circle

Answers

1. Answer: c
2. Answer: d
3. Answer: a
4. Answer: c
5. Answer: b
6. Answer: b
7. Answer: b
8. Answer: d
9. Answer: a
10. Answer: b
11. Answer: a
12. Answer: b
13. Answer: b
14. Answer: a
15. Answer: a
16. Answer: d
17. Answer: c
18. Answer: b
19. Answer: b
20. Answer: d
21. Answer: c
22. Answer: d
23. Answer: c

CHAPTER 2: ORTHOGRAPHIC PROJECTIONS

2.1 INTRODUCTION OF PROJECTIONS

An orthographic projection enables us to represent a 3-D object in 2-D. An orthographic projection is a system of drawings that represent different sides of an object. These drawings are formed by projecting the edges of the object perpendicular to the desired planes of projection.

Orthographic projections allow us to represent the shape of an object using two or more views. These views together with dimensions and notes are sufficient to manufacture the part.

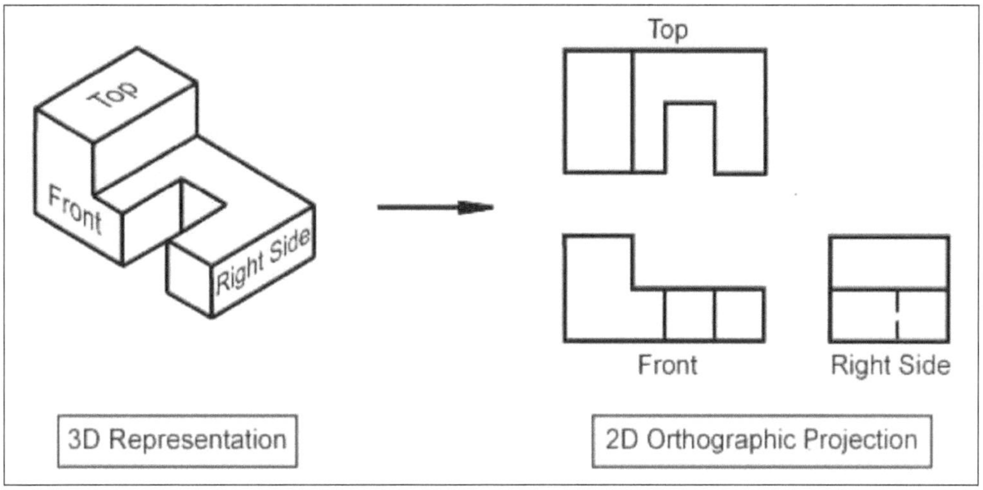

Figure 2.1

Reference Lines

Studying the projections independently, it can be seen that while considering the front view in figure 2.2, which is the view as seen from the front, the H.P. coincides with the line xy. In other words, xy represents the H.P.

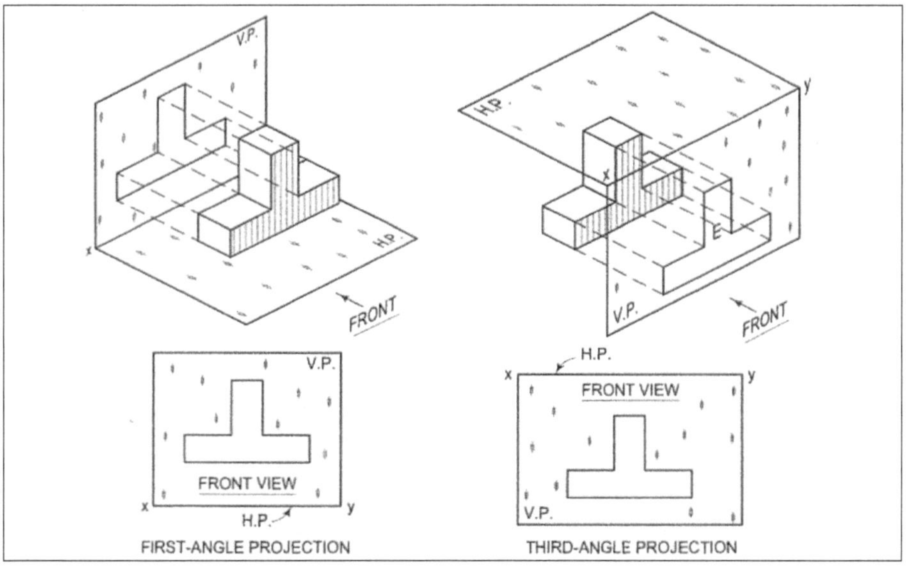

Figure 2.2

Similarly, while considering the top view (figure 2.3), which is the view obtained by looking from above, the same line xy represents the V.P. Hence, when the two projections are drawn in correct relationship with each other (figure 2.3), xy represents both the H.P. and the V.P.

This line xy is called the reference line. The squares or rectangles for individual planes are thus unnecessary and are therefore discarded. Further, in first-angle projection method, the H.P. is always assumed to be so placed as to coincide with the ground on or above which the object is situated. Hence, in this method, the line xy is also the line for the ground.

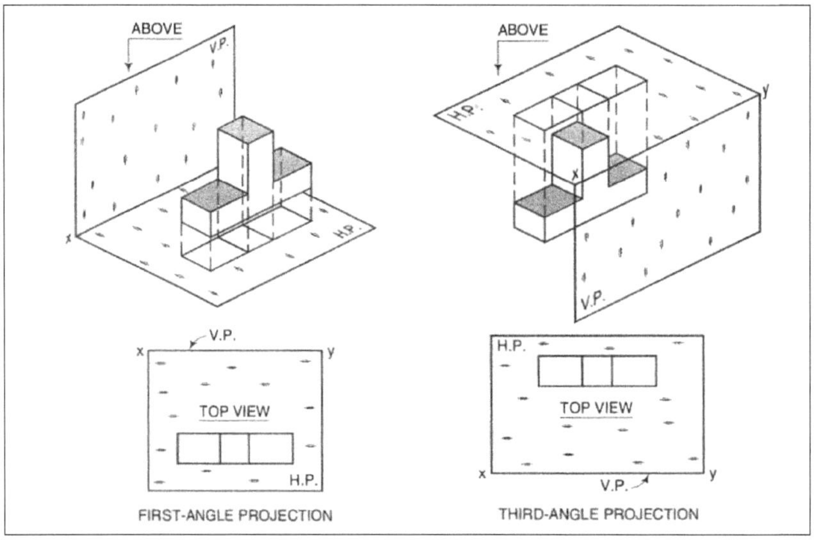

Figure 2.3

In third-angle projection method, the H.P. is assumed to be placed above the object. The object may be situated on or above the ground.

Hence, in this method, the line xy does not represent the ground. The line for the ground, denoted by letters GL, may be drawn parallel to xy and below the front view [figure 2.4].

In brief, when an object is situated on the ground, in first-angle projection method, the bottom of its front view will coincide with xy in third-angle projection method, it will coincide with CL, while xy will be above the front view and parallel to Ground line.

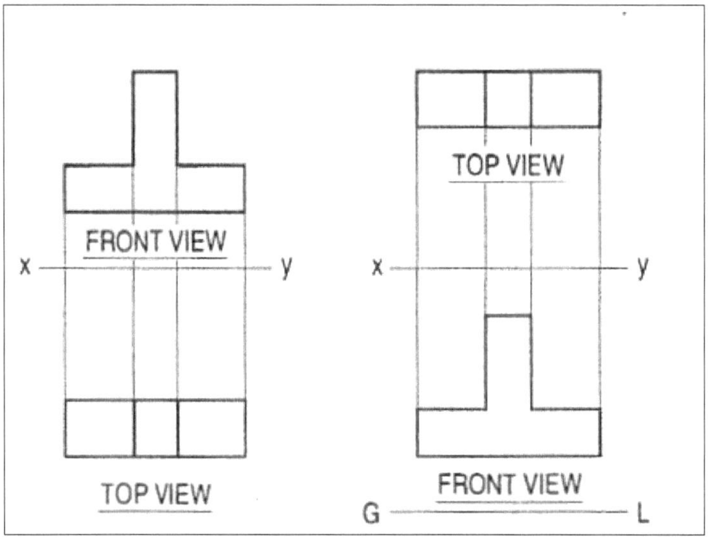

Figure 2.4

There are Six Principle Views in which six principle views of an orthographic projection are shown in Figure 2.5. Each principle view is created by looking at the object in the directions indicated in Figure 2.5 and drawing what is seen as well as what is hidden from view.

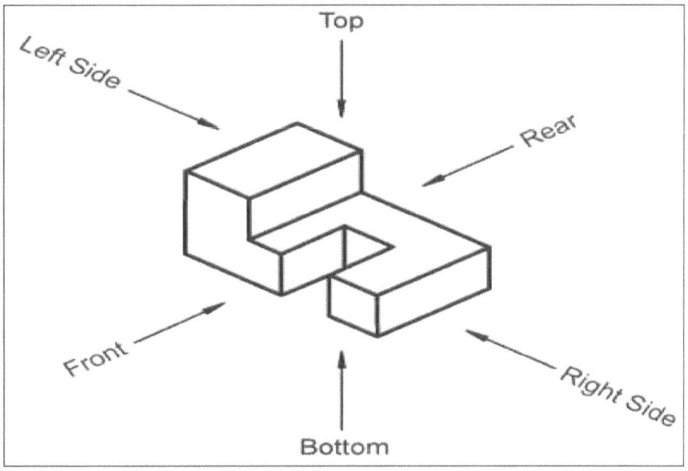

Figure 2.5

Generally, the following three views are prepared in the orthographic drawing.
1. FRONT VIEW:
This view is prepared by placing the object in front. The length and height of an object are shown in this view.
2. TOP VIEW:
This view is prepared by looking to the object from the upper side. The length and breadth of the object are shown in it.
3. SIDE VIEW:
This view is prepared by looking to the object from the right side or left side. The breadth and height of the object are shown in it.
Orthographic Projections is a technical drawing in which different views of an object are projected on different reference planes observing perpendicular to respective reference plane.

- Different Reference planes are:
  - Horizontal Plane (HP)
  - Vertical Plane (VP)
  - Side or Profile Plane (PP)

- Different views are:
  - Front View (FV) – Projected on VP
  - Top View (TV) – Projected on HP
  - Side View (SV) – Projected on PP

DRAWING OF ORTHOGRAPHIC PROJECTION
For drawing Orthographic Projection, different planes are placed in a particular order. Then a specific view is drawn through every plane.
A plane is placed in the following two methods:
1. Dihedral Angle
2. Trihedral Angle

1. DIHEDRAL ANGLE
In this method, two principal planes are kept perpendicular to each other. One of these planes is Frontal, while other is horizontal. Four right angles are obtained in this way which is called Dihedral Angle. Each right angle is called a quadrant. The object is placed in any of these right angles to take orthographic projection. The front view is taken on the frontal plane, and the top view is taken on the horizontal plane. Profile plane is used to take side views. This plane is held perpendicular at the ends of the other two planes.

2. TRIHEDRAL ANGLE
In this method, all the three Principal planes are taken perpendicular to one another, and eight right angles are formed. Thus, these are called Trihedral angle**.** Each right angle is called an Octant. An object is placed in any of these right

angles to take orthographic projection. The front view is taken on the frontal plane, and the top view is taken on the horizontal plane. Profile plane is used to take side views.

LINE TYPES USED IN AN ORTHOGRAPHIC PROJECTION

Line type and line weight provide valuable information to the print reader. For example, the type and weight of a line can answer the following questions: Is the feature visible or hidden from view? Is the line part of the object or part of a dimension? Is the line indicating symmetry?

There are four commonly used line types: continuous, hidden, centre and phantom. The standard recommends using, no less than, 2- line widths. Important lines should be twice as thick as the less important thin lines. Common thicknesses are 0.6 mm for important lines and 0.3 mm for the less important lines.

However, to further distinguish line importance, it is recommended to use four different thicknesses or weights: thin, medium, thick, and very thick. The actual line thickness should be chosen such that there is a visible difference between the line weights; however, they should not be too thick or thin making it difficult to read the print. The thickness of the lines should be adjusted according to the size and complexity of the part. The following is a list of common line types and widths used in an orthographic projection:

1. Visible lines: Visible lines represent visible edges and boundaries. The line type is continuous and the line weight is thick (0.5 - 0.6 mm).

2. Hidden lines: Hidden lines represent edges and boundaries that cannot be seen. The line type is dashed and the line weight is medium thick (0.35 - 0.45 mm).

3. Centre lines: Centre lines represent axes of symmetry and are important for interpreting cylindrical shapes. Crossed centre lines should be drawn at the centres of circles. They are also used to indicate circle of centres and paths of motion. The line type is long dash or short dash and the line weight is thin (0.3 mm).

4. Phantom lines: Phantom lines are used to indicate imaginary features. For example, they are used to indicate the alternate positions of moving parts, and adjacent positions of related parts. The line type is long dash or short dash and the line weight is usually thin (0.3 mm).

5. Dimension and Extension lines: Dimension and extension lines are used to show the size of an object. In general, a dimension line is placed between two extension lines and is terminated by arrowheads, which indicates the direction and extent of the dimension. The line type is continuous and the line weight is thin (0.3 mm).

6. Cutting plane lines: Cutting plane lines are used to show where an imaginary cut has been made through the object in order to view interior features. The line type is phantom and the line weight is very thick (0.6 to 0.8 mm). Arrows are placed at both ends of the cutting plane line to indicate the direction of sight.

7. Section lines: Section lines are used to show areas that have been cut by the cutting plane. Section lines are grouped in parallel line patterns and usually drawn at a 45° angle. The line type is usually continuous and the line weight is thin (0.3 mm).

8. Break lines: Break lines are used to show imaginary breaks in objects. A break line is usually made up of a series of connecting arcs. The line type is continuous and the line weight is usually thick (0.5 – 0.6 mm).

Using the line type definitions, match each line type name with the appropriate line type.

- Visible Line
- Hidden Line
- Center Line
- Phantom Line
- Dimension and Extension Lines
- Cutting Plane Line
- Section Lines
- Break Line

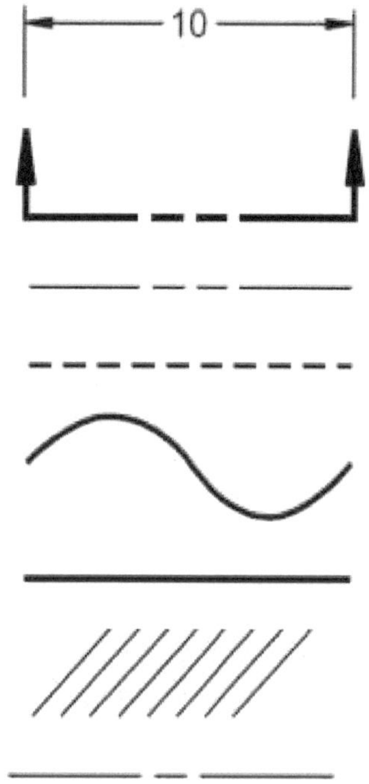

Figure 2.6

CREATING AN ORTHOGRAPHIC PROJECTION

The steps presented in this section are meant to help you create a technically correct orthographic projection using the 3rd angle projection standard. To understand and visually see how views are created using the 3rd angle projection standard, put your right hand on a table palm up. You are looking at the front view of your hand. Now rotate your hand so that your thumb points up and your little finger is touching the table. This is the right sided view of your hand. Put your hand back in the front view position. Now rotate your hand so that your finger tips are pointing up and your wrist is touching the table. This is the top view of your hand. The following steps will take you through the creation of an orthographic projection. Once you become experienced and proficient at creating orthographic projections, you will develop short cuts and may not need to follow the steps exactly as written.

1. Choose a front view. This is the view that shows the most about the object.
2. Decide how many views are needed to completely describe the object. If you are unable to determine which views will be needed, draw the
   standard views (front, top and right side).
3. Draw the visible features of the front view.
4. Draw projectors off of the front view horizontally and vertically in order to create the boundaries for the top and right sided views.
5. Draw the top view. Use the vertical projectors to fill in the visible and hidden features.
6. Project from the top view back to the front view. Use the vertical projectors to fill in any missing visible or hidden features in the front view.
7. Draw a 45° projector off of the upper right corner of the box that encloses the front view.
8. From the top view, draw projectors over to the 45° line and down in order to create the boundaries of the right sided view.
9. Draw the right sided view.
10. Project back to the top and front view from the right sided view as needed.
11. Draw centre lines where necessary.

Following the forementioned steps will ensure that the orthographic projection is done correctly. That is, it will insure that:

√ The front and top views are vertically aligned.
√ The front and right sided views are horizontally aligned.
√ Every point or feature in one view is aligned on a projector in any adjacent view (front and top, or front and right side).
√ The distance between any two points of the same feature in the related views (top and right side) are equal.

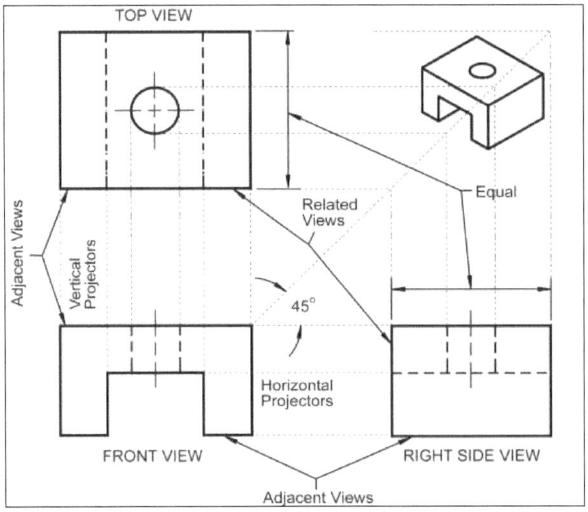

Figure 2.7

TYPES OF VIEWS
There are different types of projections:
TYPES OF PROJECTIONS:
Parallel projections
Perspective (Converge) projections

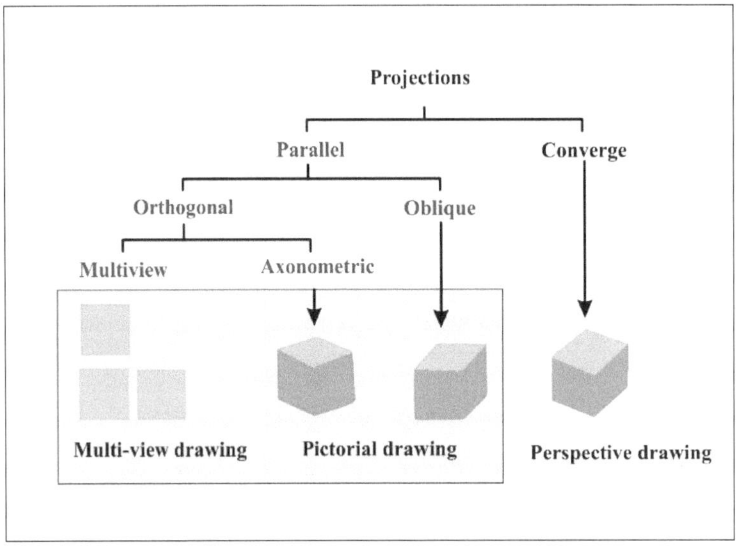

Figure 2.8

PARALLEL PROJECTION
By drawing parallel projections from each object vertex until they cross the screen's plane, a parallel projection is created. An item is transformed to the view plane along parallel lines by parallel projection.

If the centre of the projection is at an infinite distance from the projected plane, the projection is said to be parallel. A parallel projection obtains precise views of an object's multiple sides while maintaining the relative proportion of items.

The projection lines extend from the object in a straight line, cross the view plane, and are parallel to one another. It is used in drafting to create scale drawings of 3D objects and preserves relative propositions of objects. The point of intersection is a projection and is not a realistic representation.

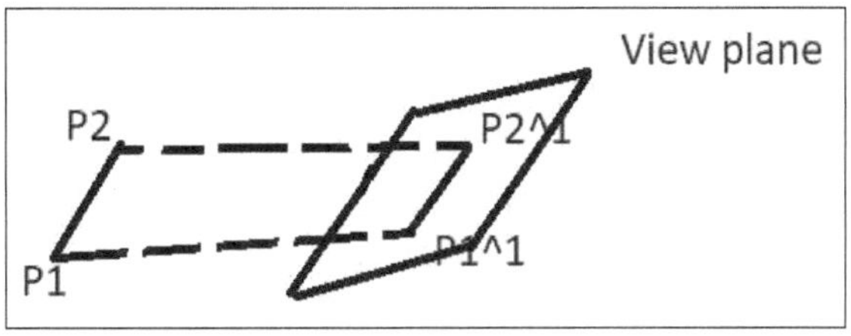

Figure 2.9

Parallel projection is divided into two parts and these two parts are sub divided into many.

ORTHOGRAPHIC PROJECTIONS

In an orthographic projection, the projection direction is parallel to the plane's projection. Orthographic lines are perpendicular to one another and form a 90° angle with the view plane. By projecting points along parallel, perpendicular to the projection line, lines, orthographic parallel projections are created.

Evaluations are what orthographic projections are used for when processing an object's front, side and top perspectives. These orthographic projections are often used in engineering and architectural designs.

Orthographic parallel projection transformation formulae are simple. Plan view and side elevations are two examples of unique orthographic parallel projections. We can also do monometric orthographic projections, which are orthographic views that show more than one phase of an object.

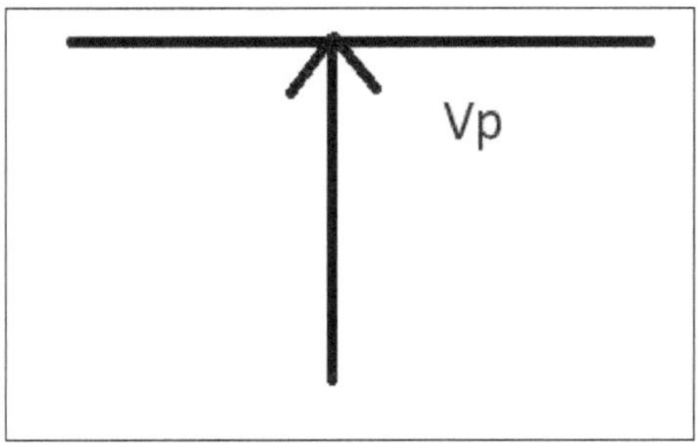

Figure 2.10

OBLIQUE PROJECTIONS

Oblique projections are obtained by projectors along parallel lines that are not perpendicular to the projection plane. An oblique projection shows the front and top surfaces that include the three dimensions of height, width and depth. The front or principal surface of an object is parallel to the plane of projection. Effective in pictorial representation.

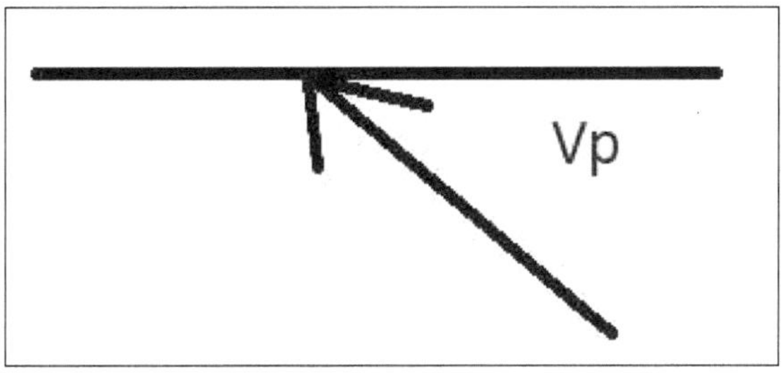

Figure 2.11

Isometric Projections: Orthographic projections that show more than one side of an object are called axonometric orthographic projections. The most common axonometric projection is an isometric projection. In this projection parallelism of lines are preserved but angles are not preserved.
Dimetric projections: In these two projectors have equal angles with respect to principal axis.
Trimetric projections: The direction of projection makes unequal angle with their principal axis.

CAVALIER PROJECTIONS

All lines perpendicular to the projection plane are projected with no change in length. If the projected line making an angle 45 degrees with the projected plane, as a result the line of the object length will not change.

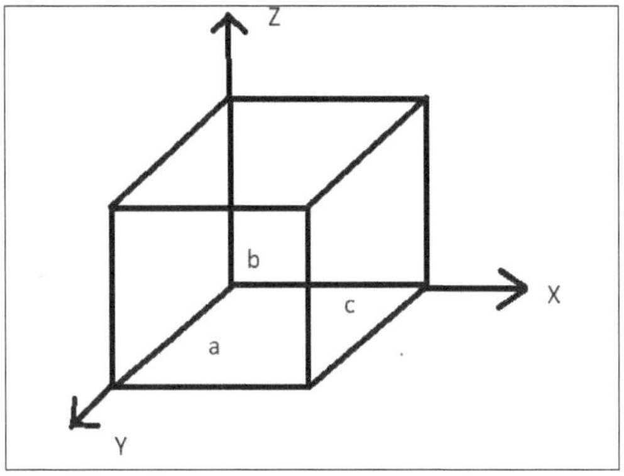

Figure 2.12

CABINET PROJECTIONS

All lines perpendicular to the projection plane are projected to one half of their length. These gives a realistic appearance of object. It makes 63.4 degrees angle with the projection plane. Here lines perpendicular to the viewing surface are projected at half their actual length.

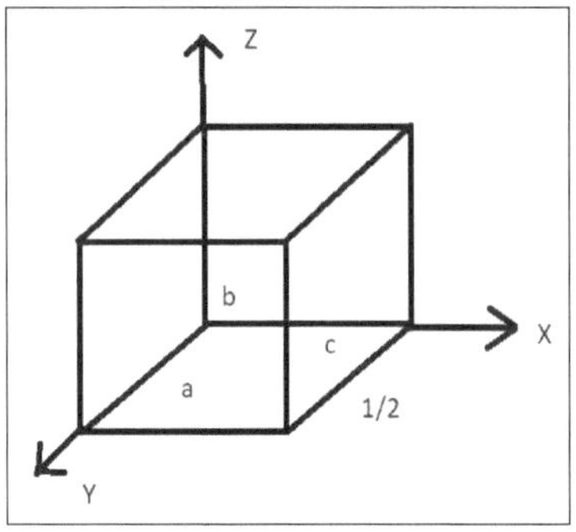

Figure 2.13

PERSPECTIVE PROJECTIONS

A perspective projection is the one produced by straight lines radiating from a common point and passing through point on the sphere to the plane of projection.

Perspective projection is a geometric technique used to produce a 3-dimensional graphic image on a plane, corresponding to what person sees.

Any set of parallel lines of object that are not parallel to the projection plane are projected into converging lines. A different set of parallel lines will have a separate vanishing point.

Coordinate positions are transferred to the view plane along lines that converge to a point called projection reference point.

The distance and angles are not preserved and parallel lines do not remain parallel. Instead, they all converge at a single point called centre of projection there are three types of perspective projections.

Two characteristics of perspective are vanishing point and perspective force shortening. Due to fore shortening objects and lengths appear smaller from the centre of projections. The projections are not parallel and we specify a centre of projection cop.

Different types of perspective projections:
One point perspective projections: In this, principal axis has a finite vanishing point. Perspective projection is simple to draw.

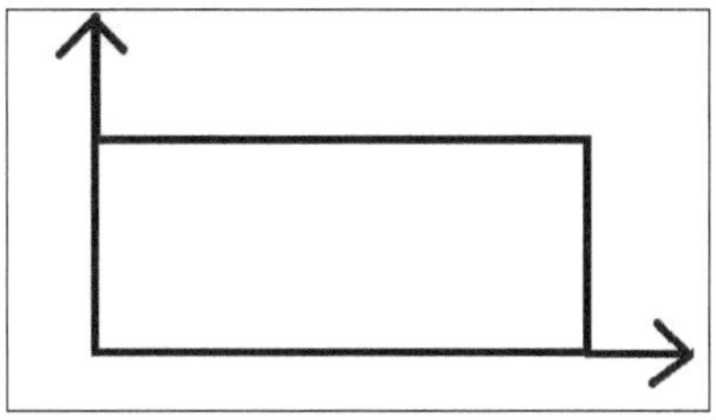

Figure 2.14

Two-point perspective projections: Exactly two principals have vanishing points. Perspective projection gives better impression of depth.

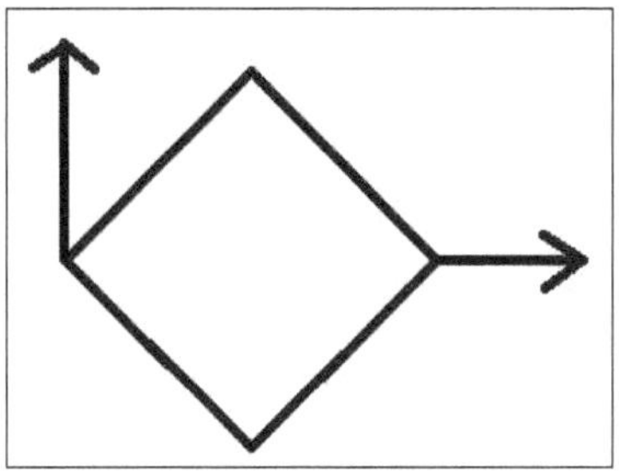

Figure 2.15

Three-point perspective projections: All the three principal axes have finite vanishing point. Perspective projection is most difficult to draw.

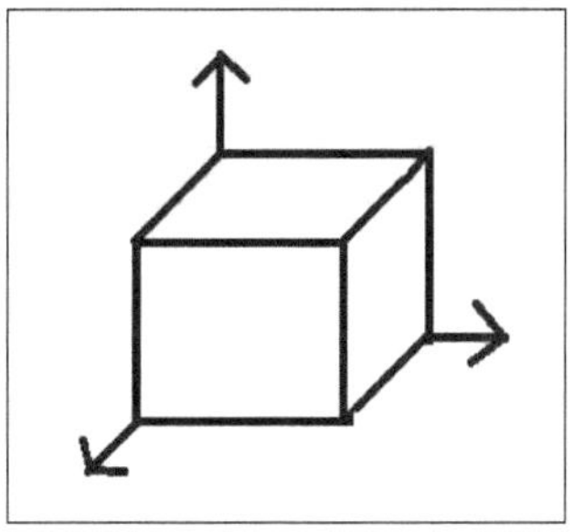

Figure 2.16

View comparison

Type		
Multi-view drawing	• Accurately presents object's details, i.e. size and shape.	• Require training to visualization.
Pictorial drawing	• Easy to visualize.	• Shape and angle distortion Circular hole becomes ellipse Right angle becomes obtuse angle.
Perspective drawing	• Object looks more like what our eyes perceive.	• Difficult to create • Size and shape distortion Distorted width

PROJECTION SYSTEMS

In the United States, we use 3rd angle projection to create an orthographic projection. This is the method of creating orthographic projections that is described in this chapter. In some parts of Europe and elsewhere 1st angle projection is used.

To inform the print reader what projection method was used, the projection symbol should be placed in the bottom right hand corner of the drawing. If the drawing uses metric units, the text "SI" is placed in front of the projection symbol. The projection symbols and projection symbol's proportions are shown in Figures 2.17

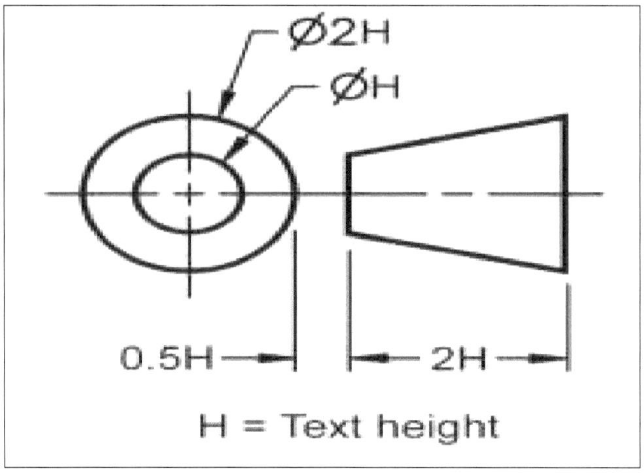

Figure 2.17

In this way the following four systems are formed:

1. First Angle System
2. Second Angle System
3. Third Angle System
4. Fourth Angle System

Generally, First Angle System and Third Angle System are used. It is because in the Second Angle System and Fourth Angle System, the lines of views of the object overlap.

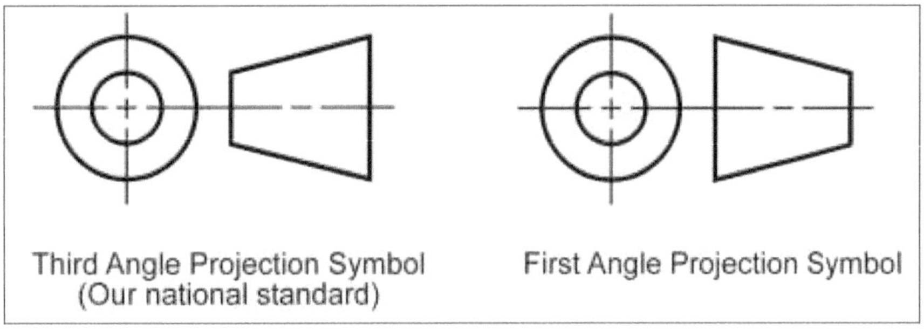

Figure 2.18

2.1.1 FIRST ANGLE PROJECTION METHOD

The views of an object should be taken by placing it in the first quadrant of dihedral and or first octant of a trihedral angle. Such a system is called **First Angle System.** By taking the Front view on the frontal plane, top view on the horizontal plane and side view on the Profile plane, the planes are then straightened by rotation. In this way, the front view comes over the top view; side view comes beside the front view.

1. FIRST ANGLE SYSTEM

- European countries
- ISO standard

The object is positioned in the first quadrant of the first angle projection. The object is situated at the top of the horizontal plane and at the front of a vertical plane. India and European nations frequently employ first-angle projection. Between the observer and projection planes is the object. In the first angle projection, the projection plane is assumed to be solid.

Symbol – Graphical Representation of 1st Angle Projection method

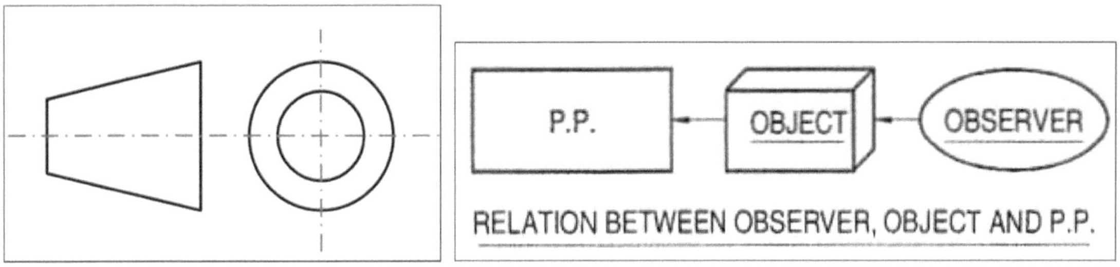

Figure 2.19

Symbols for methods of projection: For every drawing it is absolutely essential to indicate the method of projection adopted. This is done by means of a symbolic figure drawn within the title block on the drawing sheet.

The symbolic figure for the first-angle projection method is shown in figure 2.19 and 2.20, while that for the third-angle projection method is shown in figure 2.19 which is self explanatory. These symbolic figures are actually the projections of a frustum of cone of convenient dimensions according to the size of drawing.

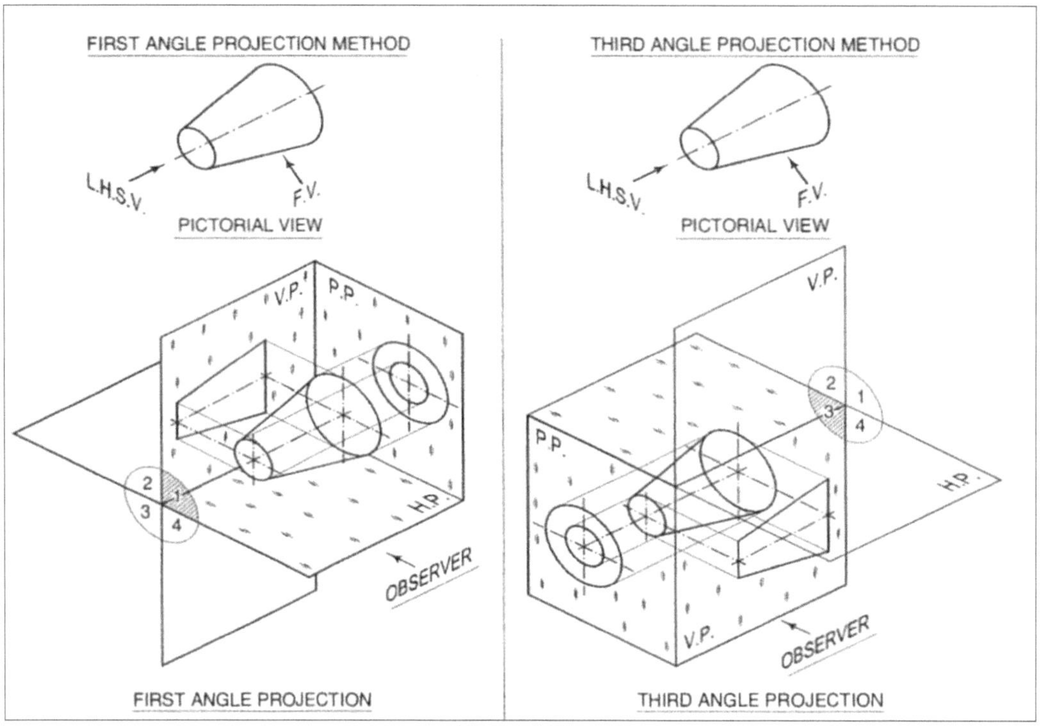

Figure 2.20

CHARACTERISTICS OF FIRST ANGLE PROJECTION:

1. Front view always comes over the top view.
2. Top view always comes under the front view.
3. Right side view always comes to the left at the front view.
4. Left side view always comes to the right of the front view.
5. The view is always in opposite direction to the observer.
6. The object is always in the middle of the view and the observer.

2. THIRD ANGLE SYSTEM

If the views of an object are taken by placing it in the third quadrant of dihedral or third octant of a trihedral tingle. Such a system is called **Third Angle System.** Here, front view forms on the frontal plane, top view forms on the horizontal plane and side view forms on the profile plane. After making the views, the planes are set straight by rotation. In this way, the top view comes over the front view, while side view forms by the side of front view.

Third angle system

- Canada, USA
- Japan, Thailand

The object is positioned in the third quadrant in the third angle projection. The item is positioned below the horizontal plane and behind the vertical planes. In the US, third angle projection is frequently used. Between the object and the observer are the projection planes. In third-angle projection, the projection plane is assumed to be transparent.

Symbol – Graphical Representation of 3rd Angle Projection method

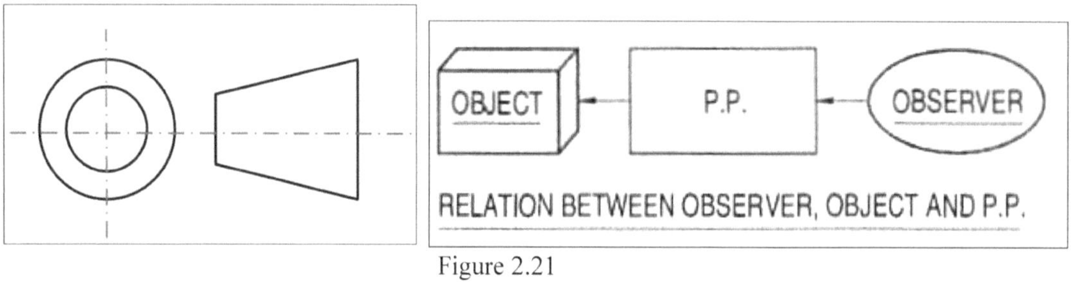

Figure 2.21

CHARACTERISTICS OF THIRD ANGLE PROJECTION:

1. Top view always comes over the Front view.
2. Front view always comes under the top view.
3. Right side view always comes to the right of the front view.
4. Left side view always comes to the left of the front view.
5. The view is always formed to the side of the observer.
6. The view is always in the middle of the object and the observer.

RULES OF ORTHOGRAPHIC DRAWING

Following rules should be followed while forming orthographic drawing.

1. Front view and top view always form over/under each other.
2. The front view shows the length and height of an object.
3. The side view shows the breadth and height of an object.
4. The top view shows the length and breadth of an object.
5. Side view always forms beside the front view.
6. Projection line always forms by the meeting of two surfaces.
7. The hidden detail of an object is always shown by dotted line.

SELECTION OF VIEW

Following points should be kept in mind at the selection of view,

1. The more detailed view should be selected.
2. The selection of view with maximum detail should be made.
3. The selection of view with more internal details should be made.
4. The selection of view of required details should be made.

Problem 1: AL-shaped solid object having dimensions of length width (W) and height (HJ in the figure 2.22) Assuming that this object is lying in the first quadrant. Draw its front view, top view and side view.
Solution:
When the given object is in the first quadrant, its front view appears in the imaginary vertical plane V.P. behind it while its top view appears in an imaginary horizontal plane, H.P. below. The side view appears to the right or left of the front view depending on from which side the object is being viewed.
(i) Mark the visible corners of the given block as shown front view in figure2.22. Assume that you are viewing the object in the direction of the arrow towards the imaginary V.P. What you will see is a rectangle of height 'H' and width 'W' on V.P. This will be the front view.
(ii) To draw the above view: Draw a reference line xy, which represents the intersecting line of the planes V.P. and H.P. Draw a rectangle as shown W and H, above xy make sure that the width is parallel to the Iine xy. The rectangle is the front orthogonal view of the object.
(iii) Draw a line parallel to and thickness of h to the line 1-2. The rectangle 1-2-4-3 is the front view of the horizontal L-shaped stem of the object.
(iv) Draw vertical projectors from 1 and 2 and extend them beyond the line xy. Draw a line 9-10 below and parallel to the reference line xy. Draw the lines 9-3 and 10-4 equal to the length L of the object. Join line 3-4. The rectangle 9-10-3-4 is the top view of the object.
(v) Draw a line 11-12 parallel to and below 9-10 of thickness h. The rectangle 9-11-12-10 is the view of the vertical stem of the object.
(vi) Draw a reference vertical line MN at right angles to xy cutting it at P. From P draw a construction line at 45° in the fourth quadrant.
(vii) Project lines from the points 10, 12 and 4 of the top view to meet this inclined line at 1 O', 12' and 4'.
(viii) Project lines from points 2, 4 and 12 from the front view parallel to line xy. From points 1 O', 12' and 4' project lines vertically upwards to meet these horizontal projections.
(ix) Join points 5-1-3-7-11-9. This will be the side view of the object.
(x) Finally draw the symbol of first angle projection at the right bottom corner of the drawing.

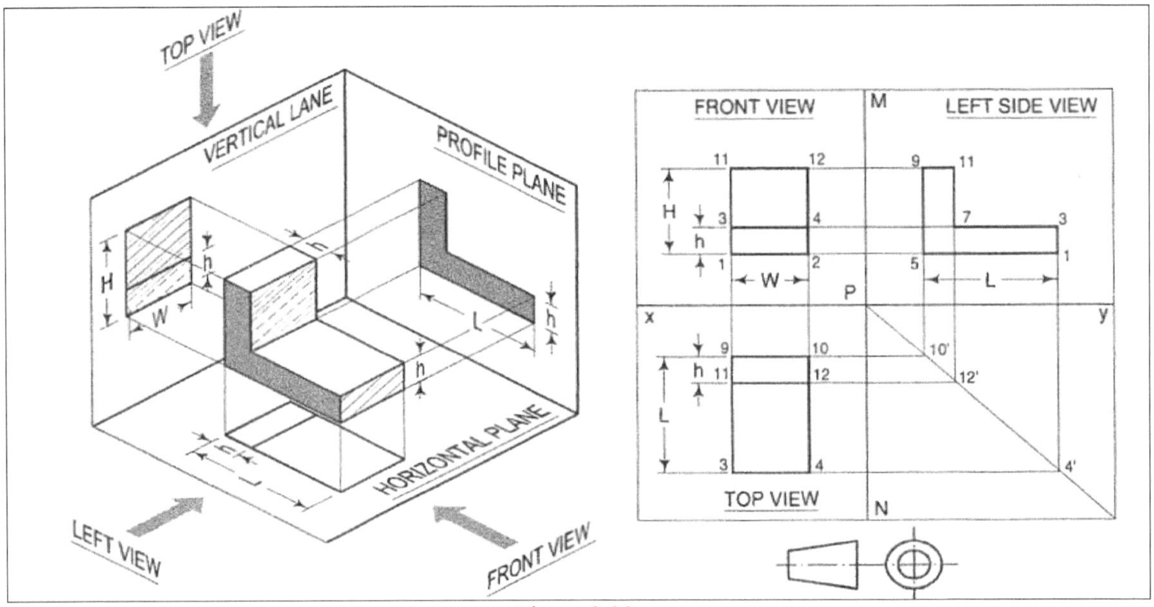

Figure 2.22

2.2 CONVERSION OF PICTORIAL VIEW TO ORTHOGRAPHIC VIEW

For converting a pictorial view of an object into orthographic views, the direction from which the object is to be viewed for its front view is generally indicated by means of an arrow.

When this is not done, the arrow may be assumed to be parallel to a sloping axis.

What are pictorial views?

A view of an object (actual or imagined) as it would be seen by an observer who looks at the object either in a chosen direction or from a selected point of view. Pictorial sketches often are more readily made and more clearly understood than are front, top, and side views of an object.

BASIC TYPES OF PICTORIAL VIEWS

There are three basic types of pictorial drawings: axonometric, oblique and perspective. Figure 2.23 also shows a front view in each case for comparison, and brings out the fact the axonometric projection is a special case of orthographic projection.

A Pictorial drawing provides a 3D image to help understand the shape of an object or to assist in interpreting a drawing. There are three main ways to draw a pictorial drawing:

1. Isometric
2. Oblique
3. Other methods of Pictorial drawings include dimetric drawing, trimetric drawing

1. ISOMETRIC DRAWING

Definition – An Isometric drawing is a pictorial representation of an object in which all three dimensions are drawn at full scale. The term isometric means “equal measurement”.

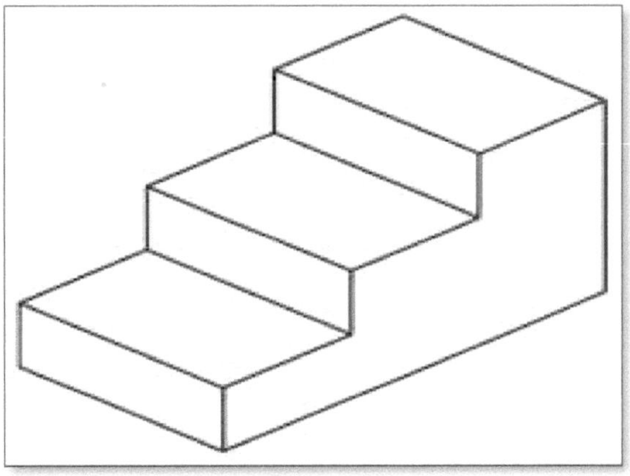

Figure 2.23

PICTORIAL DRAWING ISOMETRIC

Isometric steps

Isometric drawing is the most commonly used method of pictorial drawing. Isometric drawings are built on three lines, called isometric axes. One is drawn vertically and the other two with the 30° set square either side of it. An Isometric drawing is a way of presenting designs in three dimensions (3D). They can be completed using the 30/60 set square or freehand.

An Isometric drawing is one of a family of three-dimensional views called pictorial drawings. In an isometric drawing, the object's vertical lines are drawn vertically, and the horizontal lines in the width and depth planes are shown at 30 degrees to the horizontal.

The true dimension of the object is used to construct the drawing. These dimensions can be taken from either drawing completed in orthographic or by measuring an object. Isometric drawings or images have become the industry standard for parts manuals, technical proposals, patent illustrations and maintenance publications due to their use of true length and the ability for untrained people to understand. An isometric projection results if the plane is oriented so that it makes equal angles with the three principal planes of the object. The representation of the object seen below is an isometric drawing of a cube.

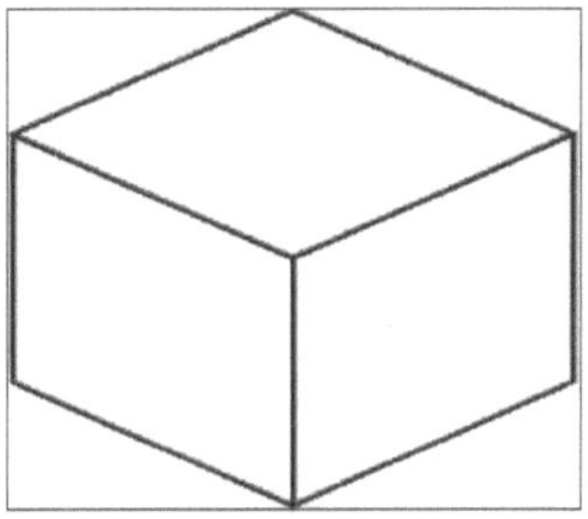

Figure 2.24

ISOMETRIC CUBE

In this drawing the three visible faces appear as equilateral parallelograms; that is, while all of the parallel edges of the cube are projected as parallel lines, the horizontal edges are drawn at an angle (usually 30°) from the normal horizontal axes, and the vertical edges, which are parallel to the principal axes, appear in their true proportions.

2. OBLIQUE PROJECTION

Oblique projection is probably the simplest method of producing a pictorial drawing because surfaces are directly in front of the viewer will look very similar to the front view in an orthogonal drawing.

It is a method of drawing objects in three dimensions and may be drawn at any angle although 45^0 is generally used. It is quite a simple technique compared to isometric or even perspective drawing. However, to draw accurately in oblique projection traditional drawing equipment is needed. Oblique projections illustrate the general 3-dimensional shape of the object. However only faces of the object parallel to the plane of projection are shown at their true size and shape. In fact, the oblique projection of these faces is the same as an orthographic front view.

Cavalier and Cabinet pictorial drawings of a cube:

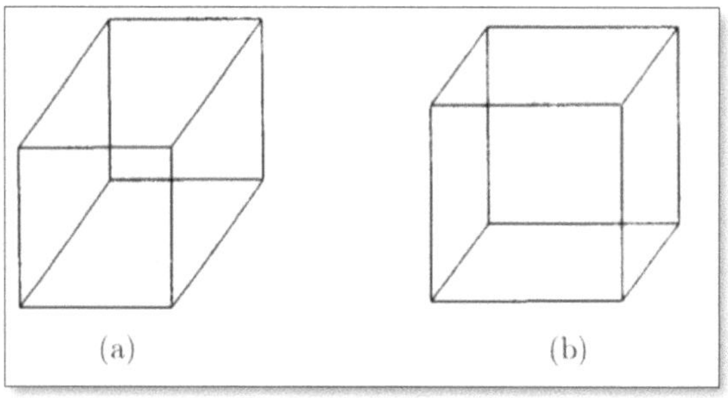

Figure 2.25

Figure 2.25 (a) Cavalier (b) Cabinet

There are two specific forms of Oblique drawings. (Figure 2.25) These are Cavalier and Cabinet projection. The difference between the two is how the depth of the object is represented.

A cavalier projection is obtained when the angle between oblique projectors and the plane of projection is 45° and the rear projections are true to size. The resulting figure appears too thick and distorted. A cabinet projection is used to correct this deficiency. Cabinet projection is an oblique projection where the length of the lines projecting backwards are halved. Figure 2.25 shows a drawing of a cube. Diagram (a) has been drawn in Cavalier and diagram (b) is in Cabinet.

Problem 2: From the given figure 2.26 of machine element. Draw the first view, right hand view and top view according to the first angle projection.

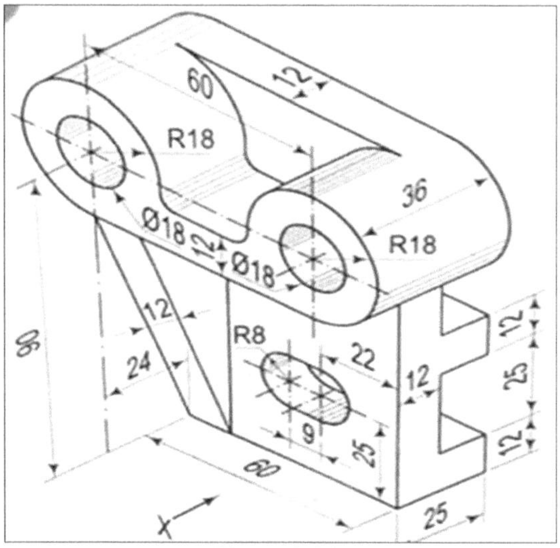

Figure 2.26

Solution:

Step 1: Draw the front view of the machine element.

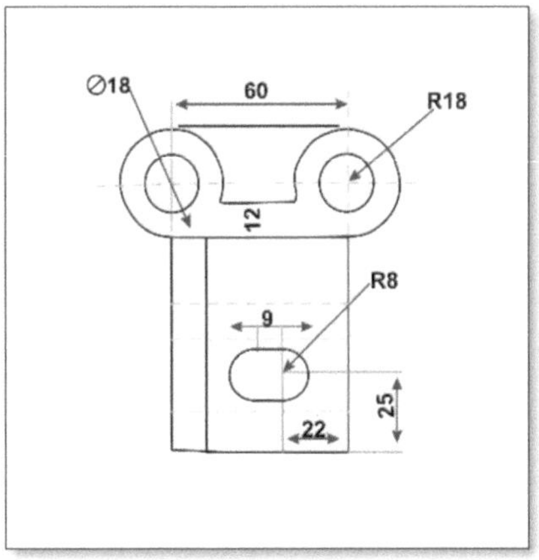

Figure 2.27

Step 2: Draw the right hand side view of the machine element

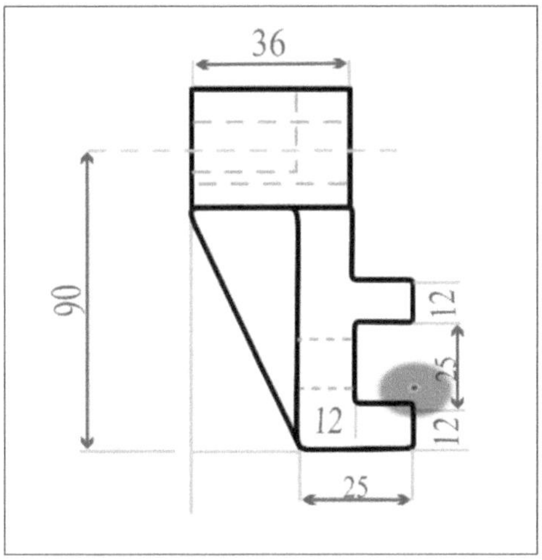

Figure 2.28

Step 3: Draw the Top view of the machine element.

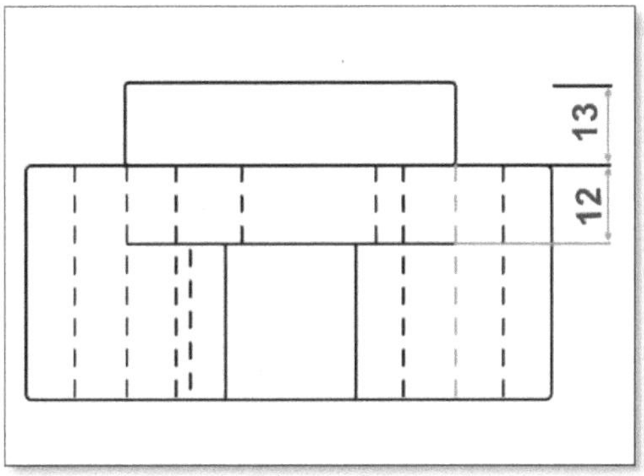

Figure 2.29

Problem 3: Draw the pictorial view of the given figure 2.30

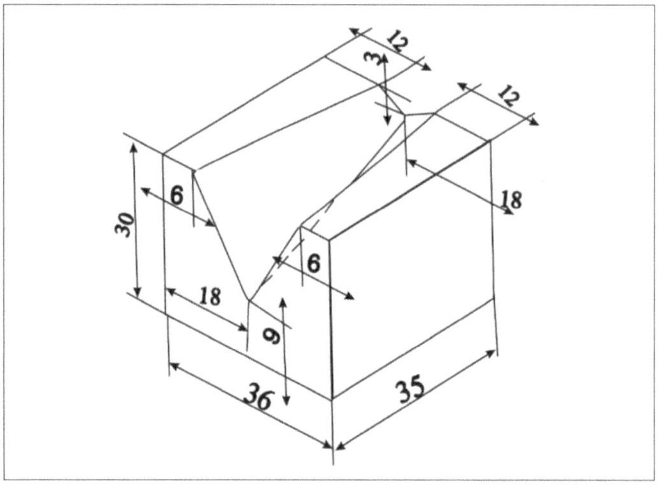

Figure 2.30

Solution: The Conversion of the view is given below:

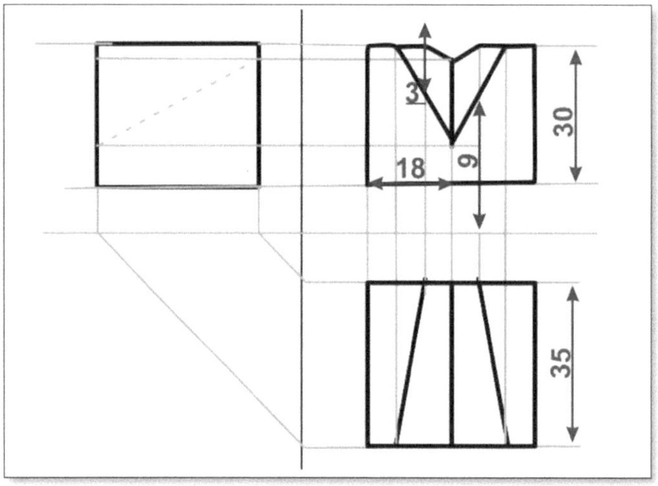

Figure 2.31

2.3 PROJECTION OF POINT, LINE, PLANE, SOLID

The projections are shown on a flat surface in their respective positions either above or below or in xy. Define its position with respect to the coordinates. With respect to the VP, HP, & P.

2.3.1 INTRODUCTION

A point may be situated, in space, in any one of the four quadrants formed by the two principal planes of projection or may lie in any one or both of them. Its projections are obtained by extending projectors perpendicular to the planes. One of the planes is then rotated so that the first and third quadrants are opened out.

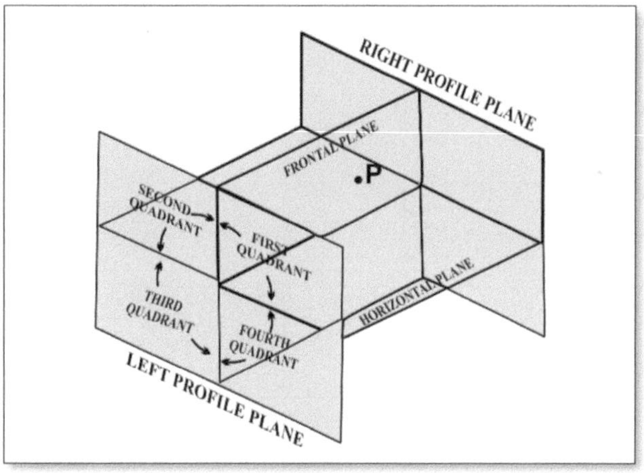

Figure 2.32

PROJECTION OF A POINT IN FIRST QUADRANT

Consider a point A placed in the first quadrant. This is at a height h mm above HP, at a distance d mm in front of VP. Its front view a' is projected onto VP and the top view a is projected onto HP. Now the HP is rotated in the clockwise direction for 900 and is obtained in vertical position. The projections will be seen as given in Figure 2.33

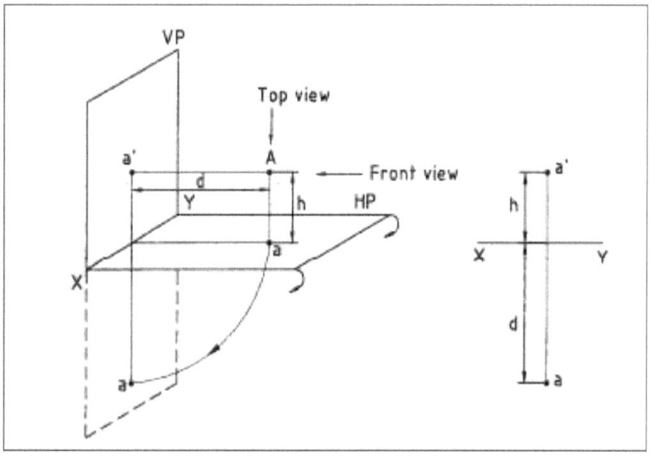

Figure 2.33

It is drawn with reference to XY line. Mark a point a' at a height h mm above XY, and at a distance d mm below XY. The projector joining a' and a is always perpendicular to XY.

PROJECTION OF POINT IN THIRD QUADRANT

Consider a point C placed in the third quadrant. The point is at a height h mm below HP and a distance d mm behind VP. Its front view c' is projected onto VP and the top view c is projected onto HP.

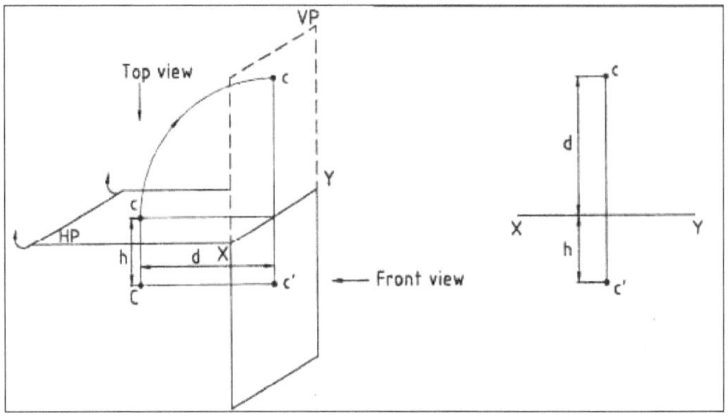

Figure 2.34

Now the HP is rotated in the clockwise direction for 900 and is obtained in vertical position. The projections will be seen as given in Figure 2.34. It is drawn with reference to XY line. Mark a point c' at a height h mm below XY and c at a distance d mm above XY.

Problem 4: A point A is 20 mm above HP and 30 mm in front of VP. Draw its projection.

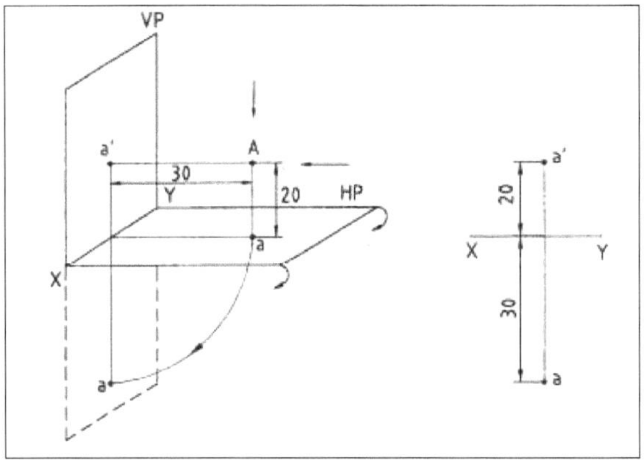

Figure 2.35

Solution:

To draw the projections:

1. Draw the reference line XY.
2. Mark a point a' at a distance of 20 mm above XY. Through this point draw a perpendicular line to XY and mark the top view at a distance of 300 mm below XY.
3. A point C is 20 mm below HP and 30 mm behind VP.
4. Draw its projection.

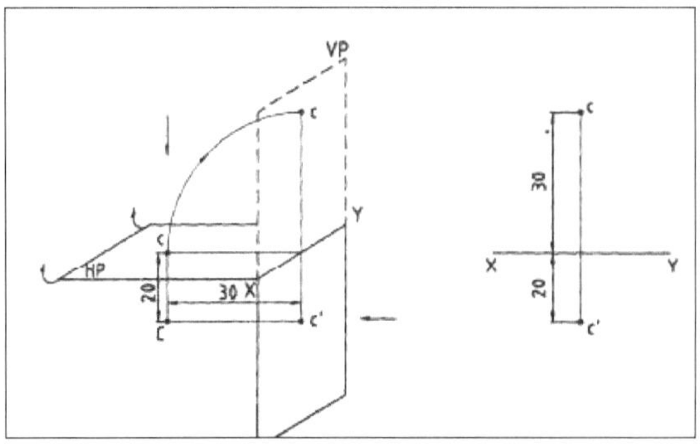

Figure 2.36

2.3.2 PROJECTION OF STRAIGHT LINES

A straight line is the shortest route to join any two given points. It is a one-dimensional object having only length (l). The projection of straight line is obtained by joining the top and front views of the respective end points of the line. The actual length of the straight line is known as true length (TL).

Example:

Hold a pencil in your hand and open a note-book in such a way that some pages are on the table and some pages are held perpendicular to the table top. Consider these as two planes HP and VP.

Put one end of the pencil on the HP and let it make some angle with HP. Make the pencil parallel to VP (imagine the pencil to be a line).

Under the condition if the line is projected on HP and VP, its projections will be as shown in following figure 2.37

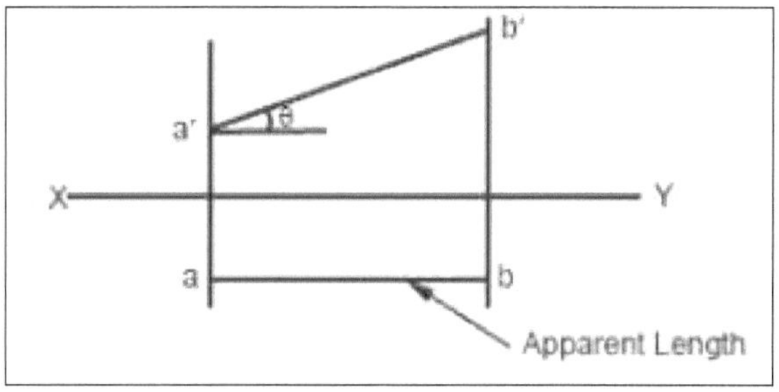

Figure 2.37

The elevation of the line a′b′ will be true length, the angle θ will also be true angle but plan ab will be shorter than true length. The length ab is a function of angle θ. If θ is increased, ab will be reduced further. Now consider the pencil parallel to HP and making some angle φ with VP. Its projections will be as shown in following figure.

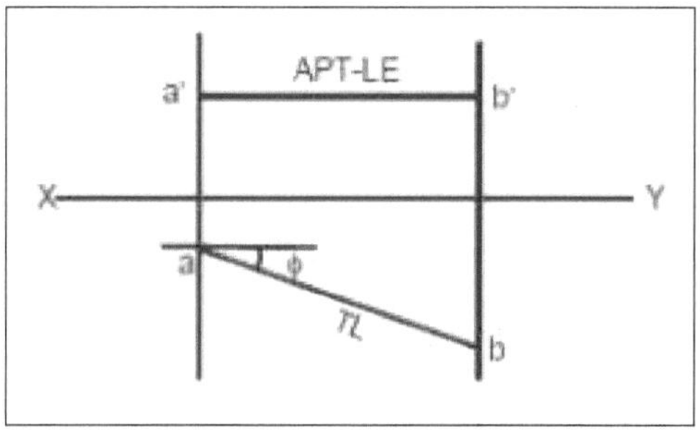

Figure 2.38

The length of plan is True length TL, the angle made by this plan with VP, i.e., with XY line is true angle φ, and the length of elevation is a function of angle φ. Elevation is parallel to XY (in second case, the line is given parallel to HP, while in this case the elevation is a line parallel to XY and the plan of the line is TL, the angle made φ is true angle).

From the example, we see that in first case when the line was parallel to VP, the elevation of the line is TL, the angle made, i.e., θ, it is also true angle and the plan is a line parallel to X-Y.

Hence, learn a rule that if the projections of the line are given and if one of the views (plan or elevation) is parallel to XY, the other view MUST BE true length and angle made by this TL will be true angle. Its reverse is also true, i.e., from among the views of a line, if one view is TL, the other view MUST BE parallel to X-Y.

A straight line is placed with reference to the planes of projections in the following positions:

1. Line perpendicular to HP and parallel to VP.
2. Line perpendicular to VP and parallel to HP.
3. Line parallel to both HP and VP.
4. Line inclined to HP and parallel to VP.
5. Line inclined to VP and parallel to HP.
6. Line inclined to both HP and VP.

In first angle projection method, the line is assumed to be placed in the first quadrant. The projections of the line in the above mention positions are discusses below.

Projection of a Line kept Perpendicular to HP and Parallel to VP

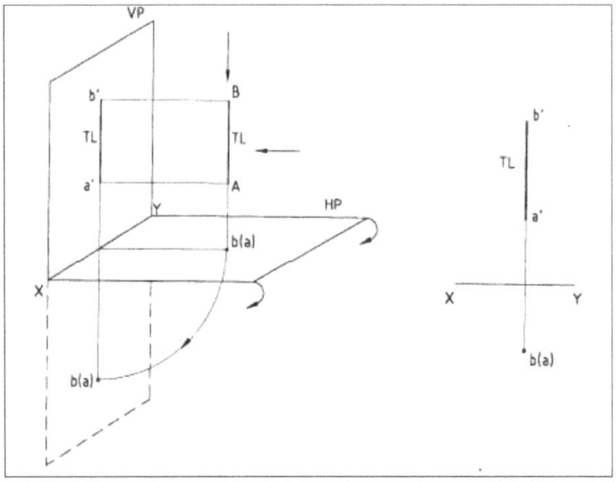

Figure 2.39

Consider a straight line AB kept perpendicular to HP and parallel to VP as shown in Figure 2.39. Its front view is projected onto VP which is having true length. The top view is projected onto HP which is a point end b which is visible while the other end a is invisible and is enclosed within ( ).

Now the HP is rotated in the clockwise direction through 900 and is obtained in the vertical position. The projections obtained are seen as given in Figure 2.39. It is drawn with reference to the XY line as follows:

1. Draw the XY line.
2. Draw the front view a'b'. which is a line perpendicular to XY and having true length (TL).
3. Projected the top view ab. The end b is visible and the invisible end a is marked inside ( ).

Projection of a Line kept Perpendicular to VP and Parallel to HP

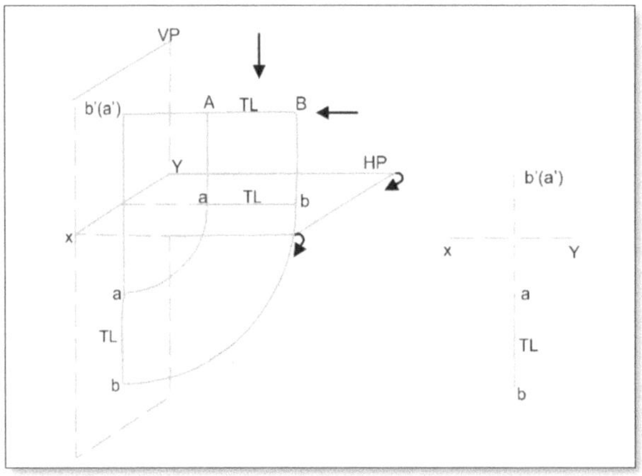

Figure 2.40

Consider a straight line AB kept perpendicular to VP and parallel to HP Figure 2.40. Its top view is projected onto HP which is a line having true length (TL).

The front view is projected onto VP which is point, the end b' of which is visible and the other end a' is invisible which is shown enclosed in ( ).

Now the HP is rotated in the clockwise direction through 900 and is obtained in the vertical position. The projections obtained are seen as given in Figure 2.40. It is drawn with reference to the XY line as follows:

1. Draw the XY line.
2. Draw the top view ab, a line perpendicular to XY and having true length (TL).
3. Projected the front view a'b'. The end b' is visible ad the invisible end a' is marked inside ( ).

Projection of a Line kept Parallel to Both HP and VP

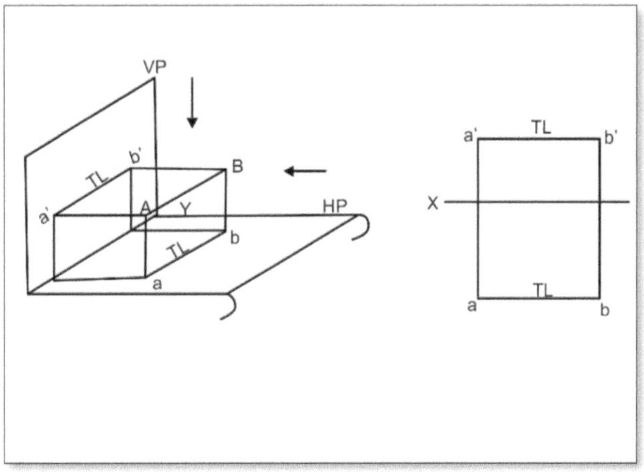

Figure 2.41

Consider a straight line AB kept parallel to both HP and VP Figure 2.41. Its front view is projected onto VP which is a line having true length (TL). The top view is projected onto HP which is also a line having true length.
Now the HP is rotated in the clockwise direction through an angle of 900 and is obtained in the vertical position. The projections obtained are seen as given in Figure 2.41. It is drawn with references to the XY line as follows:
1. Draw the XY line.
2. Draw the front view a'b', a line parallel to XY and having true length (TL)
3. Project the top view ab which is also a line parallel to XY having true length (TL).

Projection of a Line kept inclined to HP and Parallel to VP
Consider a straight line AB kept inclined to HP and parallel to VP Figure 2.42. Its front view is projected onto VP which is an inclined line at an angle θ to XY and having true length (TL). The top view is projected onto HP which is also a line but smaller than the true length and parallel to XY. The inclination of the line with HP is always represented by the symbol θ.

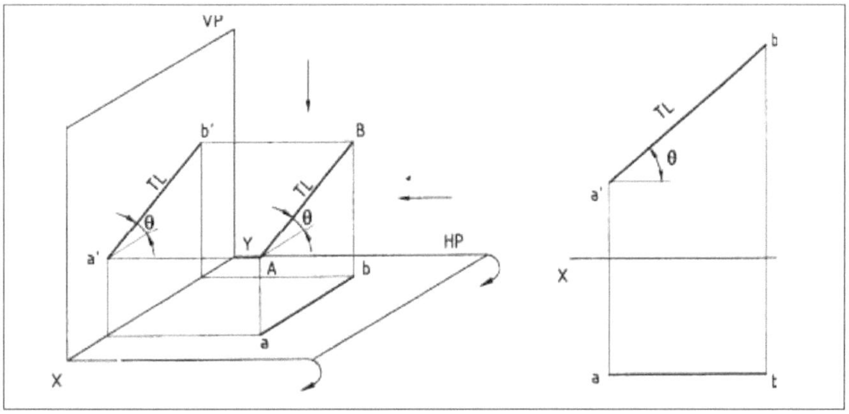

Figure 2.42

Now the HP is rotated in the clockwise direction through 900 and is obtained in the vertical position. The projections obtained are seen as given in Figure 2.42. It is drawn with reference to the XY line as follows:
1. Draw the XY line.

2. Draw the front view a'b', a line inclined at an angle θ to XY having true length (TL).
3. Project the top view ab which is also a line parallel to XY and smaller than true length.

Projection of a Line kept Inclined to VP and Parallel to HP

Consider a straight line AB kept inclined to VP and parallel to HP Figure 2.37. Its top view is projected onto HP which is a line inclined at an angle ϕ to XY and having true length (TL). The front view is projected onto VP which is also a line but smaller than true length and is parallel to XY. The inclination of the line with VP is always represented by the symbol ϕ.

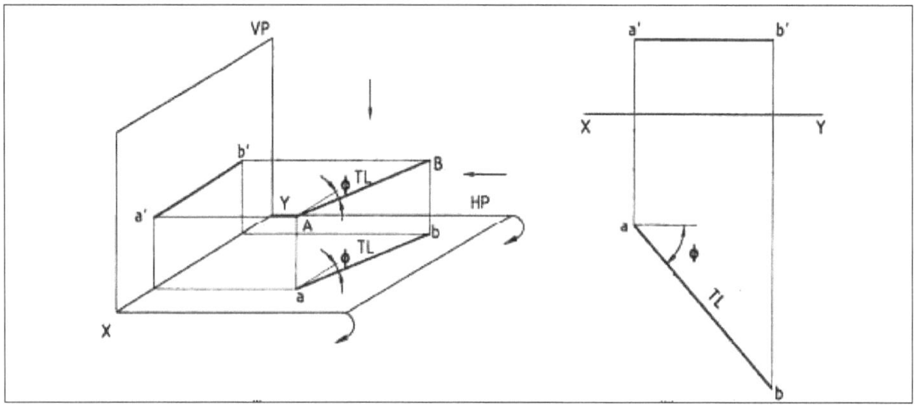

Figure 2.43

Now the HP is rotated in the clockwise direction through an angle of 900 and is obtained in the vertical position. The projections obtained are seen as given in Figure 2.43. It is drawn with reference to the XY line as follows:

1. Draw the XY line
2. Draw the top view ab, a line inclined at an angle ϕ to XY and having true length (TL).
3. Project the front view a'b', which is also a line parallel to XY but smaller than true length.

2.3.2.1 CLASSIFICATION OF LINES

The sorts of lines suggested by the Bureau of Indian Standards and displayed in figure 2.44 must be used for general engineering drawings. From any of the six groups listed in Table 1, the line thickness must be selected based on the kind and size of the drawing.

The conventions adopted must be explained in notes on the relevant drawing in cases where other types of thickness of lines are used for special cases, such as electrical and piping drawings, or if the lines specified in table 2 are used for applications other than those listed in the last column.

TYPES OF LINES USED IN ENGINEERING DRAWING

FOLLOWING ARE THE DIFFERENT **TYPES OF LINES** USED IN ENGINEERING DRAWING:

A type – Continuous Thick
B type – Continuous THIN
C type – Continuous THIN Freehand
D type – Continuous THIN Zig-Zag
E type – Dashes THICK
F type – Dashes THIN
G type – Chain Thin
H type – Chain THIN and THICK
J type – Chain THICK
K type – Chain THIN Double Dash

TYPE A

Outline of Parts

Outline of Parts type lines represent the visible edges of the objects, hence should be outstanding in appearance and therefore, are drawn as bold i.e., thick continuous lines.

TYPE B

Dimension, Projection, Leader, Hatching Lines

Dimension, Projection, Leader, hatching type lines must be drawn thin and continuous. The extension lines for dimensioning should run from the outlines without leaving a gap and extend beyond the dimension lines. This type is also used to draw outlines of adjacent and revolved sections.

TYPE A

Outline of Parts

Outline of Parts type lines represent the visible edges of the objects, hence should be outstanding in appearance and therefore, are drawn as bold i.e., thick continuous lines.

TYPE B

Dimension, Projection, Leader, Hatching Lines

Dimension, Projection, Leader, hatching type lines must be drawn thin and continuous. The extension lines for dimensioning should run from the outlines without leaving a gap and extend beyond the dimension lines. This type is also used to draw outlines of adjacent and revolved sections.

TYPE A

Outline of Parts

Outline of Parts type lines represent the visible edges of the objects, hence should be outstanding in appearance and therefore, are drawn as bold i.e., thick continuous lines.

TYPE B

Dimension, Projection, Leader, Hatching Lines

Dimension, Projection, Leader, hatching type lines must be drawn thin and continuous. The extension lines for dimensioning should run from the outlines without leaving a gap and extend beyond the dimension lines. This type is also used to draw outlines of adjacent and revolved sections.

TYPE J

Lines to indicate Surfaces which require Additional Treatment

These are long thick lines with alternatively long and short dashes of proportion ranging from 6:1 to 4:1 and evenly spaced. These lines are drawn to indicate surfaces, to receive additional treatment like anodizing, plating, etc.

TYPE K

Lines to indicate Outlines of Adjacent parts, Extreme positions of Movable Parts, Centroidal lines, Parts situated in front of the Cutting Planes, Initial outlines prior to forming.

	Line	Description	General Application
A		Continuous thick	A1 Visible outlines. A2 Visible edges.
B		Continuous thin (straight or curved)	B1 Imaginary lines of intersection. B2 Dimension lines. B3 Projection lines. B4 Leader lines. B5 Hatching lines. B6 Outlines of revolved sections in place. B7 Short centre lines
C		Continuous thin free hand	C1 Limits of partial or interrupted views and sections, If the limit is not a chain thin.
D		Continuous thin (straight) with zigzags	D1 Long break line
E		Dashed thick	E1 Hidden outlines. E2 Hidden edges.
F		Dashed thin	F1 Hidden outlines. F2 Hidden edges.
G		Chain thin	G1 Center lines. G2 Lines of symmetry. G3 Trajectories
H		Chain thin, thick at ends and changes of direction	H1 Cutting planes.
J		Chain thick	J1 Indication of lines or surfaces to which a special requirement applies
K		Chain thin double dashed	K1 Outlines of adjacent parts. K1 Alternative or extreme position of movable parts. K3 Centroidal lines. K4 Initial outlines prior to forming K5 Parts situated in front of the cutting plane

Figure 2.44

These are thin lines with a long and two short dashes alternately and evenly placed in the proportion ranging from 6:1 to 4:1. These lines are used to represent the outlines of adjacent parts, extreme positions of movable parts in the assembly drawings, parts situated in front of the cutting planes, initial outlines prior to forming, centroidal lines etc.

2.3.2.2 TRACE OF A LINE

The point of intersection or meeting of a line with the reference plane, extended if necessary, is known as the trace of a line. The point of intersection of a line with the HP is known as the horizontal trace, represented by HT and that with the VP is known as the vertical trace, represented by VT. No trace is obtained when a line is kept parallel to a reference plane.

If the line is given parallel to a plane, it will never intersect that plane and, therefore, no trace of the line on that plane.

If the line is given parallel to VP and inclined to HP, only HT will be obtained and no VT.
If the line is given parallel to HP and inclined to VP, only VT will be obtained and no HT. If the line is given parallel to both the planes, neither HT nor VT will be obtained.

Rule For obtaining HT of an inclined line AB:
For obtaining HT of an inclined line AB, inclined to HP, extend the elevation a′b′ towards XY line till it intersects X-Y (In some problems, depending upon the position of the given line, the elevation might intersect X-Y line in natural course and it need not be extended. This may happen when one end of the line is above HP and the other end below HP. In this also the point of intersection of elevation with X-Y will be h′. The point of intersection of the extended elevation with X-Y is h′.

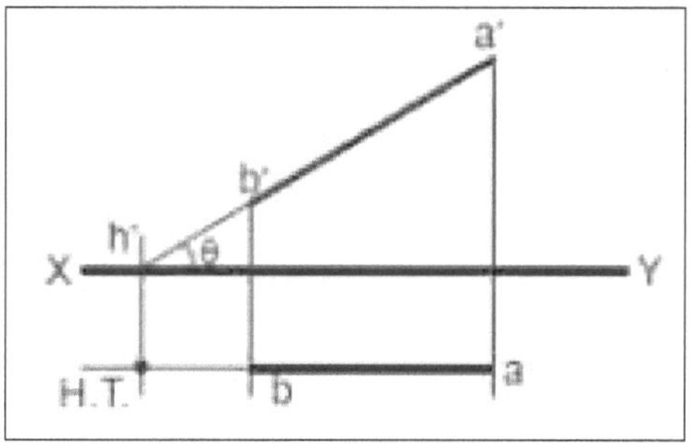

Figure 2.45

Now from h′, draw a line perpendicular to X-Y. Then extend the plan ab in its own direction, till it intersects the perpendicular from h′.
Rule For obtaining VT of an inclined line AB:
The point of intersection of extended plan ab (in some cases the plan need not be extended but the perpendicular from h′ will intersect the plan ab) and the perpendicular from h′ will be HT.
For obtaining the VT the procedure is reversed:

(a) Extend the plan towards X-Y line till it intersects X-Y. The point of intersection is v.
(b) Draw a line perpendicular to X-Y from h.
(c) Extend the elevation in its own direction, so that it intersects the perpendicular drawn from h.

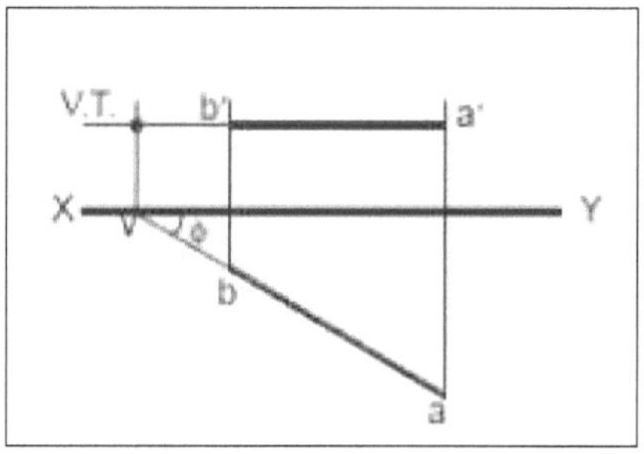

Figure 2.46

(d) The point of intersection is VT. (In some cases, the elevation need not be extended and the perpendicular from h will intersect the elevation. This point of intersection is VT)

Problem 5: A line AB 60 mm long has its end 20 mm above HP and 30 mm in front of VP. The line is kept perpendicular to HP and parallel to VP. Draw its projections. Also mark the traces.
Solution: Assume that end A of the line is nearer to HP. The front view a’ b’ is a line having true length. The top view is a point, the end b of which is visible and a is invisible which is enclosed in (). The line is extended to meet HP to obtained the horizontal trace (HT). No vertical trace (VT) is obtained because the line is kept parallel to VP.

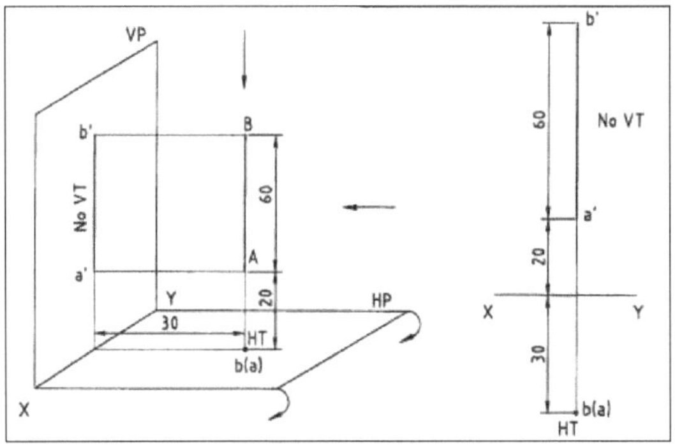

Figure 2.47

The projections obtained are drawn with reference to XY line as shown in Figure 2.47(You should not drawn diagram (i) in the examination. This diagram is given for explanation only.)

1. Mark the projections of the end A by considering it as a point. Its front view a' is 20 mm above XY and the top view a is 30 mm below XY.
2. The front view of the line a' b' is obtained by drawing a line perpendicular to XY from a' and having a length of 60 mm.
3. Top view of the line is obtained by projecting the other end b which coincides with a. The invisible end a is enclosed in ( ).
4. The horizontal trace (HT) is marked coinciding with the top view of the line. NO vertical trace (VT) is obtained.

Problem 6: A line AB 60 mm long has its end A 20 mm above HP and 30 mm in front of VP. The line is inclined at 400 to VP and parallel to HP. Draw its projections. Also mark the traces.

Solution: The top view ab is a line inclined at an angle of 400 to XY and having true length. Its front view a' b' is parallel to XY and smaller than true length. The line is extended to meet VP to obtained the vertical trace (VT). No horizontal trace (HT) is obtained because the line is kept parallel to HP. The projections obtained are drawn with reference to XY line as shown in Figure 2.48

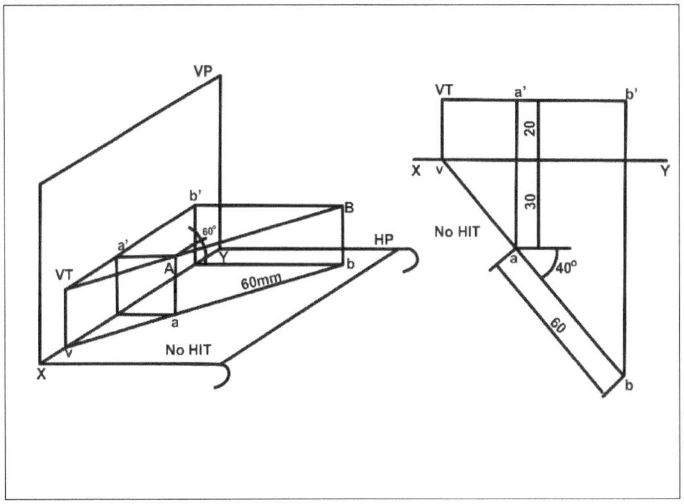

Figure 2.48

1. Mark the projections of end A by considering it as a point. Its front view a' is 20 mm above XY and top view a is 30 mm below XY.
2. The top view of the line ab is obtained by drawing a line inclined at an angle 400 to XY from a and having a length of 60 mm.
3. The front view of the line a' b' is obtained by drawing a line parallel to XY from a' and drawing a vertical line (projector) from b. It is parallel to XY and smaller than the true length.

4. To mark the vertical trace (VT) the top view of the line ab is extended to intersect with XY line at v. Then by drawing a vertical line from v and a horizontal line from a' b', the VT is located. No horizontal trace (HT) is obtained.

Problem 7: A line AB 70 mm long has its end A 15 mm above HP and 25 mm in front of VP. Its top view (plan) has a length of 40 mm. Draw its projections and find the inclination of the line with HP.

Solution: The projections of the line are drawn with reference to the XY lien as follows:

1. Mark the projections of end A by considering it as a point. Its front view a' is 15 mm above XY and top view a is 25 mm below XY.
2. The top view of the line ab is drawn parallel to XY to the given length of 40 mm.
3. Draw a vertical line (projector) from b.
4. Using true length 70 mm as radius and a' as centre, mark a point in the vertical line to get b'.
5. The inclination of a' b' with XY is measured to get the angle.

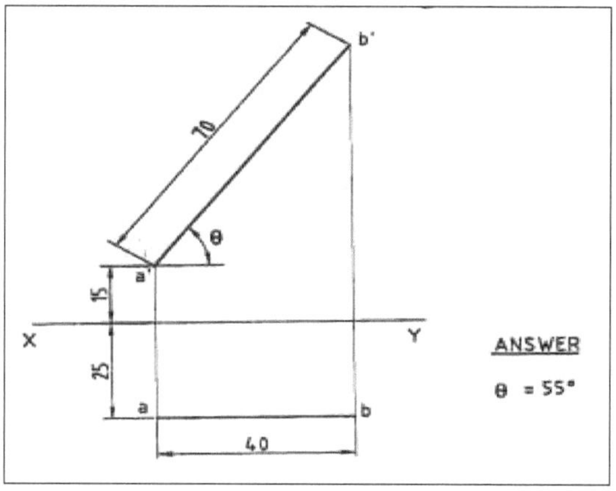

Figure 2.49

Projection of a Line kept Inclined to Both HP and VP

When a line is placed inclined to both HP and VP, its projections obtained in top and front views are smaller than the true length of the line and inclined to the XY line. So, it is impossible to project and draw the top or front view of the line directly. Any one of the following methods may be used to draw the projections:

1. Rotating line method
2. Rotating trapezoidal plane method
3. Auxiliary plane method

PROJECTIONS OF PLANES

A plane is a 2-dimensional object having length and breadth only. Its thickness is always neglected. Various shapes of plane figures are considered such as square, rectangle, circle, pentagon, hexagon, triangle, etc.

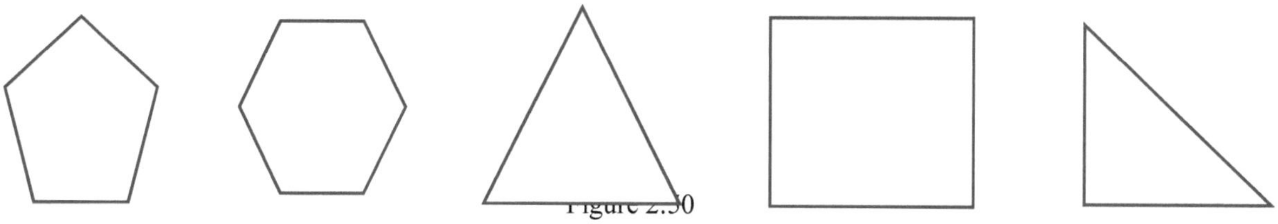

Figure 2.50

In this topic various plane figures are the objects.

What is usually asked in the problem?

To draw their projections means F.V, T.V. & S.V

What will be given in the problem?

1. Description of the plane figure. 2. It's position with HP and VP.

CASE OF A RECTANGLE – OBSERVE AND NOTE ALL STEPS

SURFACE PARALLEL TO HP PICTORIAL PRESENTATION, SURFACE INCLINED TO HP PICTORIAL PRESENTATION and SURFACE INCLINED TO HP PICTORIAL PRESENTATION

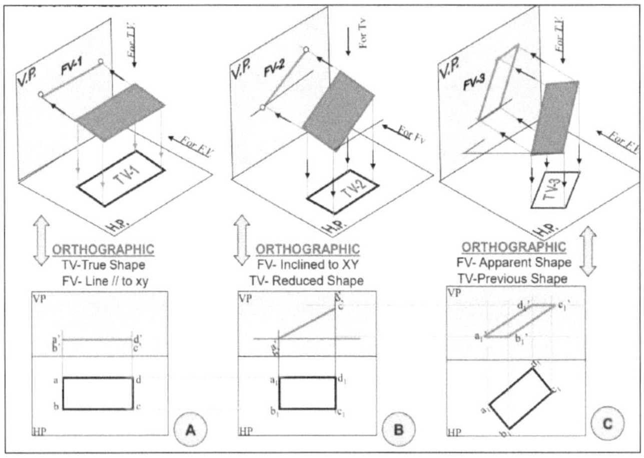

Figure 2.51

Problem 8: Rectangle 30 mm and 50 mm sides is resting on HP on one small side which is 30 inclined to VP, while the surface of the plane makes 45 inclination with HP. Draw it's projections.

Solution:

Read problem and answer following questions:

1. Surface inclined to which plane?

Ans. Towards HP

2. Assumption for initial position?

Ans Towards HP

3. So which view will show True shape?

Ans. Towards TV

4. Which side will be vertical?

Ans. One small side. Hence begin with TV, draw rectangle below X-Y drawing one small side vertical.

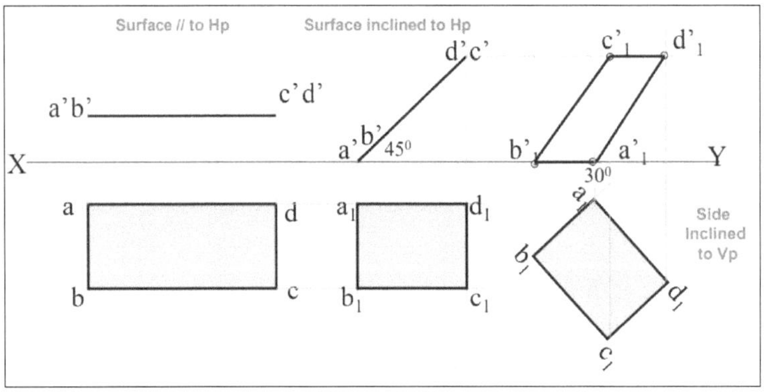

Figure 2.52

Problem 9: A regular pentagon of 30 mm sides is resting on HP on one of its sides with its surface 45 inclined to HP. Draw it's projections when the side in HP makes 30 angle with VP.

Read problem and answer following questions:

1. Surface inclined to which plane?

Ans. Towards HP

2. Assumption for initial position?

Ans. Towards HP

3. So which view will show True shape?

Ans. Towards TV

4. Which side will be vertical?
Ans. any side.

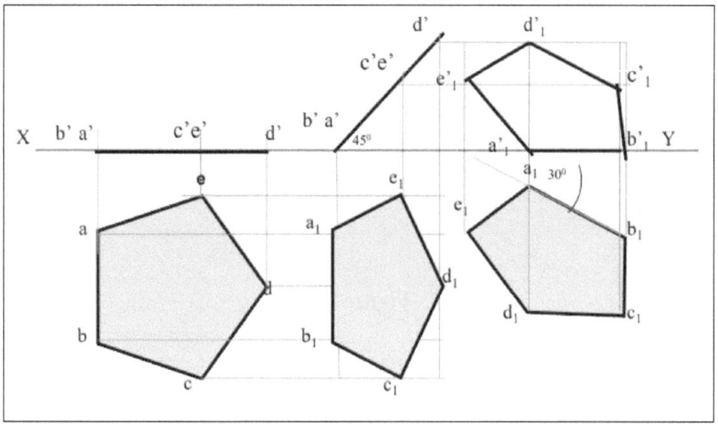

Figure 2.53

2.3.3 TYPES OF PLANES

Planes may be divided into two main types:
(1) Perpendicular planes.
(2) Oblique planes.

(1) PERPENDICULAR PLANES

These planes can be divided into the following sub-types:

(i) Perpendicular to both the reference planes.
(ii) Perpendicular to one plane and parallel to the other.
(iii) Perpendicular to one plane and inclined to the other

(i) Perpendicular to both the reference planes: A square ABCD is perpendicular to both the planes. Its H.T and V.T are in a straight line perpendicular to xy.

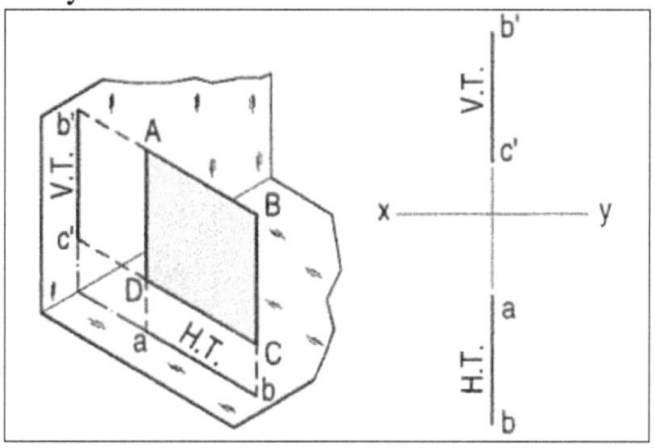

Figure 2.54

The front view b'c' and the top view ab of the square are both lines coinciding with the V.T. and the H.T. respectively.

(ii) Perpendicular to one plane and parallel to the other
(a) Plane, perpendicular to the H.P. and parallel to the V.P. A triangle PQR is perpendicular to the H.P. and is parallel to the V.P. Its H.T. is parallel to xy. It has no V.T. The front view p'q'r' shows the exact shape and size of the triangle. The top view pqr is a line parallel to xy. It coincides with the H.T.
(b) Plane, perpendicular to the V.P. and parallel to the H.P. A square ABCD is perpendicular to the V.P. and parallel to the H.P. Its V.T. is parallel to xy. It has no H.T. The top view abed shows the true shape and true size of the square. The front view a'b' is a line, parallel to xy. It coincides with the V.T.

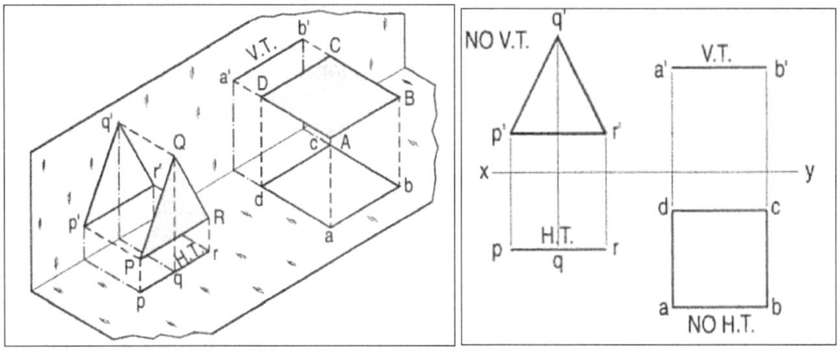

Figure 2.55

(iii) Perpendicular to one plane and inclined to the other

(a) Plane, perpendicular to the H.P. and inclined to the V.P in Figure 2.38. A square ABCD is perpendicular to the H.P. and inclined at an angle 0 to the V.P. Its V.T. is perpendicular to xy. Its H.T. is inclined at 0 to xy. Its top view ab is a line inclined at 0 to xy. The front view a'b'c'd' is smaller than ABCD.

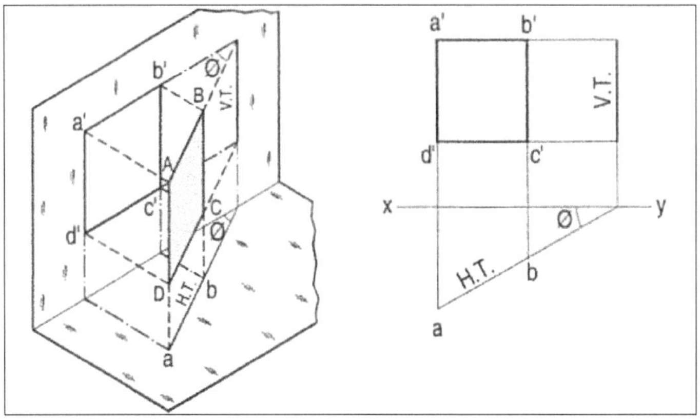

Figure 2.56

(b) Plane, perpendicular to the V.P. and inclined to the H.P. (figure 2.56) A square ABCD is perpendicular to the V.P. and inclined at an angle e to the H.P. Its H.T. is perpendicular to xy. Its V.T. makes the angle e with xy. Its front view a' b' is a line inclined at xy. The top view abed is a rectangle which is smaller than the square ABCD.

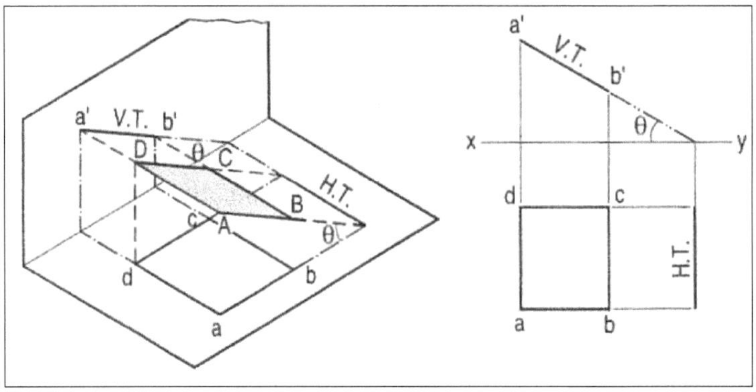

Figure 2.57

(2) OBLIQUE PLANES

Planes which are inclined to both the reference planes are called oblique planes. Representation of oblique planes by their traces is too advanced to be included in this book. A few problems on the projections of plane figures inclined to both the reference planes are however, illustrated at the end of the chapter. They will prove to be of great use in dealing with the projections of solids.

When a plane has its surface inclined to one plane and an edge or a diameter or a diagonal parallel to that plane and inclined to the other plane, its projections are drawn in different stages.

(1) If the surface of the plane is inclined to the H.P. and an edge (or a diameter or a diagonal) is parallel to the H.P. and inclined to the V.P.,

(i) in the initial position the plane is assumed to be parallel to the H.P. and an edge perpendicular to the V.P.

(ii) It is then tilted so as to make the required angle with the H.P. As already explained, its front view in this position will be a line, while its top view will be smaller in size.

(iii) In the final position, when the plane is turned to the required inclination with the V.P., only the position of the top view will change. Its shape and size will not be affected. In the final front view, the corresponding distances of all the corners from xy will remain the same as in the second front view.

If an edge is in the H.P. or on the ground, in the initial position, the plane is assumed to be lying in the H.P. or on the ground, with the edge perpendicular to the V.P. If a corner is in the H.P. or on the ground, the line joining that corner with the centre of the plane is kept parallel to the V.P.

(2) Similarly, if the surface of the plane is inclined to the V.P. and an edge (or a diameter or a diagonal) is parallel to the V.P. and inclined to the H.P.,

(i) in the initial position, the plane is assumed to be parallel to the V.P. and an edge perpendicular to the H.P.

(ii) It is then tilted so as to make the required angle with the V.P. Its top view in this position will be a line, while its front view will be smaller in size.

(iii) When the plane is turned to the required inclination with the H.P., only the position of the front view will change. Its shape and size will not be affected. In the final top view, the corresponding distances of all the corners from xy will remain the same as in the second top view.

Problem 10: A square ABCD of 50 mm side has its corner A in the H.P., its diagonal AC inclined at 30° to the H.P. and the diagonal BO inclined at 45° to the V.P. and parallel to the H.P. Draw its projections.

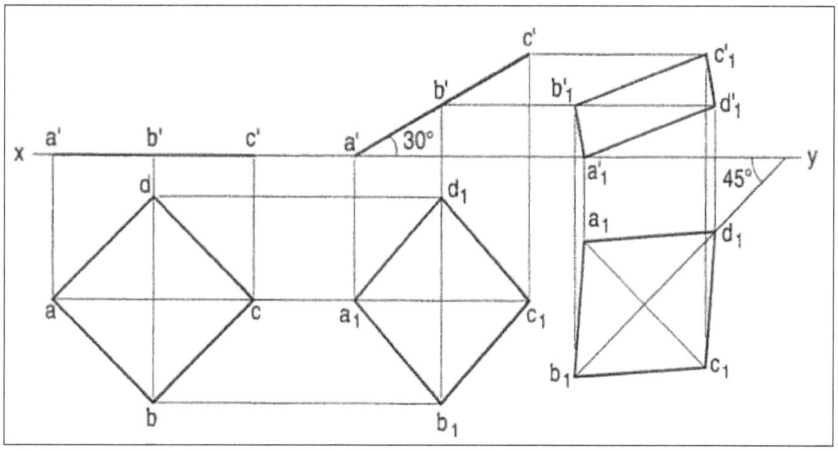

Figure 2.58

Solution:

In the initial stage, assume the square to be lying in the H.P. with AC parallel to the V.P.

(i) Draw the top view and the front view. When the square is tilted about the corner A so that AC makes 30° angle with the H.P., BO remains perpendicular to the V.P. and parallel to the H.P.

(ii) Draw the second front view with a'c' inclined at 30° to xy, keeping a' or c' in xy. Project the second top view. The square may now be turned so that BO makes 45° angle with the V.P. and remains parallel to the H.P. Only the position of the top view will change. Its shape and size will remain the same.

(iii) Reproduce the top view so that b1d1 is inclined at 45° to xy. Project the final front view upwards from this top view and horizontally from the second front view.

Problem 11: A thin 30°-60° set-square has its longest edge in the V.P. and inclined at 30° to the H.P. Its surface makes an angle of 45° vvith the V.P. Draw its projections.
Solution:
In the initial stage, assume the set-square to be in the V.P. with its hypotenuse perpendicular to the H.P.

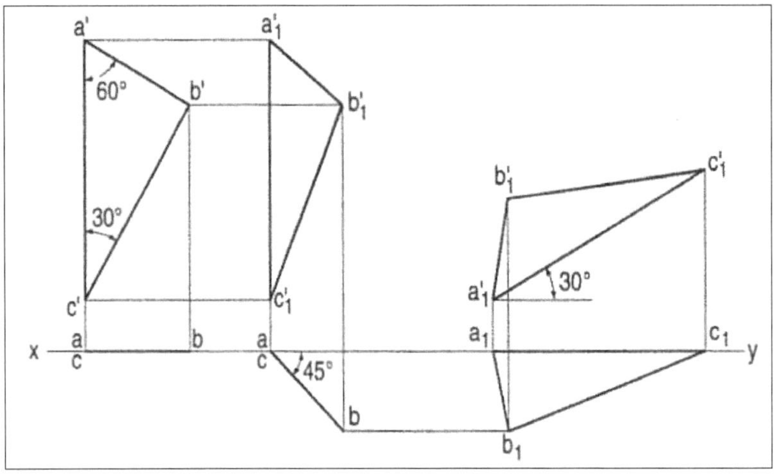

Figure 2.59

(i) Draw the front view a' b' c' and project the top view ac in xy.
(ii) Tilt ac around the end a so that it makes 45° angle with xy and project the front view a'1b'1c'.
(iii) Reproduce the second front view a'1b'1c'1 so that a'1b'1 makes an angle of 30° with xy. Project the final top view a1b1c1.

TRACES OF PLANES

A plane, extended if necessary, will meet the reference planes in lines, unless it is parallel to any one of them. These lines are called the traces of the plane. The line in which the plane meets the H.P. is called the horizontal trace or the H.T. of the plane. The line in which it meets the V.P. is called its vertical trace or the V.T. A plane is usually represented by its traces.

(1) TRACES

(a) When a plane is perpendicular to both the reference planes, its traces lie on a straight line perpendicular to xy.
(b) When a plane is perpendicular to one of the reference planes, its trace upon the other plane is perpendicular to xy (except when it is parallel to the other plane).
(c) When a plane is parallel to a reference plane, it has no trace on that plane. Its trace on the other reference plane, to which it is perpendicular, is parallel to xy.
(d) When a plane is inclined to the H.P. and perpendicular to the V.P., its inclination is shown by the angle which its V.T. makes with xy. When it is inclined to the V.P. and perpendicular to the H.P., its inclination is shown by the angle which its H.T. makes with xy.
(e) When a plane has two traces, they produce if necessary intersect in xy (except when both are parallel to xy as in case of some oblique planes).

(2) PROJECTIONS

(a) When a plane is perpendicular to a reference plane, its projection on that plane is a straight line.
(b) When a plane is parallel to a reference plane, its projection on that plane shows its true shape and size.
(c) When a plane is perpendicular to one of the reference planes and inclined to the other, its inclination is shown by the angle which its projection on the plane to which it is perpendicular, makes with xy. Its projection on the plane to which it is inclined, is smaller than the plane itself.

PROJECTION OF SOLIDS

A solid has three dimensions, viz. length, breadth and thickness. To represent a solid on a flat surface having only length and breadth, at least two orthographic views are necessary. Sometimes, additional views projected on auxiliary planes become necessary to make the description of a solid complete. This chapter deals with the following topics:
1. Types of solids.
2. Projections of solids in simple positions.
(a) Axis perpendicular to the H.P.
(b) Axis perpendicular to the V.P.
(c) Axis parallel to both the H.P. and the V.P. 3.
Projections of solids with axes inclined to one of the reference planes and parallel to the other.
(a) Axis inclined to the V.P. and parallel to the H.P.
(b) Axis inclined to the H.P. and parallel to the V.P. 4. Projections of solids with axes inclined to both the H.P. and the V.P. 5. Projections of spheres.

TYPES OF SOLIDS:
Solids may be divided into two main groups:
1. Polyhedra
2. Solids of revolution

POLYHEDRA
A polyhedron is defined as a solid bounded by planes called faces. When all faces are equal and regular, the polyhedron is said to be regular.
There are seven regular polyhedra which may be defined as stated below:
(i) Tetrahedron: It has four equal faces, each an equilateral triangle.
(ii) Cube or hexahedron: It has six faces, all equal squares.
(iii) Octahedron: It has eight equal equilateral triangles as faces.

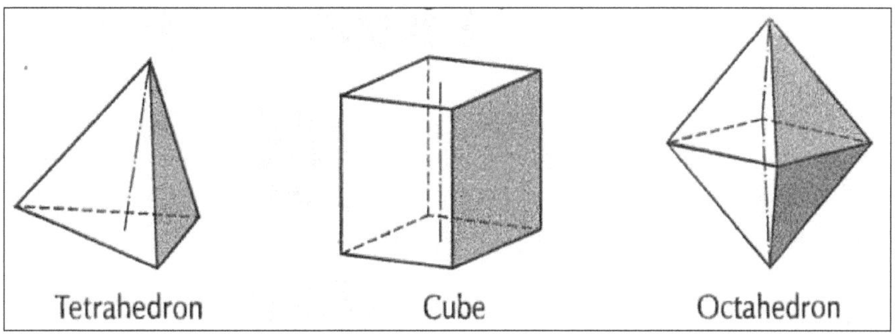

Figure 2.60

(iv) Dodecahedron: It has twelve equal and regular pentagons as faces.
(v) Icosahedron: It has twenty faces, all equal equilateral triangles.

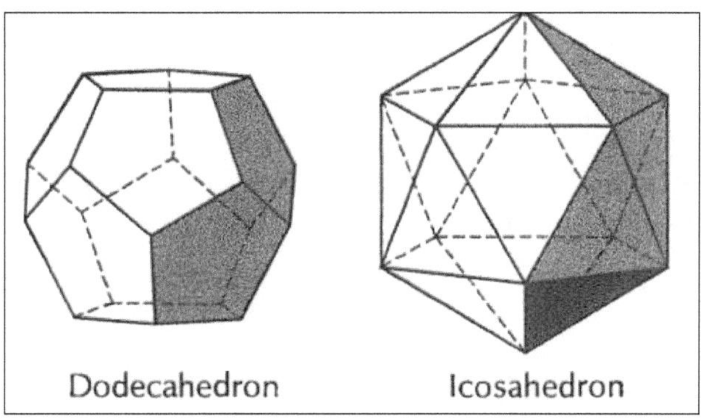

Figure 2.61

(vi) Prism: This is a polyhedron having two equal and similar faces called its ends or bases, parallel to each other and joined by other faces which are parallelograms. The imaginary line joining the centres of the bases is called the axis.

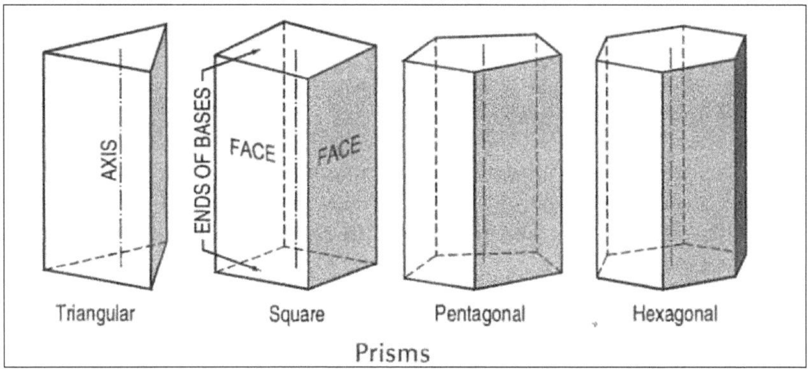

Figure 2.62

(vii) Pyramid: This is a polyhedron having a plane figure as a base and a number of triangular faces meeting at a point called the vertex or apex. The imaginary line joining the apex with the centre of the base is its axis. Prisms and pyramids are named according to the shape of their bases, as triangular, square, pentagonal, hexagonal etc.

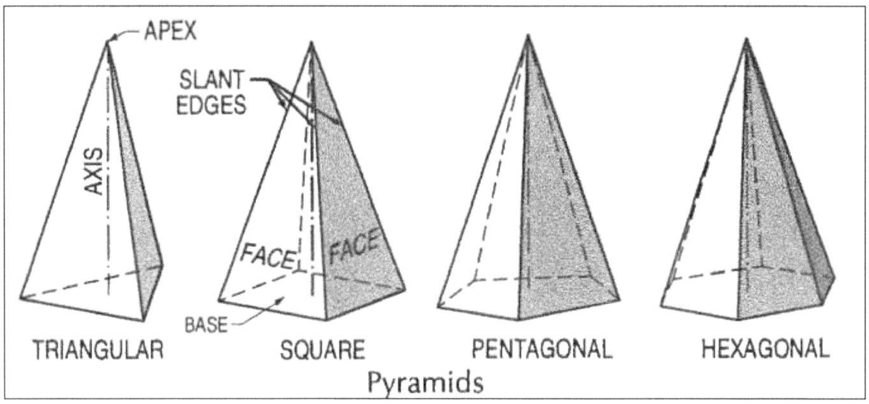

Figure 2.63

2.3.4 SOLIDS OF REVOLUTION

(i) Cylinder: A right circular cylinder is a solid generated by the revolution of a rectangle about one of its sides which remains fixed. It has two equal circular bases. The line joining the centres of the bases is the axis. It is perpendicular to the bases.

(ii) Cone: A right circular cone is a solid generated by the revolution of a right-angled triangle about one of its perpendicular sides which is fixed. It has one circular base. Its axis joins the apex with the centre of the base to which it is perpendicular. Straight lines drawn from the apex to the circumference of the base-circle are all equal and are called generators of the cone. The length of the generator is the slant height of the cone.

(iii) Sphere: A sphere is a solid generated by the revolution of a semi-circle about its diameter as the axis. The mid-point of the diameter is the centre of the sphere. All points on the surface of the sphere are equidistant from its centre.

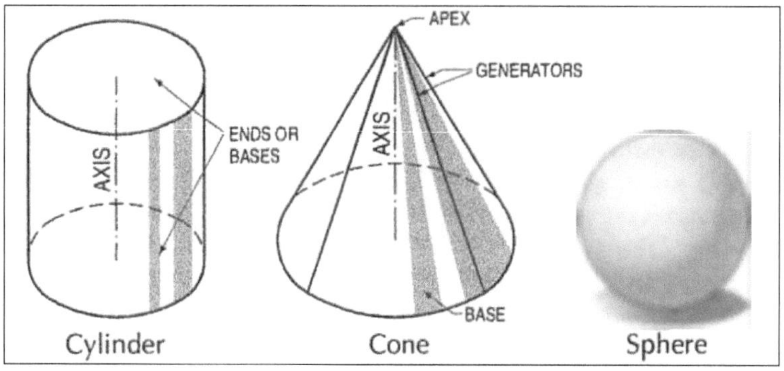

Figure 2.64

(iv) Frustum: When a pyramid or a cone is cut by a plane parallel to its base, thus removing the top portion, the remaining portion is called its frustum.

(v) Truncated: When a solid is cut by a plane inclined to the base it is said to be truncated. In this book mostly right and regular solids are dealt with. Hence, when a solid is named without any qualification, it should be understood as being right and regular.

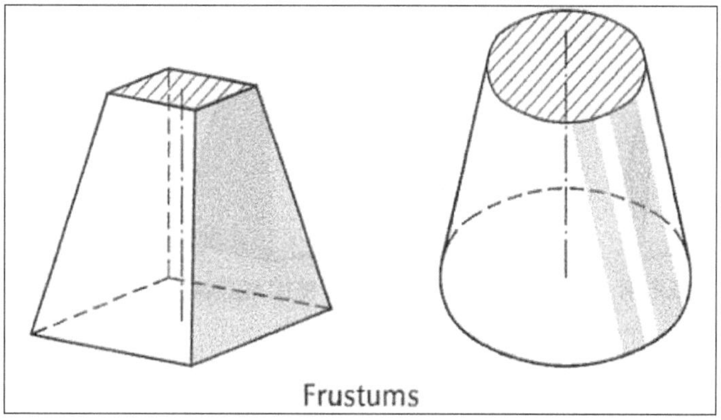

Figure 2.65

PROJECTION OF SOLID IN SIMPLE POSITION

A solid in simple position may have its axis perpendicular to one reference plane or parallel to both. When the axis is perpendicular to one reference plane, it is parallel to the other. Also, when the axis of a solid is perpendicular to a plane, its base will be parallel to that plane.

We have already seen that when a plane is parallel to a reference plane, its projection on that plane shows its true shape and size.

Therefore, the projection of a solid on the plane to which its axis is perpendicular, will show the true shape and size of its base. Hence, when the axis is perpendicular to the ground, i.e., to the H.P., the top view should be drawn first and the front view projected from it.

When the axis is perpendicular to the V.P., beginning should be made with the front view. The top view should then be projected from it.

When the axis is parallel to both the H.P. and the V.P., neither the top view nor the front view will show the actual shape of the base. In this case, the projection of the solid on an auxiliary plane perpendicular to both the planes, viz. the side view must be drawn first. The front view and the top view are then projected from the side view. The projections in such cases may also be drawn in two stages.

(1) AXIS PERPENDICULAR TO THE H.P.

Problem 12: Draw the projections of a triangular prism, base 40 mm side and axis 50 mm long, resting on one of its bases on the H.P. with a vertical face perpendicular to the V.P.

Solution:
(i) As the axis is perpendicular to the ground i.e., the H.P. begin with the top view. It will be an equilateral triangle of sides 40 mm long, with one of its sides perpendicular to xy. Name the corners as shown, thus completing the top view. The corners d, e and fare hidden and coincide with the top corners a, b and c respectively.
(ii) Project the front view, which will be a rectangle. Name the corners. The line b' e' coincides with a'd'.

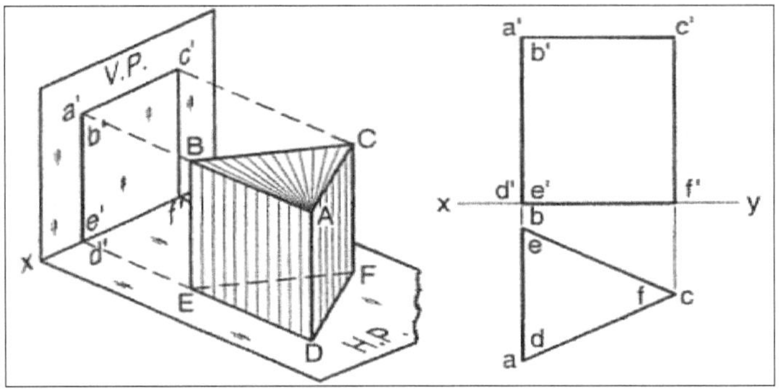

Figure 2.66

(2). AXIS PERPENDICULAR TO THE V.P.
Problem 13: A hexagonal prism has one of its rectangular faces parallel to the H.P. Its axis is perpendicular to the V.P. and 3.5 cm above the ground.
Draw its projections when the nearer end is 2 cm in front of the V.P. Side of base 2.5 cm long; axis 5 cm long.
Solution:
(i) Begin with the front view.
Construct a regular hexagon of 2.5 cm long sides with its centre 3.5 cm above xy and one side parallel to it.
(ii) Project down the top view, keeping the line for nearer end, viz. 1-4, 2 cm below xy.

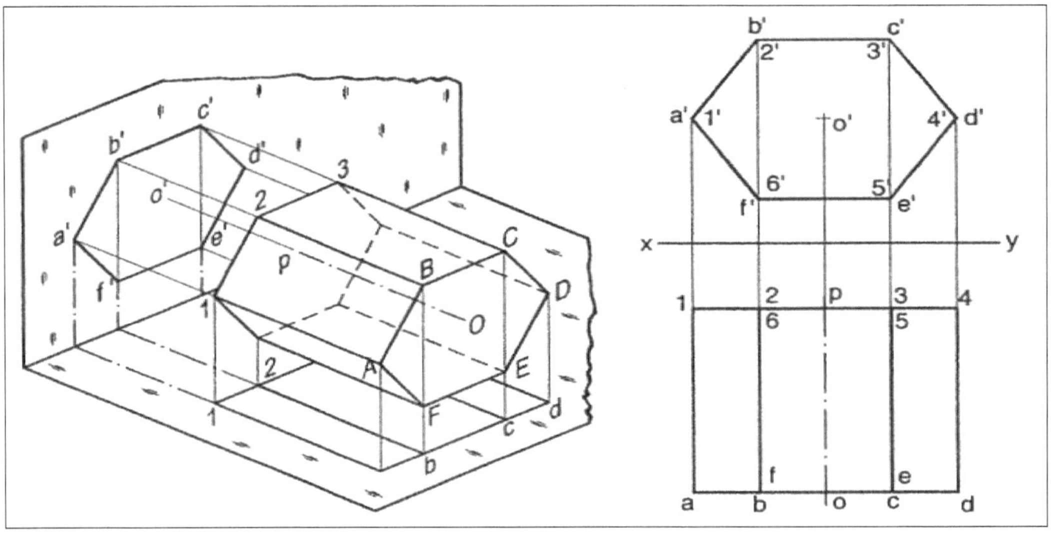

Figure 2.67

(3). AXIS PARALLEL TO BOTH THE H.P. AND THE V.P.
Problem 14: A triangular prism1 base 40 mm side and height 65 mm is resting on the H.P. on one of its rectangular faces with the axis parallel to the V.P. Draw its projections.
Solution:
As the axis is parallel to both the planes, begin with the side view.
(i) Draw an equilateral triangle representing the side view, with one side in xy.
(ii) Project the front view horizontally from this triangle.
(iii) Project down the top view from the front view and the side view, as shown.

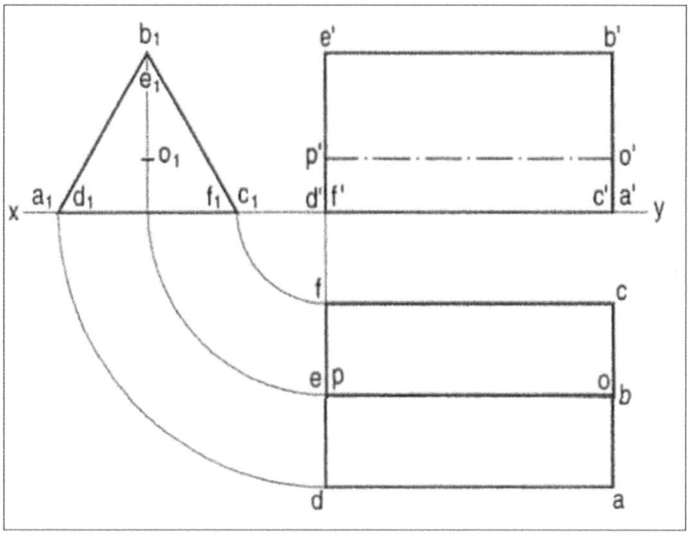

Figure 2.68

SOLID WITH AXIS INCLINED TO BOTH THE RPs

Two Methods of drawing the projections of solids:

1. Change of position method - the solids are placed first in the simple position and then tilted successively in two or three stages to obtain the final position.
2. Auxiliary plane method (Change of reference-line method) – the solids are placed initially in the simple position and then one or two auxiliary planes are setup to obtain the views in the required position.

Problem 15: A cube of 30 mm side rests with one of its edges on HP such that one of the square faces containing that edge is inclined at 300 to HP and the edge on which it rests being inclined to 600 to VP. Draw its projections.

Solution:

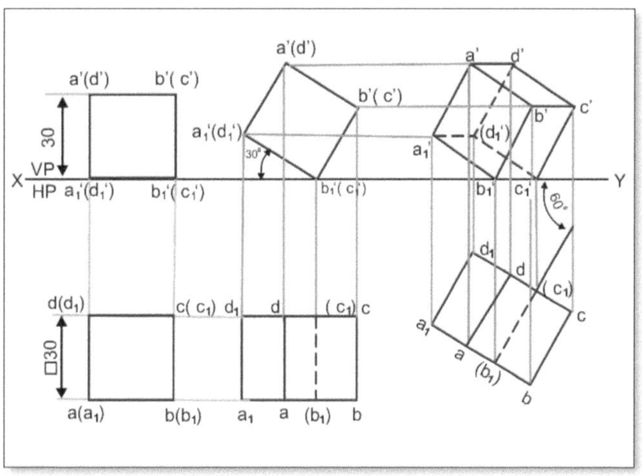

Figure 2.69

Problem 16: Draw the top and front views of a rectangular pyramid of sides of base 40 X 50 mm and height 70 mm when it lies on one of its larger triangular faces on HP. The longer edge of the base of the triangular face lying on HP is inclined at 600 to VP in the top view with the apex of the pyramid being nearer to VP.

Solution:

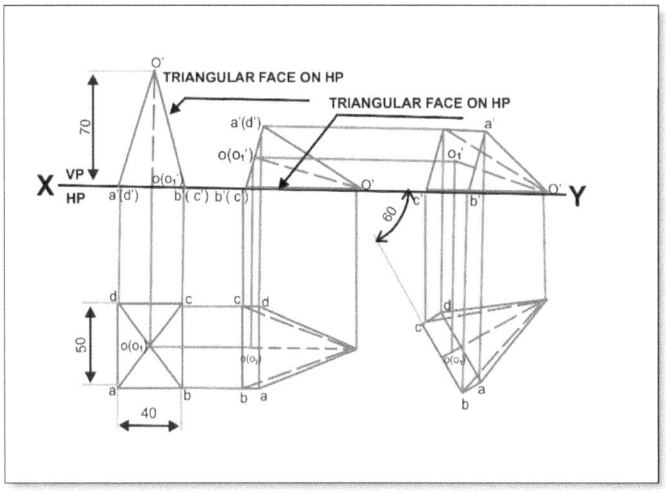

Figure 2.70

Assignment

1. In 1st angle projection the object is kept in _________
a) 1st quadrant
b) 2nd quadrant
c) 3rd quadrant
d) 4th quadrant
2. 1st angle projection is recommended by _____________
a) USA
b) ISI
c) Bureau of Indian Standards
d) ASME
3. In 1st angle projection the _________ lies between ___________ and ____________
a) object, projection plane, observer
b) projection plane, object, observer
c) reference line, side view, front view
d) reference line, left side view, right side view
4. In 1st angle projection, the front view will be below the top view.
a) True
b) False
5. In 1st angle projection the positions of front and top views are __________
a) top view lies above the front view
b) front view lies above the top view
c) front view lie left side to top view
d) top view lie left side to front view
6. In 1st angle projection, the left side view will be left side of front view.
a) True
b) False
7. The positions of right side view and front view of an object kept in 1st quadrant and projection are drawn?
a) Right side view is right side of front view
b) Right side view is left side of front view
c) Right side view is above the front view
d) Right side view is below the front view
8. The positions of reference line and top view in 1st angle projection are __________
a) reference line lies above the top view
b) reference line lies below the top view
c) reference line lie left side to top view

d) reference line lie right side to top view
9. If an object is placed in 1st quadrant such that one of the surfaces of object is coinciding with vertical plane, what is the correct position of views from the following?
a) The front view touches the reference line
b) The side view touches the reference line
c) The top view touches the reference line
d) The bottom view touches the reference line
10. If an object is placed in 1st quadrant such that one of the surfaces of object is coinciding with horizontal plane, what is the correct position of views from the following?
a) The front view touches the reference line
b) The side view touches the reference line
c) The top view touches the reference line
d) The bottom view touches the reference line
11. If an object is placed in 1st quadrant such that one of the surfaces of object is coinciding with both vertical plane and horizontal plane, what is the correct position of views from the following?
a) The top view touches the reference line
b) The top view and side view touch each other
c) Both side views touch each other
d) The top view and front touches each other at reference line
12. Where is the position of bottom view in 1st angle projection?
a) left side of right hand side view
b) right side of right hand side view
c) above the front view
d) below the top view
13. Where is the position of back view in 1st angle projection?
a) left side of right hand side view
b) right side of right hand side view
c) above the front view
d) below the top view
14. In 3rd angle projection the object is kept in _________
a) 1st quadrant
b) 2nd quadrant
c) 3rd quadrant
d) 4th quadrant
15. 3rd angle projection is recommended by _____________
a) USA b) ISI
c) Bureau of Indian Standards d) IS
16. In 3rd angle projection the _________ lies between ___________ and ____________
a) object, projection plane, observer b) projection plane, object, observer
c) reference line, side view, front view d) reference line, left side view, right side view
17. In 3rd angle projection, the front view will be below the top view.
a) True
b) False
18. In 3rd angle projection, the positions of front view and top views are?
a) Top view lies above the front view
b) Front view lies above the top view
c) Front view lie left side to top view
d) Top view lie left side to front view
19. The positions of right side view and front view of an object kept in 3rd quadrant and projection are drawn?
a) right side view is right side of front view
b) right side view is left side of front view
c) right side view is above the front view
d) right side view is below the front view
20. The positions of reference line and top view in 3rd angle projection are?
a) reference line lies above the top view

b) reference line lies below the top view
c) reference line lie left side to top view
d) reference line lie right side to top view
21. If an object is placed in 3rd quadrant such that one of the surfaces of object is coinciding with vertical plane, what is the correct position of views from the following?
a) The front view touches the reference line
b) The side view touches the reference line
c) The top view touches the reference line
d) The bottom view touches the reference line
22. If an object is placed in 3rd quadrant such that one of the surfaces of object is coinciding with horizontal plane, what is the correct position of views from the following?
a) The front view touches the reference line
b) The side view touches the reference line
c) The top view touches the reference line
d) The bottom view touches the reference line
23. If an object is placed in 3rd quadrant such that one of the surfaces of object is coinciding with both vertical plane and horizontal plane, what is the correct position of views from the following?
a) The top view touches the reference line
b) The top view and side view touch each other
c) Both side views touch each other
d) The top view and front touches each other at reference line
24. Where is the position of bottom view in 3rd angle projection?
a) left side of right hand side view
b) right side of right hand side view
c) above the front view
d) below the top view
25. Where is the position of back view in 3rd angle projection?
a) left side of right hand side view
b) right side of right hand side view
c) above the front view
d) below the top view
26. In 3rd angle projection, the left side view will be left side of front view.
a) True
b) False
27. A plane surface has ___ dimension(s).
a) one
b) two
c) three
d) no dimension
28. Which of the following position is not possible for a plane?
a) plane perpendicular to both VP and HP
b) plane parallel to both VP and HP
c) plane perpendicular to HP, parallel to VP
d) plane perpendicular to VP, parallel to HP
29. A plane perpendicular to HP and parallel to VP, will make
a) vertical trace (VT)
b) horizontal trace (HT)
c) both vertical (VT) and horizontal trace (HT)
d) none of the above
30. A regular hexagonal lamina has
a) equal sides
b) equal angles
c) both (a) and (b)
d) none of the above
31. The front view of a regular pentagonal lamina perpendicular to HP, is
a) a point

b) a line
c) a regular pentagon
d) a circle
32. The top view of a circular disc perpendicular to HP, is
a) a point
b) a line
c) an ellipse
d) a circle
33. If a lamina is perpendicular to both HP and VP, then its true shape can be obtained in
a) front view (elevation)
b) top view (plan)
c) side view
d) none of the above
34. If a regular hexagonal lamina is perpendicular to VP, its top view will be a
a) a point
b) a line
c) regular hexagon
d) a rectangle
35. The true shape of an oblique plane is obtained in
a) horizontal plane (HP)
b) vertical plane (VP)
c) profile plane (PP)
d) none of the above
36. An oblique plane is
a) inclined to horizontal plane (HP)
b) inclined to vertical plane (VP)
c) both (a) and (b)
d) perpendicular to both vertical and horizontal plane
37. The top view of a circular disc inclined to Horizontal plane (HP) and parallel to Vertical plane (VP), is
a) a circle
b) an ellipse
c) a line
d) a point
38. Which one of the following position is not possible in projection of a line?
a) Line perpendicular to HP, parallel to VP
b) Line perpendicular to VP, parallel to HP
c) Line parallel to both VP and HP
d) Line perpendicular to both VP and HP
39. When the line is parallel to both VP and HP, we can get its True length in
a) Top view
b) Front view
c) both (a) and (b)
d) Side view
40. The top view of a line perpendicular to HP, is
a) a line of shorter length
b) a line having true length
c) a point
d) any of the above
41. The front view of a line perpendicular to HP, is
a) a line of shorter length
b) a line having true length
c) a point
d) any of the above
42. The front view of a line perpendicular to VP, is
a) a line of shorter length
b) a line of true length

c) a point
d) any of the above
43. The top view of a line perpendicular to VP, is
a) a line of shorter length
b) a line of true length
c) a point
d) any of the above
44. When the line is inclined to HP and parallel to VP, we get its true length in
a) front view
b) top view
c) side view
d) none of the above
45. When the line is inclined to VP and parallel to HP, we get its true length in
a) front view
b) top view
c) side view
d) none of the above
46. True inclination of a line with HP, is denoted by
a) α
b) β
c) Φ
d) θ
47. When the line is inclined to both VP and HP, and parallel to PP, we get its true length in
a) front view
b) top view
c) side view
d) none of the above
48. Following method(s) is (are) used to find true length of a straight line
a) Rotation method
b) Trapezoidal method
c) Both (a) and (b)
d) None of the above
49. If a line is inclined to HP and parallel to VP, it will make
a) Horizontal trace
b) Vertical trace
c) Both (a) and (b)
d) None of the above
50. If a solid is positioned that its axis is perpendicular to one of the reference plane. Which of the following is false?
a) Axis is parallel to other reference plane
b) Base is parallel to reference plane
c) Projection on that plane gives true shape of its base
d) Base is perpendicular to horizontal plane
51. If a solid's axis is perpendicular to one of the reference planes then the projection of solid on to the same plane gives the true shape and size of its ___________
a) lateral geometry
b) base
c) cross-section
d) surface
52. When the axis of solid is perpendicular to H.P, the ______view should be drawn first and ____ view then projected from it.
a) front, top
b) top, side
c) side, front
d) top, front

53. When the axis of solid is perpendicular to V.P, the ______view should be drawn first and ____ view then projected from it.
a) front, top
b) top, side
c) side, front
d) top, front
54. When the axis of solid is parallel to H.P & V.P, then ______view should be drawn first and ____ and _______view then projected from it.
a) front, top, side
b) top, side, front
c) side, front, top
d) top, front, side
55. When the axis of solid is parallel to H.P & V.P, then ______view should be drawn first and ____ and _______view then projected from it.
a) front, top, side
b) top, side, front
c) side, front, top
d) top, front, side
56. The front view, side view and top view of a regular square pyramid standing on horizontal plane base on horizontal plane.
a) triangle, triangle and square
b) square, triangle and triangle
c) square, triangle and square
d) triangle, square and triangle
57. The front view, side view and top view of a cylinder standing on horizontal plane base on horizontal plane.
a) circle, rectangle and rectangle
b) rectangle, rectangle and circle
c) rectangle, circle and rectangle
d) circle, triangle and triangle
58. The side view, top view and front view of a regular hexagonal pyramid placed base parallel to profile plane.
a) Triangle, triangle and hexagon
b) hexagon, triangle and triangle
c) hexagon, triangle and hexagon
d) triangle, hexagon and triangle
59. The side view, top view and front view of a regular cone placed base parallel to profile plane.
a) Triangle, triangle and circle
b) circle, triangle and triangle
c) rectangle, triangle and circle
d) triangle, circle and triangle
60. The side view, top view and front view of a regular pentagonal prism placed axis perpendicular to vertical plane.
a) rectangle, rectangle and pentagon
b) pentagon, rectangle and rectangle
c) pentagon, rectangle and pentagon
d) rectangle, pentagon and rectangle
61. Square pyramid, cylinder, triangular prism and cone are placed one next to other in between the reference planes in different positions given below match the following.

Sr No.	Position of Solids		Front View
1.	Square pyramid-axis perpendicular to V.P.	i.	Rectangle
2.	Cylinder- base parallel to profile plane	ii.	Circle
3.	Triangular prism – axis perpendicular to H.P.	iii.	Square
4.	Cone – base parallel to V.P.	iv.	Triangle

a) 1, i; 2, ii; 3, iii; 4, iv
b) 1, iv; 2, iii; 3, ii; 4, i
c) 1, ii; 2, iv; 3, iii; 4, i
d) 1, iii; 2, i; 3, iv; 4, ii

62. Pentagonal pyramid, tetrahedron, cuboid and cone are placed one next to other in between the reference planes in different positions given below match the following.

Sr. No.	Position of solids		Front View1
1.	Pentagonal pyramid – axis perpendicular to H.P	i.	Smallest base
2.	Tetrahedron – base parallel to vertical plane	ii.	Pentagon
3.	Cuboid – smallest side parallel to H.P	iii.	Triangle
4.	Cone – base parallel to P.P(profile plane)2	iv.	Pentagon

a) 1, i; 2, ii; 3, iii; 4, iv
b) 1, iv; 2, iii; 3, i; 4, ii
c) 1, ii; 2, iv; 3, iii; 4, i
d) 1, iii; 2, i; 3, iv; 4, ii

63. Square prism, hexagonal pyramid, cube, sphere are placed one next to other in between the reference planes in different positions given below match the following.

Sr. No.	Position of solids		Front View
1.	Square prism – axis perpendicular to V.P.	i.	Circle
2.	Hexagonal pyramid – base parallel to H.P.	ii.	Rectangle
3.	Triangular prism – axis perpendicular to P.P.	iii.	Triangle
4.	Sphere	iv.	Triangle

a) 1, i; 2, ii; 3, iii; 4, iv
b) 1, ii; 2, iii; 3, ii; 4, i
c) 1, ii; 2, iv; 3, iii; 4, i
d) 1, iii; 2, i; 3, iv; 4, ii

Answers

1. Answer: a
2. Answer: c
3. Answer: a
4. Answer: b
5. Answer: b
6. Answer: b
7. Answer: b
8. Answer: a
9. Answer: c
10. Answer: a
11. Answer: d
12. Answer: c
13. Answer: b
14. Answer: c
15. Answer: a
16. Answer: b
17. Answer: a
18. Answer: b
19. Answer: b
20. Answer: a
21. Answer: a
22. Answer: a
23. Answer: d
24. Answer: b
25. Answer: d
26. Answer: a

27. Answer: b
28. Answer: d
29. Answer: b
30. Answer: c
31. Answer: c
32. Answer: d
33. Answer: a
34. Answer: c
35. Answer: c
36. Answer: c
37. Answer: b
38. Answer: c
39. Answer: b
40. Answer: a
41. Answer: b
42. Answer: a
43. Answer: a
44. Answer: a
45. Answer: b
46. Answer: a
47. Answer: c
48. Answer: c
49. Answer: a
50. Answer: a
51. Answer: b
52. Answer: a
53. Answer: c
54. Answer: a
55. Answer: b
56. Answer: b
57. Answer: a
58. Answer: c
59. Answer: a
60. Answer: b
61. Answer: a
62. Answer: b
63. Answer: c

Chapter 3: Isometric Projections

3.1 ISOMETRIC PROJECTION

The isometric projection of an item is a one aircraft view drawn with the item so located with recognize to the aircraft of projection that each one the three foremost axes seem like willing to every different at a same perspective of 120°.

PRINCIPLE OF ISOMETRIC PROJECTIONS

It's a pictorial orthographic projection of an object where a transparent cube containing the object is tilted before one of those solid diagonals of the cube becomes perpendicular to the vertical plane along with the three axes are equally inclined to this vertical plane. The isometric projection of a cube in steps is shown in the below figure 3.1. Here ABCDEFGH is the isometric projection of the cube.

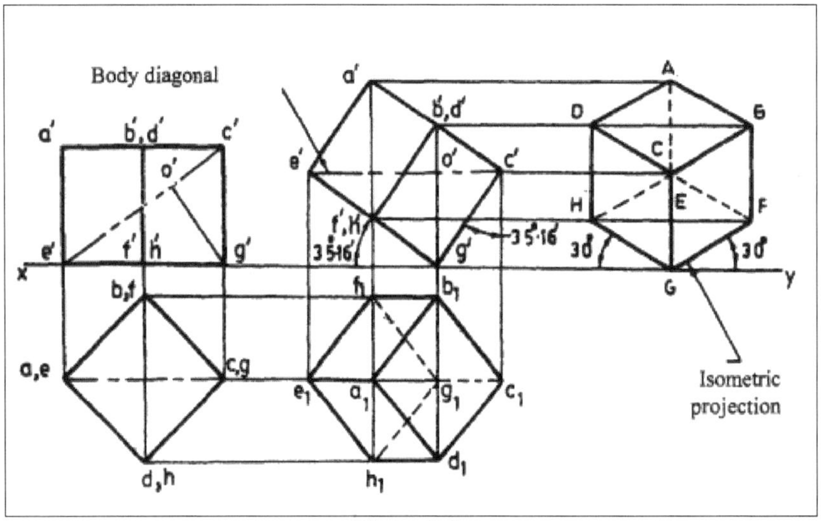

Figure 3.1

The front view of this cube, resting on one of its corners (G), is the isometric projection of the cube.

3.2 ISOMETRIC PROJECTION OF CUBE

A cube is placed with three of its perpendicular edges making equal inclinations with the vertical plane of projection and is projected on it. As shown in figure 3.2

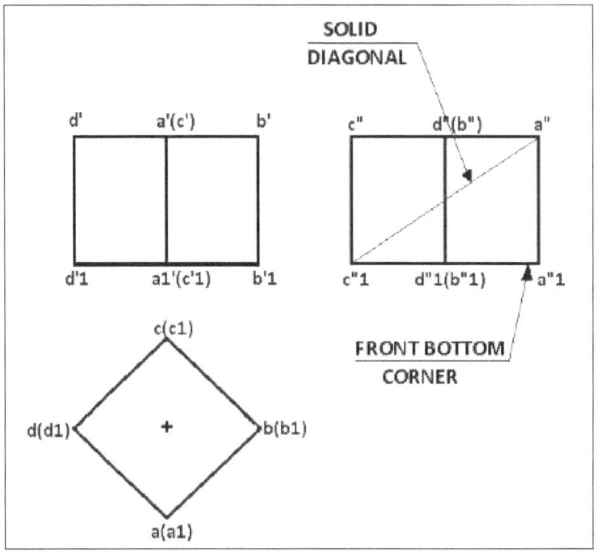

Figure 3.2

The projections of a cube when placed on HP such that two of its square faces making equal inclinations with the plane of projection.
Next, tilt the cube towards the observer such that it rests on the front bottom corner with the solid diagonal opposite to the corner in which it rests and passing through the top front corner, is perpendicular to the plane of projection.

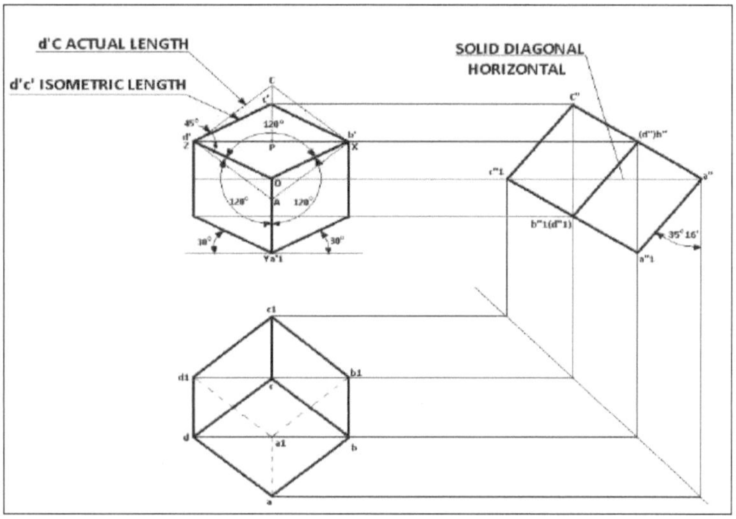

Figure 3.3

The Figure3.3 illustrates this position of the cube. As seen in the left view, the solid diagonal a "c1" is perpendicular to the plane of projection. In this position, the front and top views are projected. The three perpendicular edges AB, AD & AA1 will be equally inclined at an angle of 35° 16′ to the plane of projection.
But the projection of these edges, i.e., a'b', a'd' & a'a1′ are at an angle of 120° to each other and also they are foreshortened to the same length. The front view of the cube, therefore, represents its isometric projection.

3.3 ISOMETRIC LENGTH

To find the extent to which the lengths of the edges are foreshortened.
Draw a square d'Ab'C of sides equal to the actual length of the edges of the cube with d'b' as the common diagonal.
d'C is the actual length of the edge, whereas corresponding edge d'c' in the isometric projection is foreshortened. This foreshortened length is known as isometric length of the edge of the cube.
d'C is the actual length of the edge, whereas corresponding edge d'c' in the isometric projection is foreshortened. This foreshortened length is known as the isometric length of the edge of the cube.

In the triangle d'Pc', $\frac{d'P}{d'c} = \cos 30° = \frac{3}{2}$

In the triangle d'PC, $\frac{d'P}{d'C} = \cos 45° = \frac{1}{2}$

$$\frac{d'c'}{d'C} \quad \frac{2}{3} * \frac{1}{2}$$

$$= \frac{2}{3}$$

$$= 0.816$$

i.e., $\frac{d'c'}{d'C} = \frac{\text{Isometric Length}}{\text{Actual Lenght}} = 0.816$

i.e., Isometric Length = 0.816 Actual Length. Thus, in the isometric projection, the actual lengths of the objects are reduced in the ratio of 2:3. Or the isometric length is 0.816 of the actual length.

When an isometric projection is to be drawn, all the edges of the object parallel to the three isometric axes, have to be reduced to their corresponding isometric lengths which will be equal to 0.816 times their actual length.

3.4 ISOMETRIC SCALE

An **isometric scale** can be used to draw correct **isometric** projections. All distances in this **scale** are 2/3 × true size, or approximately 80% of true size. Figure 3.4 shows an **isometric scale.**

In the isometric projection of a cube shown in the show above figure 3.4 the top face ABCD is sloping away from the observer, and hence the edges of the top face will appear fore-shortened.

The true shape of the triangle DAB is represented by the triangle DPB. The extent of reduction of an isometric line is easily found by the construction of a diagram known as isometric scale.

For this, reproduce the triangle DPA as shown in the show below figure 3.4 Mark the divisions of true length on DP. Through these divisions, draw vertical lines to get the corresponding points on DA.

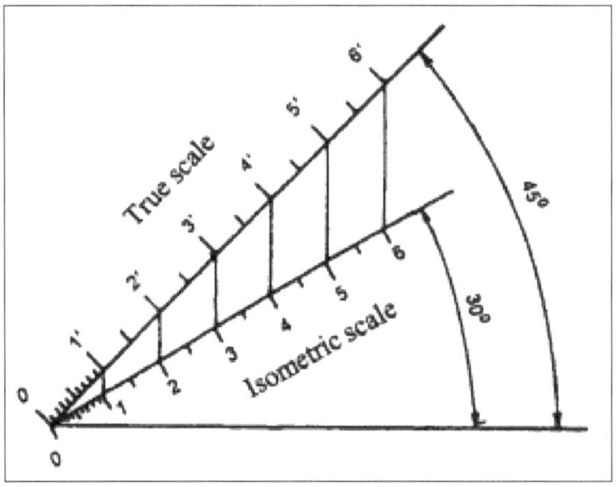

Figure 3.4

The divisions of the line DA give dimensions to isometric scale. In the triangle ADO and PDO at the show above figure 3.4(**An Isometric Cub)**, the ratio of the isometric length to the true length,

i.e., DA/DP = cos 45° /cos30° = 0.816

The isometric axes are reduced in the ratio 1:0.816, i.e., 82% approximately.

LINES IN ISOMETRIC PROJECTION

The following are the relations between the lines in isometric projection which are evident from the show above figure 3.4(**An Isometric Cub)**

1. The lines which are parallel to the object are parallel at the isometric projection.
2. Vertical lines on the object appear vertical at the isometric projection.
3. Horizontal lines on the item are drawn at an angle of 30° with the horizontal at the isometric projection.
4. A line parallel to an isometric axis is called an isometric line, and it's fore-shortened to 82 percent.
5. A line that's not parallel to any isometric axis is known as the non-isometric line, and the extent of the fore-shortening of non-isometric lines is different if their inclinations with the vertical planes are different.

3.5 ISOMETRIC PROJECTION VIEWS

Shows in below Figure **(a)** rectangular block in pictorial form and shows in below Figure **(b)**, the steps for drawing an isometric projection using the isometric scale.

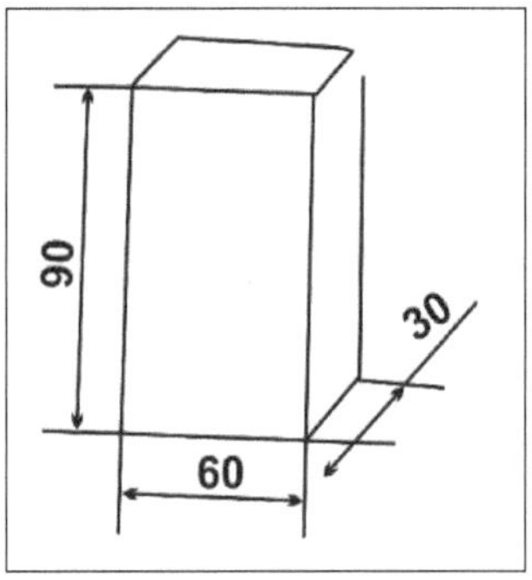

Isometric projection

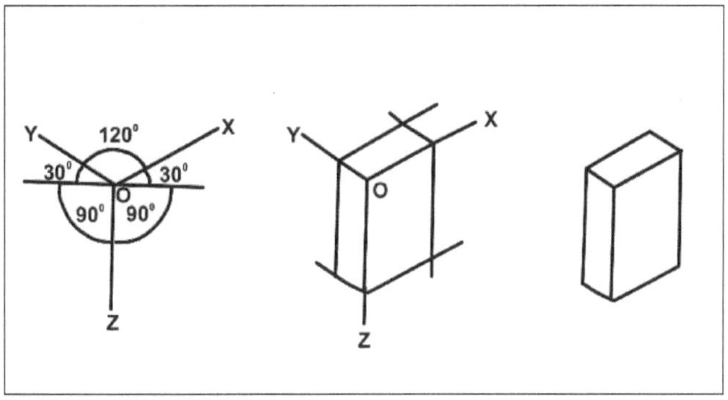

Isometric projection

ISOMETRIC VIEW OR ISOMETRIC DRAWING

To draw the isometric projection of an object, every dimension of the object along the isometric axes, have to be reduced to the isometric scales.

This has two disadvantages:

First, it is time-consuming and

It is on such a drawing when required the distances between any two points along the isometric axes cannot be measured directly with the actual scale.

To overcome these difficulties, an object can be quickly represented in the isometric form using the actual scale instead of the isometric scale. Such a view will be precisely is of the same shape but larger than that drawn to the isometric scale.

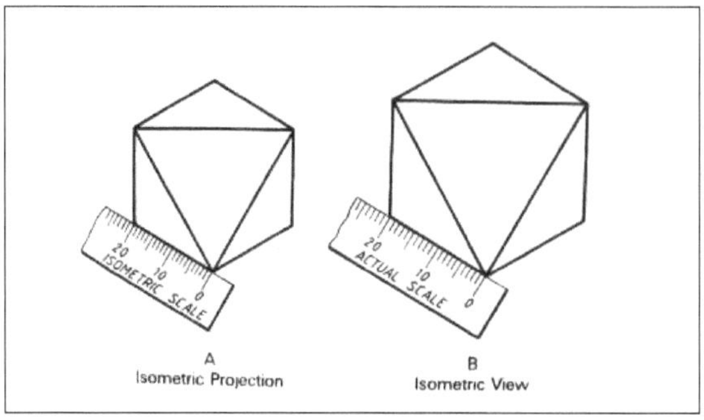

Figure 3.5

An **isometric drawing** is a type of 3D **drawing** that is set out using 30-degree angles. It's a type of axonometric **drawing** in which the same scale is used for every axis, resulting in a non-distorted image.

Drawing of objects are seldom drawn in true isometric projections, since using an isometric scale is inconvenient. Instead, a convenient way in which the foreshortening of lengths is ignored and actual or true lengths are utilized to obtain the projections, known as isometric drawing, or isometric perspective is normally utilized.

This is advantageous because the measurement could be made directly from a drawing. The isometric drawing of the figure is slightly larger (approximately 22%) than the isometric projection.
Since the proportions are the same, the increased size doesn't affect the pictorial value of this representation, and at the same time, it might be done quickly.
Shows in below a figure the difference between isometric drawing and isometric projection.

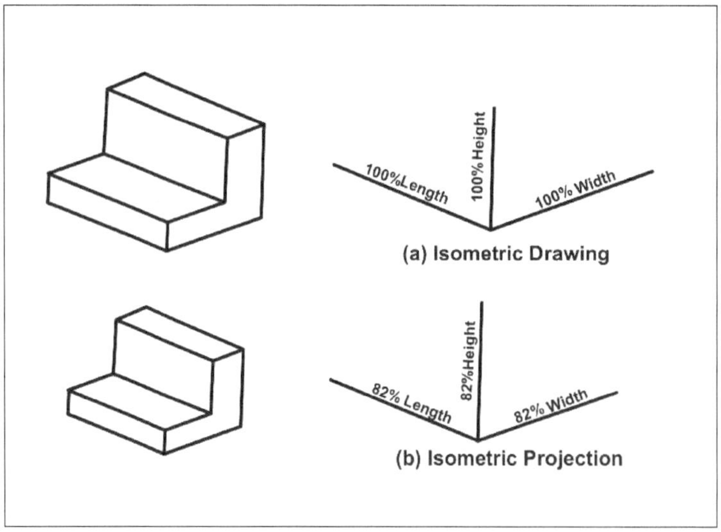

Figure 3.6

Steps to be followed to make an Isometric drawing from orthographic views are given below:
Study the given views and note the principal dimensions and other features of this object.
Draw the isometric axes (a) Orthographic view (figure 3.7)

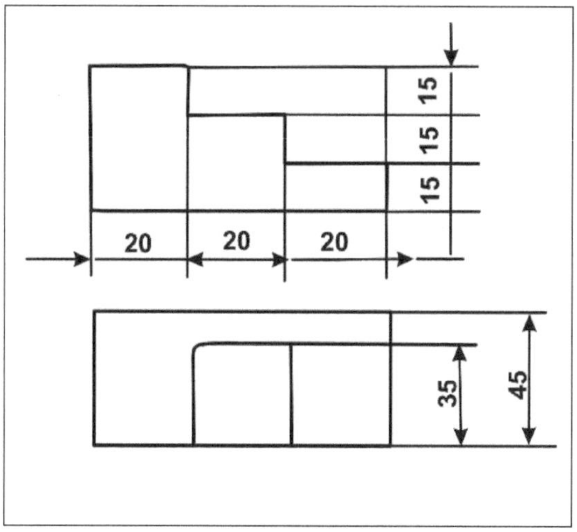

Figure 3.7

Mark the principal dimensions to-their true values along the isometric axes (b). Isometric view (figure 3.8)

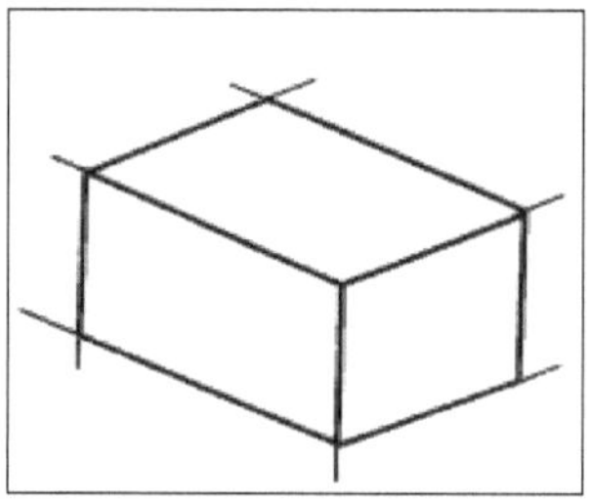

Figure 3.8

Complete the housing block by drawing lines parallel to the isometric axes and passing Through the above markings [c] isometric view (figure 3.9)

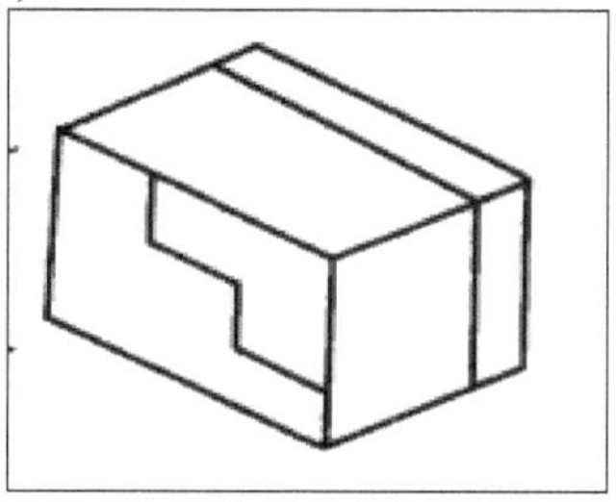

Figure 3.9

Locate the principal corners of all the features of the object on the three faces of the Housing block (d) isometric view (figure 3.10)

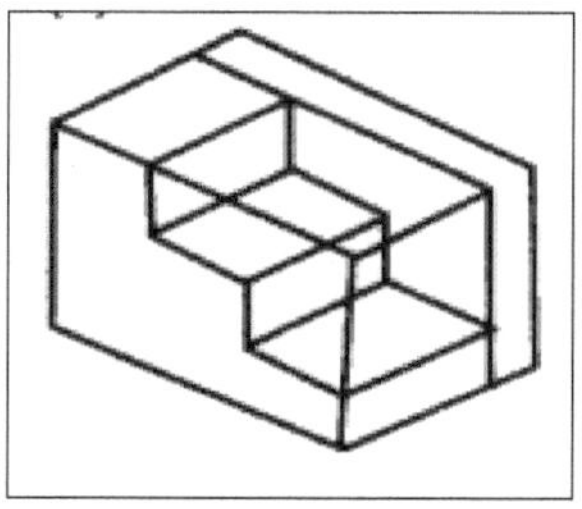

Figure 3.10

Draw lines parallel to the axes and passing through the above points and obtain the isometric Drawing of this object by darkening the visible edges (e) isometric view (figure 3.11)

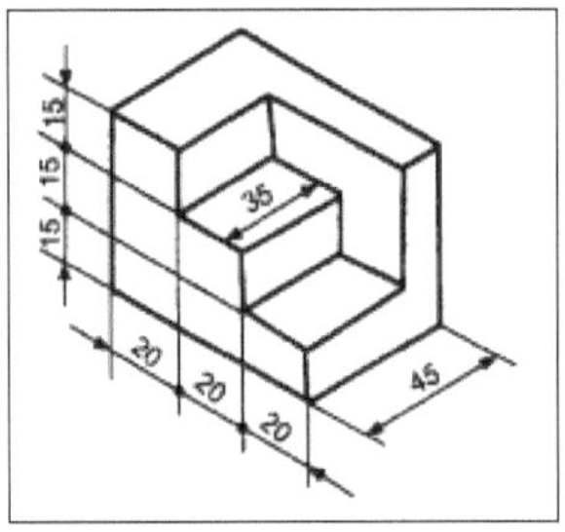

Figure 3.11

ISOMETRIC & NON-ISOMETRIC LINES

In the figure, the three perpendicular edges of the cube OX, OY, & OZ are foreshortened equally and are at equal inclinations of 120° to each other and are known as isometric axes. The lines drawn parallel to the isometric axes are known as Isometric Line.

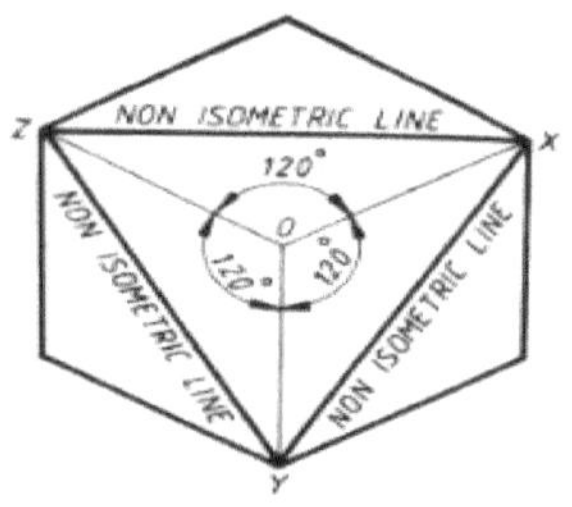

Isometric & Non-ISometric Lines

Figure 3.12

Any other line which is not parallel to any of the isometric axes is known as Non-Isometric Line. The lines XY, YZ & ZX are called non-isometric lines.

Since these lines are not parallel to the isometric axes, they are not foreshortened in the same proportion as the isometric lines. Also, those horizontal edges of the object which are non-isometric must not be drawn at 30°. To draw the non-isometric lines, their ends should be located and then joined. The surface XYZ is an oblique surface in isometric.

DIFFERENT POSITIONS OF ISOMETRIC REPRESENTATION

An object in isometric projection may be shown in eight different positions as shown in the Figure. Anyone particular position may be selected so as to illustrate all the important details of the object clearly.

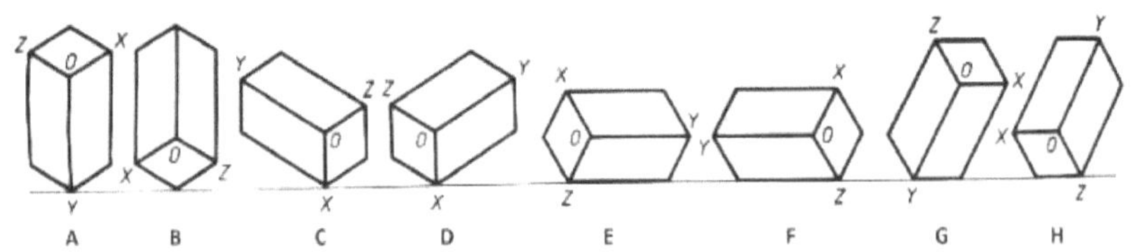

Different Positions Of Isometric Representations

Figure 3.13

However, the object will be drawn in isometric with its longest measurement either vertical or horizontal. The position in figure-B is known as an isometric projection with axis reversed is drawn when the underneath surface of the object is to be shown.

3.6 ISOMETRIC VIEW OF POLYGONS

1. Polygons are first enclosed in a rectangle.
2. The corners lie on the sides of the rectangle.
3. The distances from the corner of the rectangle to the corners of the polygon are measured.
4. These distances are plotted on the isometric axes.

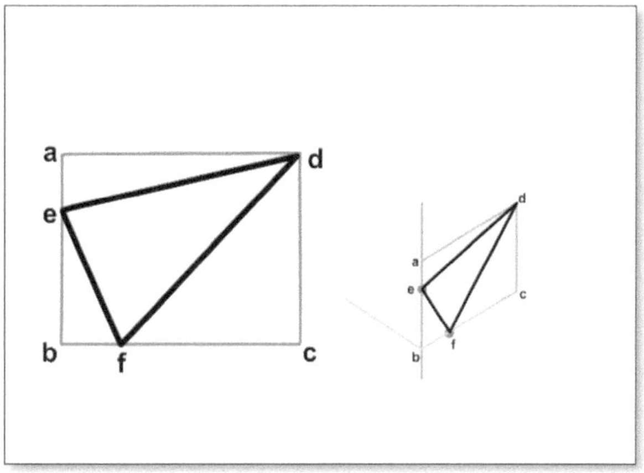

Figure 3.14

All corners of polygon not on edges of rectangle

Draw a rectangle covering as many polygon corners as possible. In this example the point i does not lie on the rectangle

From the edges of the rectangle, measure distances ki = cl and li = ck

Mark points f, g, h, j, and e in the isometric view similar to the previous example

Mark distances ck and ki on the isometric view to get point i

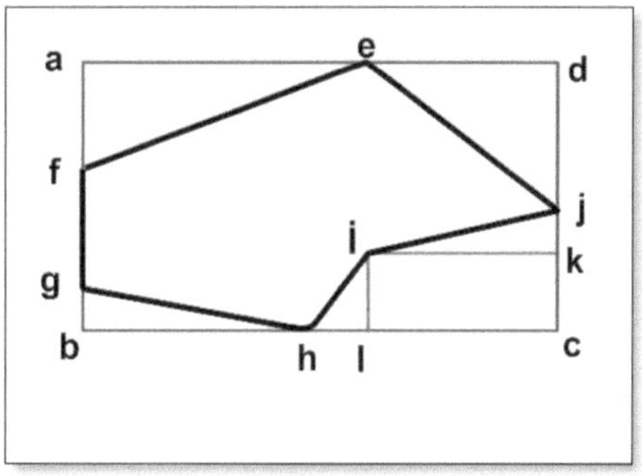

Figure 3.15

3.6.1 SQUARE

If you draw a square in isometric view, the result is a rectangle, explanation:

Any polygon that is drawn in an isometric view tends to have a base that is 30 degrees and other sides that appear to be viewed from a particular point, such as a one-dimensional perspective view. there is.

ISOMETRIC PROJECTION OF SQUARE – HORIZONTAL AND VERTICAL

Draw the isometric projection of a square lamina of side 35 mm when it is placed with its surface.

(a) Vertical and (b) Horizontal

Case (a): Square – Vertical Note: In Isometric projection/view, draw the vertical lines vertical and the horizontal lines inclined at an angle of 30° to the horizontal (shown in figure 3.16)

Case (b): Square – Horizontal Take one corner draw lines both side 30° (shown in figure 3.17)

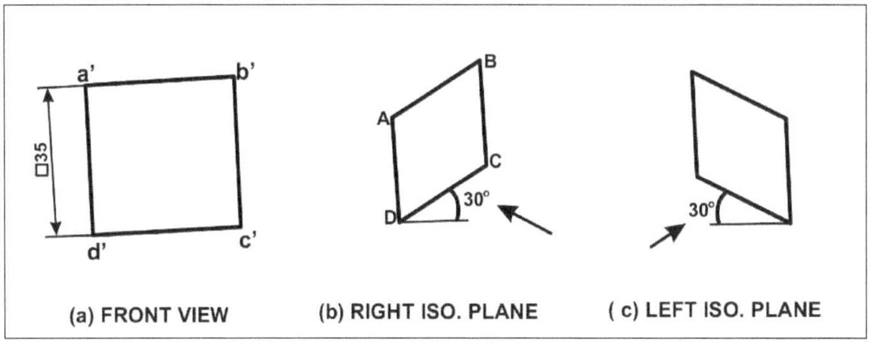

Figure 3.16

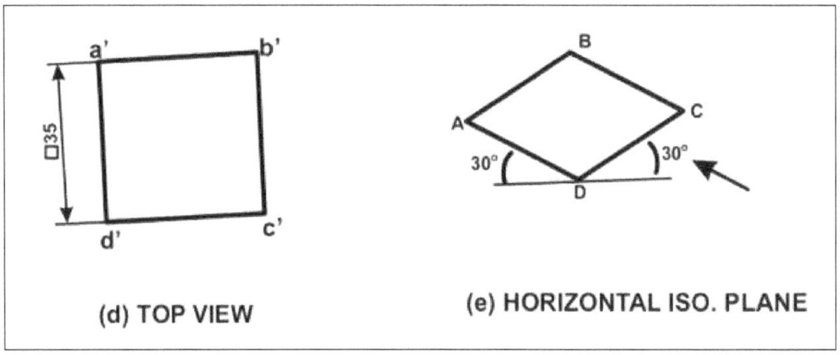

Figure 3.17

Problem 1: A square ABCD of 50 mm side has its corner A in the H.P., its diagonal AC inclined at 30° to the H.P. and the diagonal BD inclined at 45° to the V.P. and parallel to the H.P. Draw its projections.
Solution:
Keep AC parallel to the H.P. & BD perpendicular to V.P. (considering inclination of AC as inclination of the plane)
Incline AC at 30° to the H.P. i.e., incline the edge view (FV) at 30° to the HP

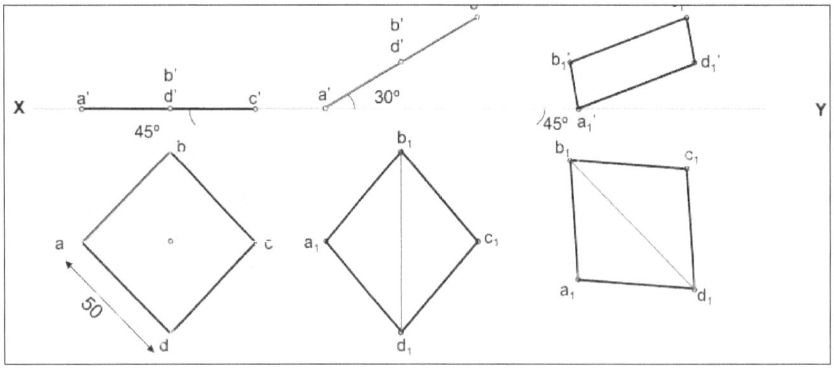

Figure 3.18

3.6.2 RECTANGLE

A rectangle appears as a parallelogram in isometric view. Three versions are possible depending on the orientation of the rectangle, i.e., right-hand vertical face, left-hand vertical face or horizontal face. The isometric view or isometric projection of a circle is an ellipse.

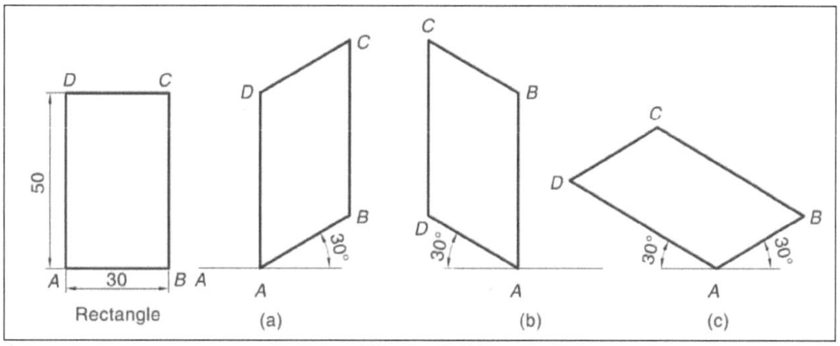

Figure 3.18

3.6.3 PENTAGON

Problem 2: A regular pentagon of 25 mm side has one side on the ground. Its plane is inclined at 40° to the HP and perpendicular to the VP. Draw its projections and show its traces.

Solution:

Hint: As the plane is inclined to HP, it should be kept parallel to HP with one edge perpendicular to VP

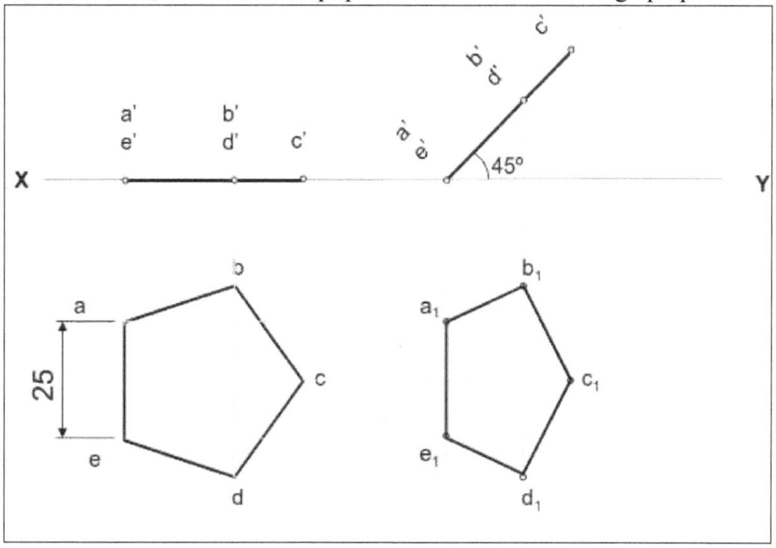

Figure 3.19

Problem 3: A regular pentagon of 30 mm sides is resting on HP on one of its sides with its surface 450 inclined to HP. Draw it's projections when the side in HP makes 300 angle with VP.

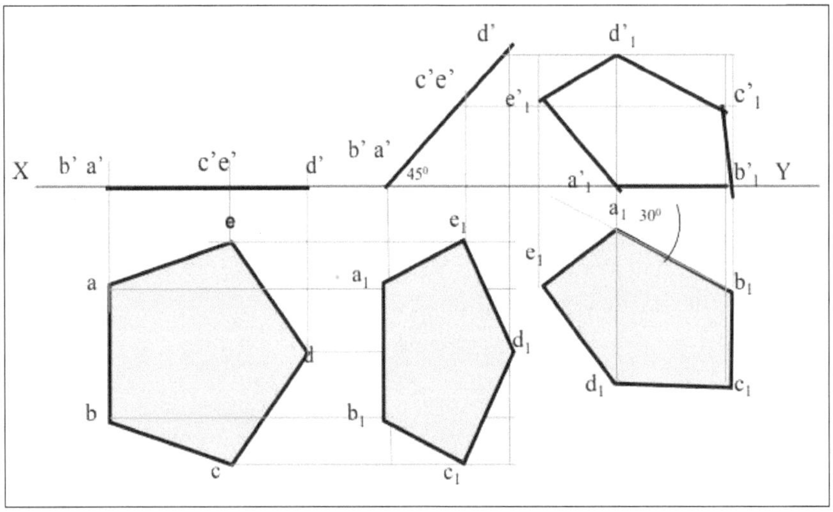

Figure 3.20

Read problem and answer following questions
1. Surface inclined to which plane?
Ans. Towards HP
2. Assumption for initial position?
Ans. Towards HP
3. So which view will show True shape?
Ans. Towards TV
4. Which side will be vertical?
Ans. Any side.
Hence begin with TV, draw pentagon below X-Y line, taking one side vertical.

3.6.4 HEXAGON

Problem 4: Draw the projections of a regular hexagon of 25 mm sides, having one of its side in the H.P. and inclined at 60 to the V.P. and its surface making an angle of 40° with the H.P.
Solution:

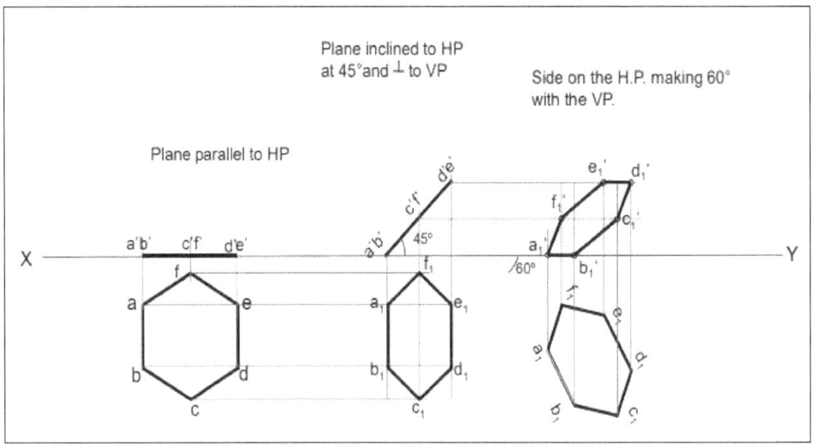

Figure 3.21

3.7 ISOMETRIC VIEW OF CIRCLE

On an isometric drawing, circles appear as ellipses and arcs as elliptical arcs. It is obtained by using four-centre method explained below.
Circles and arcs pose a particular problem when drawing in perspective. When the perspective is altered, they change in terms of both length (as lines do) and shape – they become ellipses.

This raises the question as to what ellipse should be used in order to correctly depict a circle in an isometric representation. Let's take another look at the cube. In the isometric figure on the left, a circle has been added to the top. On the right you can see the tipped cube again, viewed from the side.

Four-Centre Method: First, enclose the given circle into a square ABCD. Draw rhombus ABCD as an isometric view of the square. Join the farthest corners of the rhombus, i.e., A and C.
Obtain midpoints 3 and 4 of sides CD and AD respectively. Locate points 1 and 2 at the intersection of AC with B–3 and B–4 respectively. Now with 1 as a centre and radius 1–3, draw a small arc 3–5.
Draw another arc 4–6 with same radius but 2 as a centre. With B as a centre and radius B–3, draw an arc 3–4. Draw another arc 5–6 with same radius but with D as a centre.

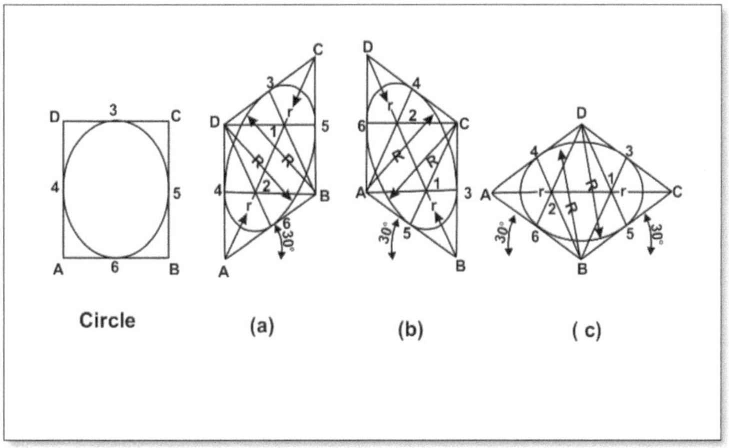

Figure 3.22

Problem 5: Draw the isometric view of a Circle with a 60 mm Diameter on all three Principle Planes Using Coordinate methods?
Solution: Construction Procedure:
1. Draw a circle with 60mm Diameter and enclose it in a square abcd.
2. Mark midpoints of the sides 1, 2, 3 and 4, where the square touches the circle tangentially
3. Draw the Diagonals of the square which cut in the circle at points 5, 6, 7 and 8 as shown in fig (a).
4. Draw a Rhombus ABCD to represent Isometric view of a square abcd.
5. Mark points 1, 2, 3 and 4 on it as the midpoint of the sides.
6. Mark points 5, 6, 7 and 8 on it, such that they are at a distance equal to Ax from the side of the square join points to obtain isometric view as shown in figures(b) (c) (d).

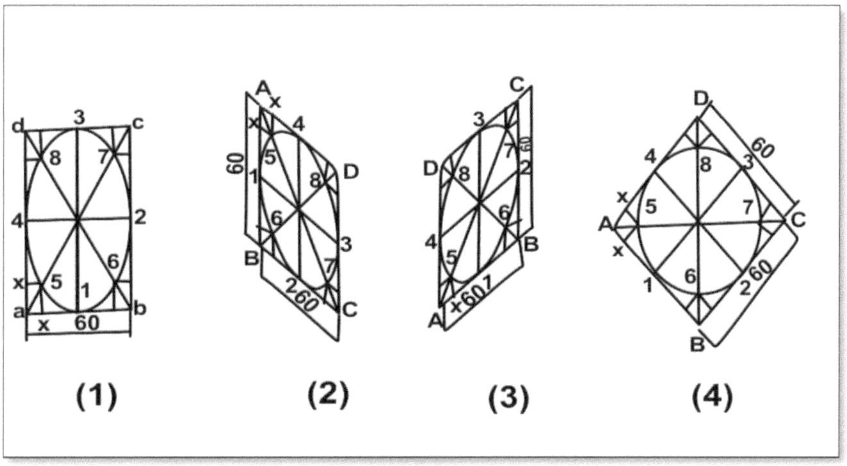

Figure 3.23

Problem 6: Draw the isometric view of a Circle lamina with a 60 mm Diameter on all three Principle Planes using for centre methods?

Solution: Construction:

1. Draw a Rhombus ABCD of 60 mm side to represent isometric view of a square
2. Mark 1, 2, 3 and 4 as a midpoints of the sides AB, BC, CD and DA respectively join (the ends of the minor diagonals) B to meet points 3 & 4 and D to meet points 1 & 2. Let B4 and D1 intersect at point E and B3 and D2 intersect at a point F. then B,E,D and F are the Four centres for drawing the ellipse
3. With centre B and radius B3 draw Arc 3-4. With centre D and Radius D1 draw Arc 1-2. With centre E and radius E1 draw Arc 1-4. With centre F and radius F2 draw Arc 2-3.
4. These Arcs join in the form of an Ellipse which represents the required isometric as shown in figure (a) (b) (c).

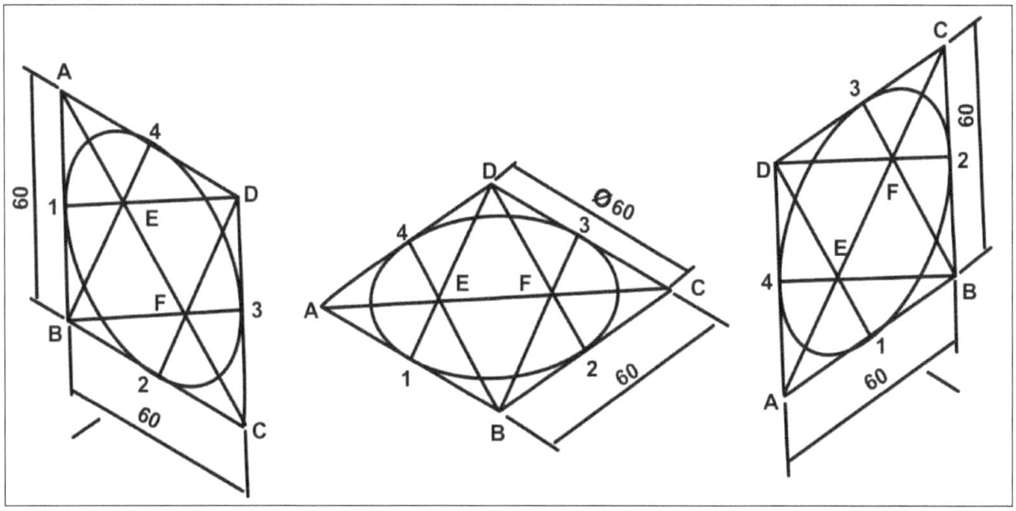

Figure 3.24

Constructing Isometric Arcs and Circles

Arcs and circles are common features on objects, especially in mechanical design and drafting. Isometric arcs are portions of isometric circles which are ellipses on isometric planes. Figure 3.25 shows a component with isometric arcs on the right face or right iso-plane.

Since the arcs are portions of isometric circles, the technique for creating iso-circles will be discussed. It is worth noting that an isometric arc can be constructed without creating a full isometric circle.

One important rule to remember when creating curves in isometric projection is that the isometric face or plane the curves lie on should be created first using guide or construction lines.

Then the curves can be created using projection of key points and intersection of projection lines from the key points.

A second rule is that true dimensions are transferred to non iso-plane.

Hence where there are inclined and oblique faces, the true sizes of features on the auxiliary views should be used during construction.

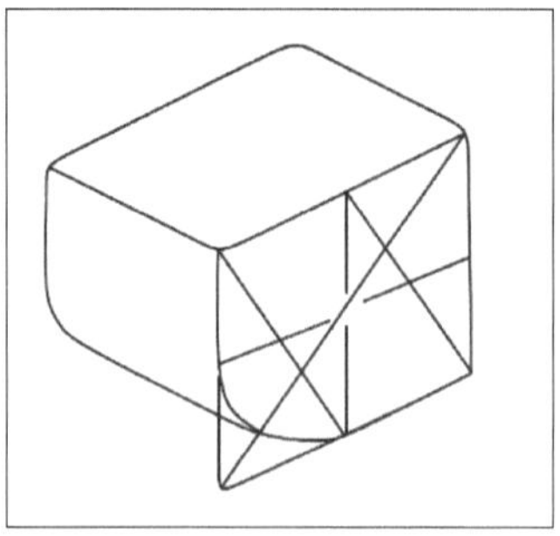

Figure 3.25

As mentioned earlier, isometric circles are ellipses and commonly called iso-circles. There are several techniques available for creating iso-circles, but an easy and more popular one is the four-centre ellipse.
This technique will be used here to create the three basic isometric circles: top iso-circle, left (front) circle, and right circle.
The four-centre ellipse is an approximate ellipse but it is usually good enough for most drafting applications. Fig. 3.26 shows in five steps, the creation of the top iso-circle.

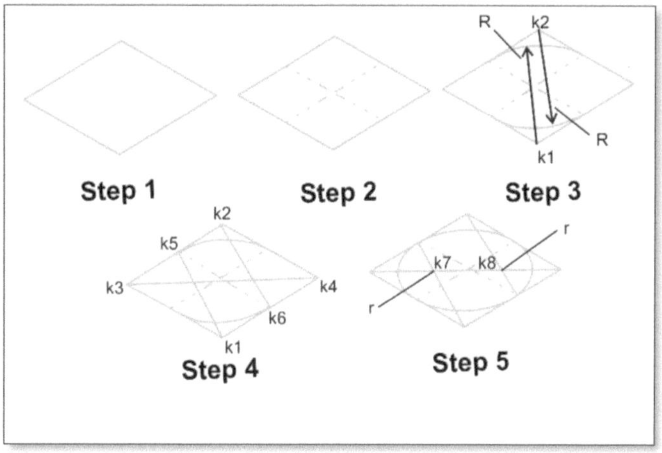

Figure 3.26

Step 1: Draw a square using the circle diameter as size for the top iso-circle, the top iso-plane is the right surface to draw the square. The top iso-plane is horizontal as can be seen in step 1 of Figure 3.26 Draw the isometric square.
Step 2: Draw the centre lines of the square. Draw the two centre lines of the square as shown in Step 2 of Figure 3.26.
Step 3: Draw the big arcs of the iso-circle. Identify the key points K1 and K2. These are two centres of the four centre ellipse technique. Notice that these centres are located at the obtuse angle corners of the isometric square.
Using the radius R, with centres at K1 and K2 draw the two big arcs for the iso-circle as shown in Step 3 of Figure 3.26
Step 4: Locate the centres of the small arcs of the iso-circle. Draw the diagonal K3-K4 between the acute angle corners of the square in Figure 3.26 Then draw lines K1-K5 and K2-K6. The intersection (K7) of the lines K3-K4 and K1-K5 in Step 5 locates one centre for a small arc.
The other small arc centre is located at K8, the intersection of lines K3-K4 and K2-K6.
Step 5: Draw the small arcs of the iso-circle. Using the centres of the small arcs K7 and K8, draw the two small arcs of radius r, as shown in Step 5 of Figure 3.26 Verify that the big and small arcs are tangent to the isometric square.
If a CAD package is used, circles could be drawn instead of arcs. The circles must then be trimmed to obtain the arcs required in the iso-circle.

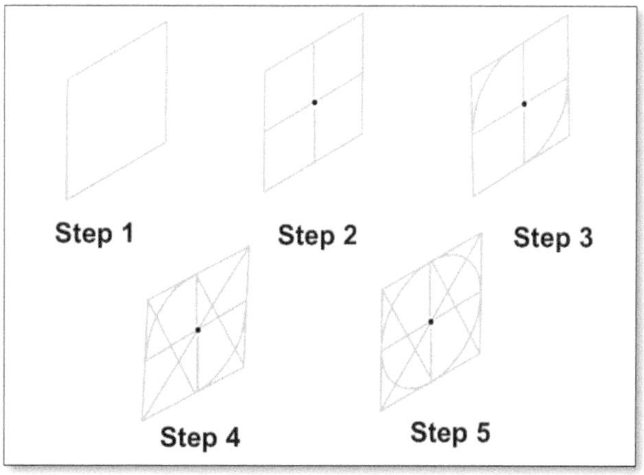

Figure 3.27

Figure 3.27 and Figure 3.28 show, respectively, in five steps how the left and right iso-circles can be created. These steps are the same as described above in Figure 3.26 for the top iso-circle, except that the iso-planes are respectively the left and right ones.

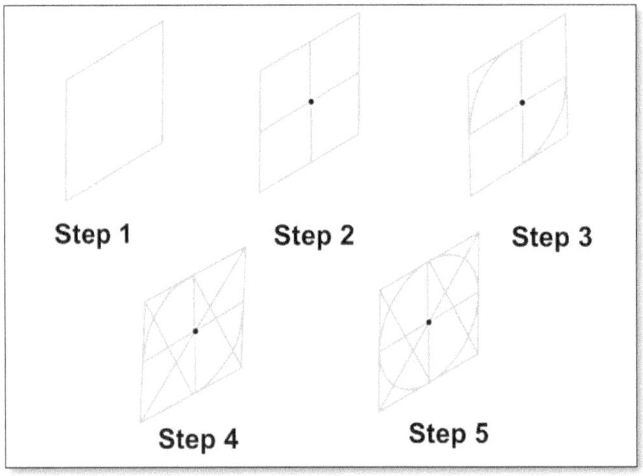

Figure 3.28

The construction of isometric arcs follows the same steps as iso-circles. However, a quick visual inspection of the arc in a problem will reveal which quadrant(s) the arc is located in.
Quarter arcs and half circle arcs are quite common in mechanical drafting.
For example, Figure 3.28 has a quarter arc on one of the acute angle corners, requiring the construction one of the small radius arcs in an iso-circle.
The five steps described above for creating iso-circles could be reduced to three as shown in Figure 3.29 by combining steps 1 and 2 as Step 1; and combining steps 3 (without drawing the large arcs) and 5 as Step 3.
This leaves Step 4 above as the new Step 2 in which all the key points K1 to K8 are created. The centres of the four arcs can then be identified as K1, K2, K7, and K8. In the last step (new Step 3), the four arcs are created as shown in Figure 3.29

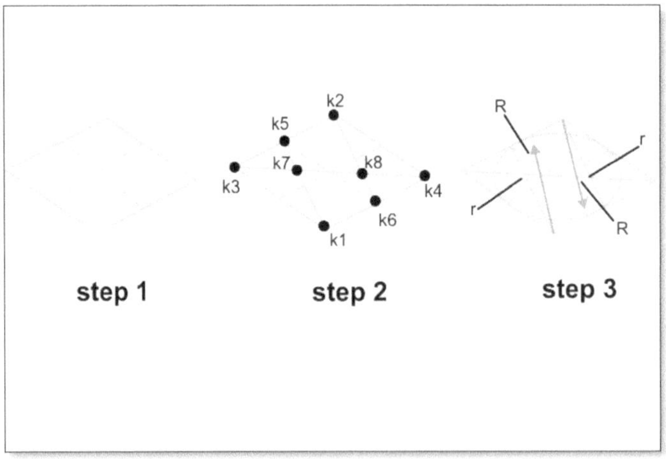

Figure 3.29

Problem 7: Shows the front view of a semi-circle whose surface is parallel to the V.P. Draw its isometric view.
(i) Enclose the semi-circle in a rectangle. Draw the isometric view of the rectangle.
(ii) Using the four-centre method, draw the half-ellipse in it which is the required view. The centre for the longer arc may be obtained as shown or by completing the rhombus.
If the view given in figure 3.30 is the top view of a horizontal semi-circle, its isometric view would be drawn as shown in figure 3.30 and figure 3.31

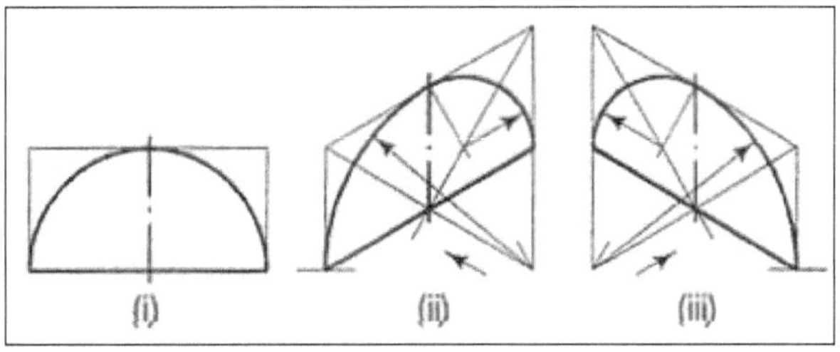

Figure 3.30

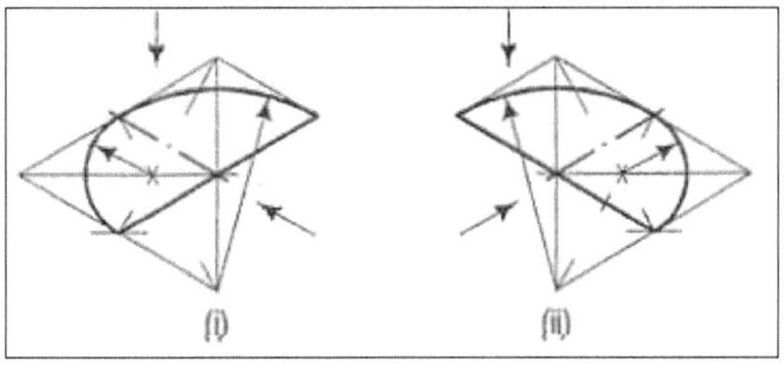

Figure 3.31

Problem 8: A circular plate of negligible thickness and 50 mm diameter appears as an ellipse in the front view, having its major axis 50 mm long and minor axis 30 mm long. Draw its top view when the major axis of the ellipse is horizontal.

Solution: A circle can be seen as an ellipse in the F.V. only when its surface is inclined to VP. So, for the first view keep the plane to VP.

Incline the T.V. till the distance between the end projectors is 30 mm.

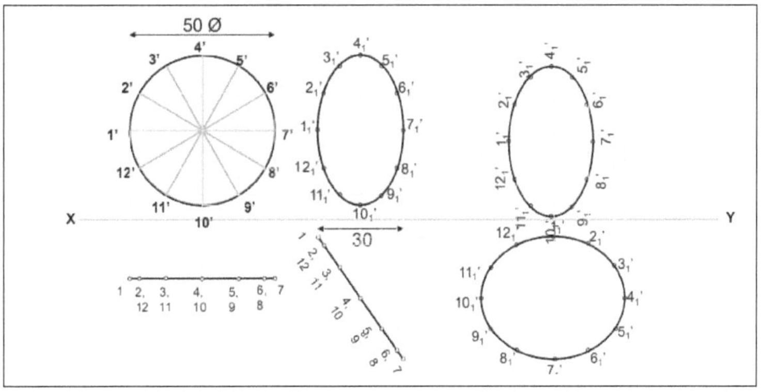

Figure 3.32

Problem 9: Shows the front view of a semi-circle whose surface is parallel to the V.P. Draw the isometric view of the semi-circle.

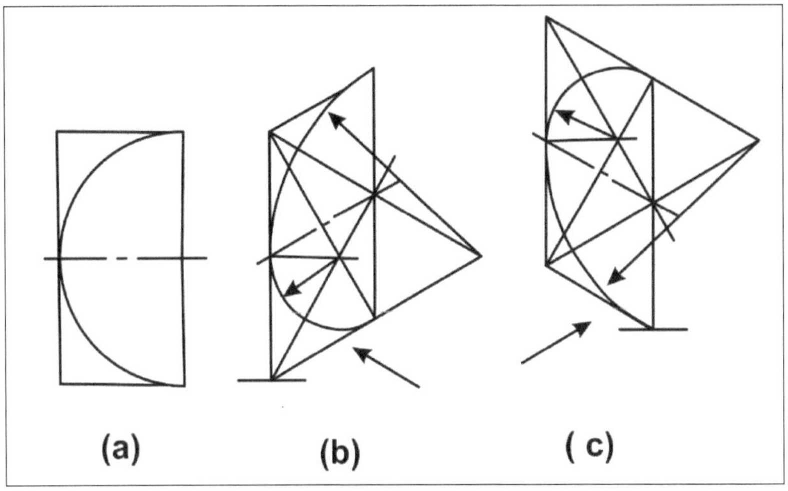

Figure 3.33

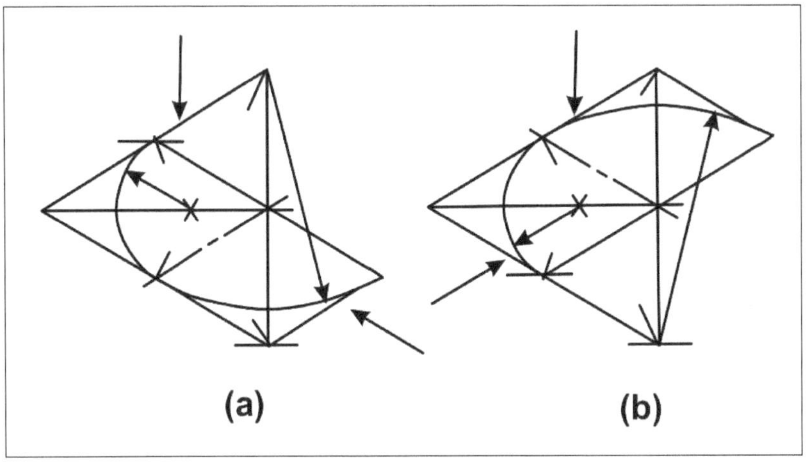

Figure 3.34

Problem 10: A semicircular plate of 80 mm diameter has its straight edge in the VP and inclined at 45 to HP. The surface of the plate makes an angle of 30 with the VP. Draw its projections.

Solution:

1. Plane in the V.P with straight edge perpendicular to H.P.
2. Plane inclined at 30 to the V.P and straight edge in the H.P.
3. Straight edge in V.P. inclined at 45 to the H.P.

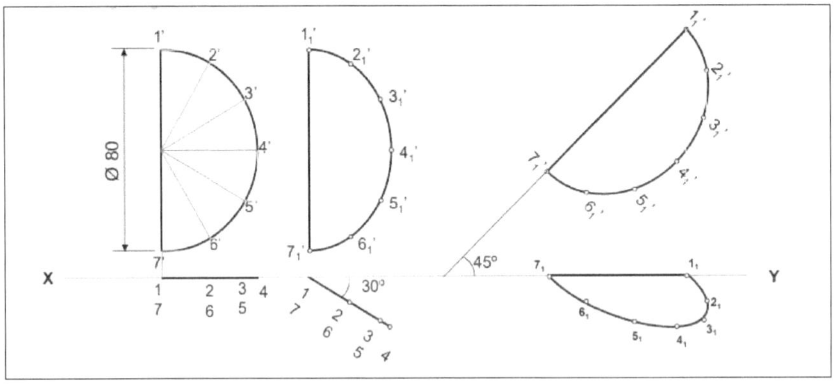

Figure 3.35

3.8 ISOMETRIC VIEW OF SOME SOLIDS

The Boxing-in Method

The four basic steps for creating an isometric drawing are: Determine the isometric viewpoint that clearly depicts the features of the object, then draw the isometric axes which will produce that view-point.

Construct isometric planes, using the overall width (W), Isometric views for solids. Construct isometric planes, using the overall width (W), height (H), and depth (D) of the object, such that the object will be totally enclosed in a box. Locate details on the isometric planes. Darken all visible lines, and eliminate hidden lines unless absolutely necessary to describe the object.

Sketch from an actual object:

1. Positioning object
2. Select isometric axis
3. Sketch enclosing box
4. Add details. Sketch from an actual object
5. Darken visible lines

Note: In isometric sketch/drawing), hidden lines are omitted unless they are absolutely necessary to completely describe the object. Sketch from an actual object.

ISOMETRIC VIEW OF SOLIDS CONATINING NON-ISOMETRIC LINES

The inclined lines of an object are represented non isometric lines in isometric projections. These are drawn by one of the following methods.

1. BOX METHOD:

In this box method, the object is assumed to be enclosed in a rectangular box and both the isometric and non-isometric lines are drawn by locating the corresponding points of contact with the surfaces and edge of the box.

This method is used when the non-isometric lines or their ends lie in isometric planes. The object is assumed to be enclosed in a rectangular box. Initially, the box is drawn in isometric. The ends of the lines for the inclined edges are then located by measuring on or from the outlines of the box.

Problem 11: Three views of a block are given. Draw its isometric view.

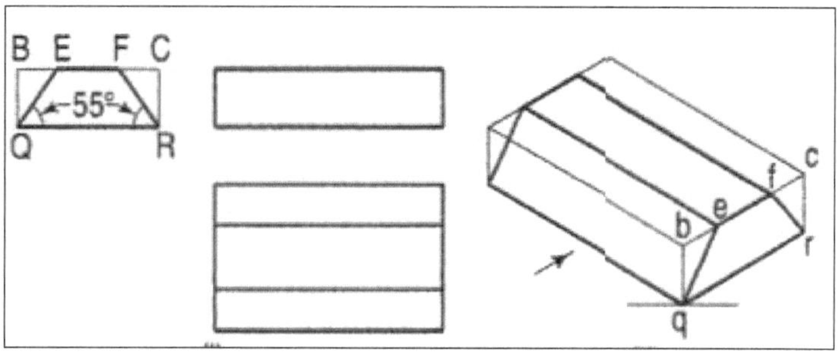

Figure 3.36

(i) Enclose the block in a rectangular box.
(ii) Draw the isometric view of the box in figure 3.37
(iii) Mark points e and f on the line be such that be = BE and fc = FC.
(iv) Complete the required view as shown.

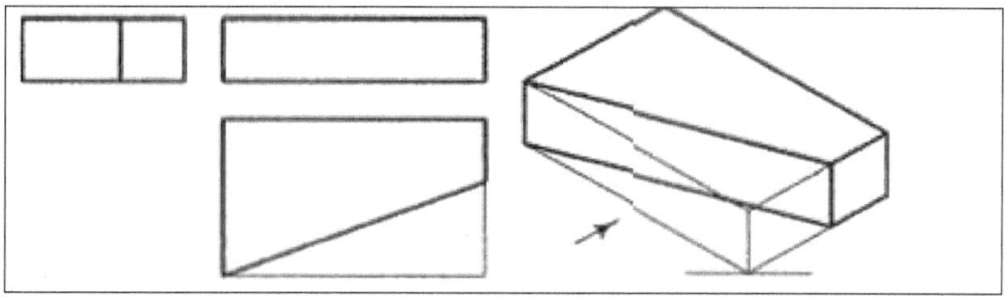

Figure 3.37

2. OFF-SET METHOD:

In this Off-set Method the lines parallel to isometric axes are drawn from every corner or reference of an end to obtain the corner or the reference point at the other end.
This method is adopted for objects in which neither non-isometric lines nor their ends lie in isometric planes. Perpendiculars are dropped from each end of the edge to a horizontal or a vertical reference plane.
The points at which the perpendiculars meet the plane, are located by drawing co-ordinates or offsets to the edges of the plane.

Problem 12: Draw the isometric view of the pentagonal pyramid, the projections of which are given in figure 3.38

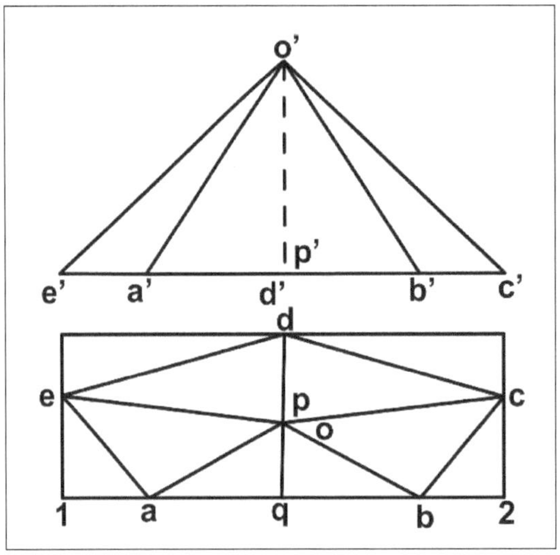

Figure 3.38

Solution:
(i) Enclose the base (in the top view) in an oblong.
(ii) Draw an offset oq (i.e. pq) on the line ab.
(iii) Draw the isometric view of the oblong and locate the corners of the base in it.
(iv) Mark a point Q on the line AB such that AQ = aq. From Q, draw a line QP equal to qo and parallel to 2C. At P, draw a vertical OP equal to o'p'.
(v) Join O with the corners of the base, thus completing the isometric view of the pyramid.

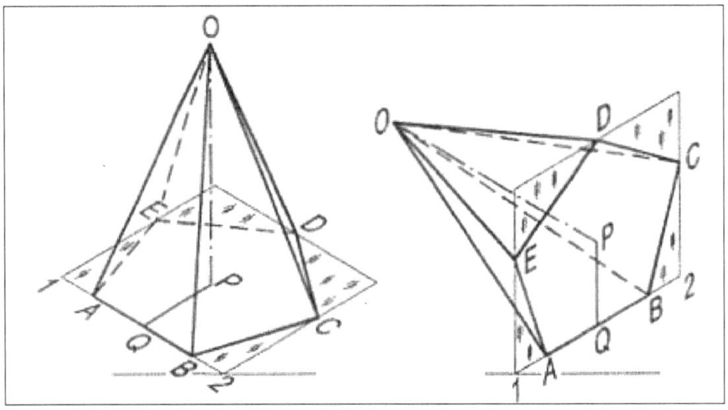

Figure 3.39

The Box Method is generally convenient for solving most of the problems

Problem 13: Draw isometric view of a hexagonal prism having a base with 30 mm side and a 70mm long axis resting on its base on the HP. With an edge of the base parallel to the VP when
(a) using Box Methods

(b) using Off-set Method
Solution:

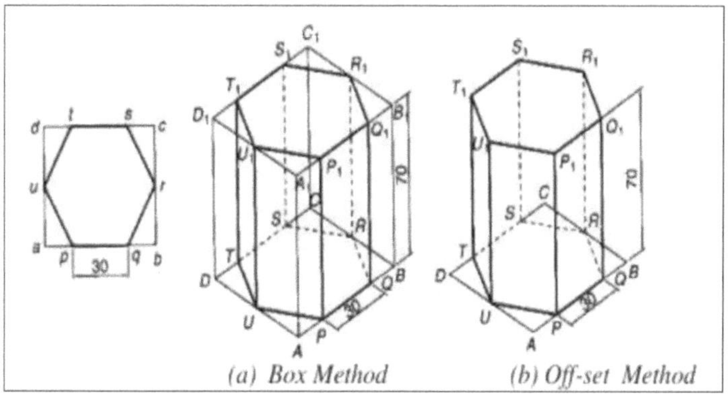

Figure 3.40

Problem 14: Draw an isometric view of a cylinder, with a 50 mm base diameter and a 70 mm long axis when
(a) The base is on the HP
(b) When one of the generators is on the HP?
Solution:

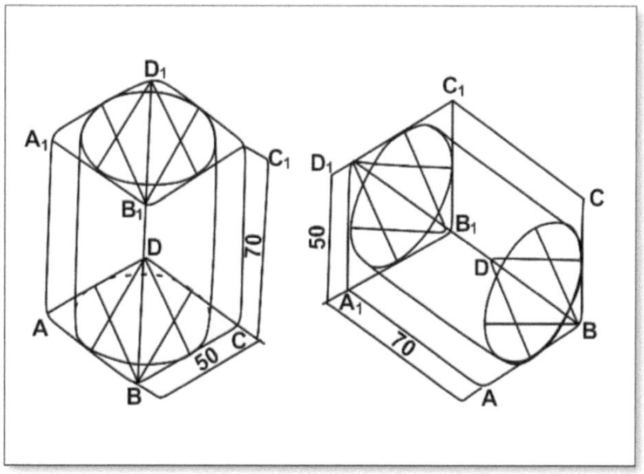

Figure 3.41

Problem 15: Draw an isometric view of a pentagonal pyramid having a base, with a 30 mm side and 50 mm long axis:
(a) When the its axis is vertical?
(b) When the its axis is horizontal?
Solution:

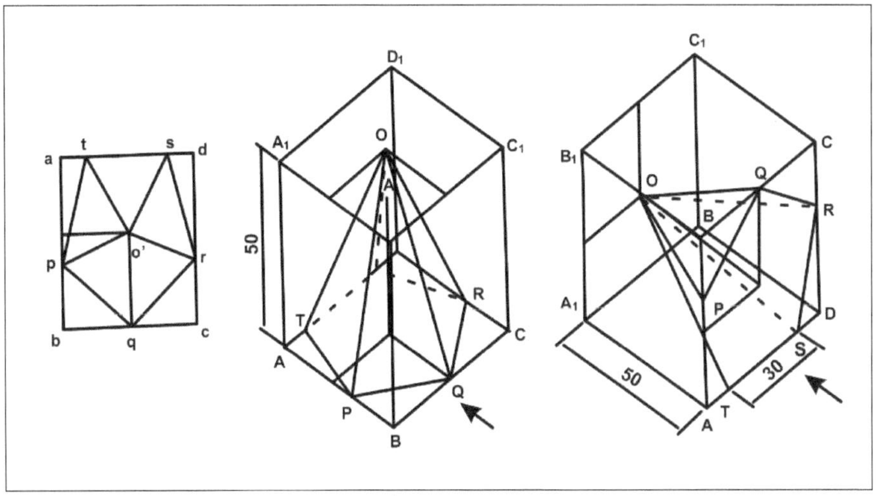

Figure 3.42

Problem 16: Draw an isometric view of Cone with a base diameter is 50 mm side and 70 mm long axis:
(a) When the base is on the HP?
(b) When the base is on the VP?
Solution:

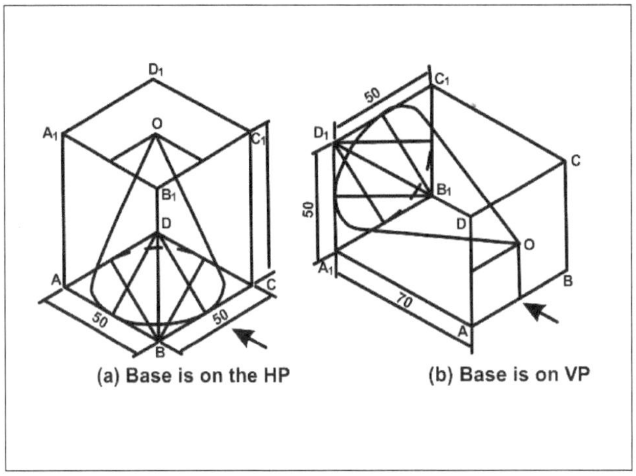

Figure 3.43

3.8.1 PRISM

Prism – a polyhedron formed by two equal parallel regular polygon, end faces connected by side faces which are either rectangles or parallelograms.

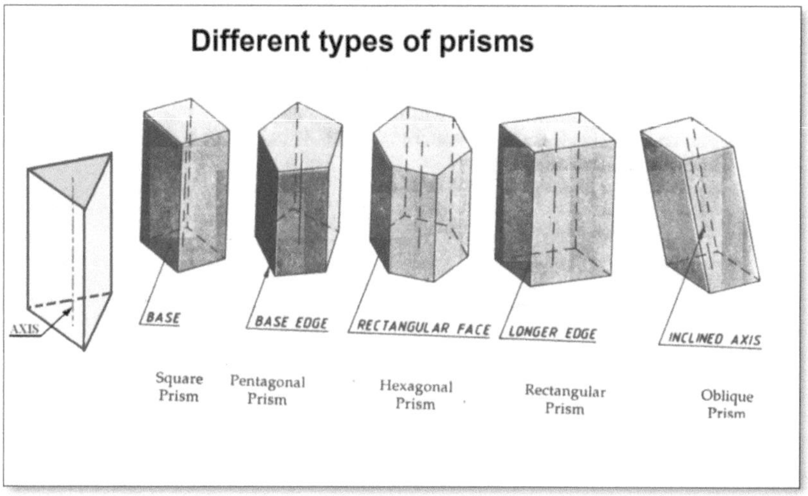

Figure 3.44

Problem 17: Draw the isometric projection of a square prism of side of base 35 mm and height 65 mm when its axis is
(i) vertical and (ii) horizontal.
Case (i) Axis of the square prism is vertical.
Case (ii) Axis of the square prism is horizontal.
Note: In isometric projection/view, hidden lines are not generally shown.
Solution:

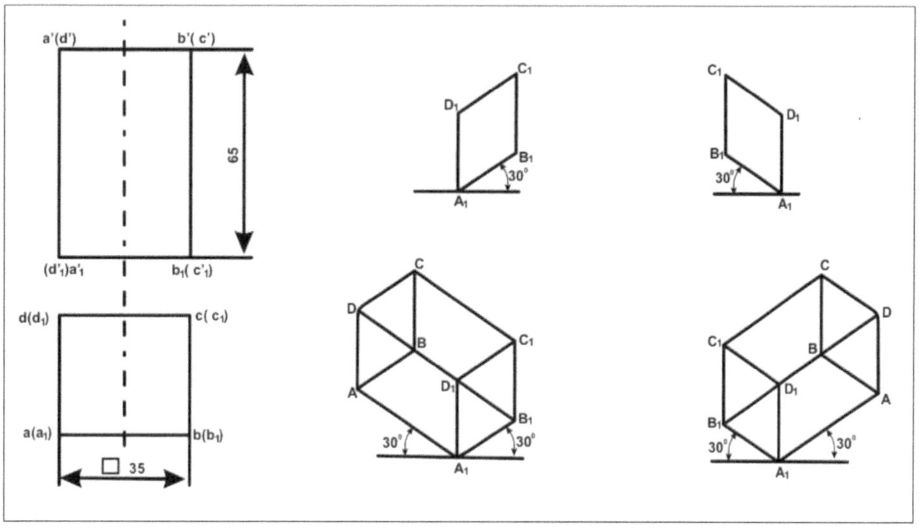

Figure 3.45

Problem 18: Draw an isometric view of a square prism having a base with 40 mm side and a 60 mm long axis, resting on the HP when:
(a) On its base with axis Perpendicular to the HP
(b) On its rectangular faces with axis perpendicular to the VP and
(c) On its rectangular face with axis parallel to VP
Solution:

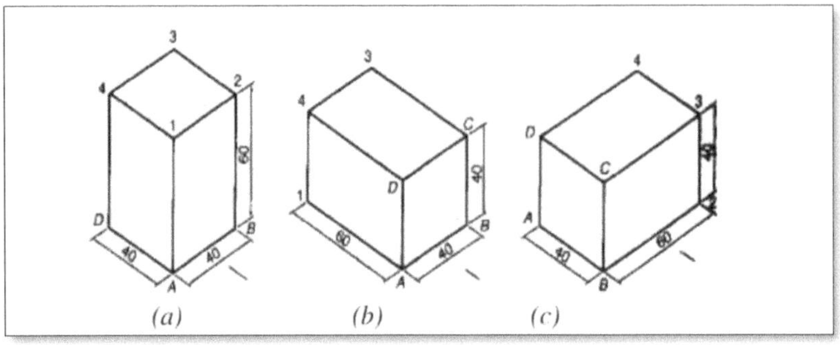

Figure 3.46

Problem 19: Draw the isometric projection of a pentagonal prism (base edge 25 mm, axial length 55 mm) resting on its face with its axis parallel to H.P. and V.P. both. Indicate the direction of viewing. Give all the dimensions.
ISOMETRIC PROJECTION OF A PENTAGONAL PRISM
Drawing helping figure.
Drawing both the isometric pentagons.
Drawing the four face edges.
Marking the axis and direction of viewing.
Dimensions
Solution:

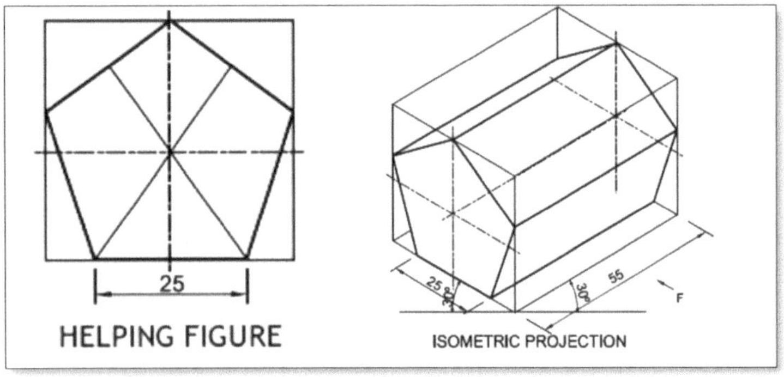

Figure 3.47

Problem 20: A hexagonal prism of 20 mm base edge and axis 45 mm long has a through square hole of 16 mm sides. The axis of the hole coincides with the axis of the prism. Two lateral faces of the hole are parallel to that of the prism.
Draw the isometric view of the solid when it is standing upright.
Solution:

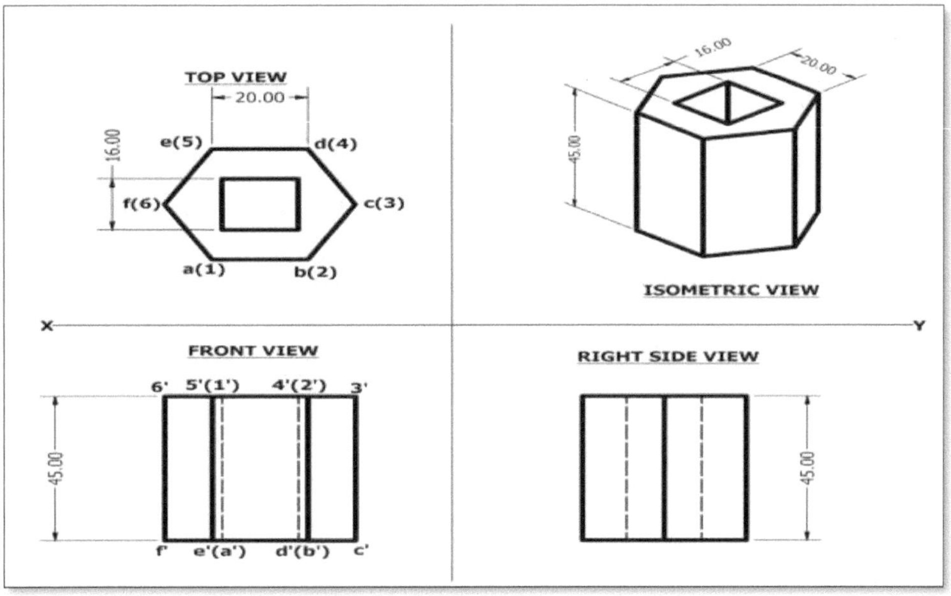

Figure 3.48

Problem 21: A thin rectangular plate of sides 60 mm X 30 mm has its shorter side in the V.P. and inclined at 30° to the H.P. Project its top view if its front view is a square of 30 mm long sides

Solution:

A rectangle can be seen as a square in the F.V. only when its surface is inclined to VP. So for the first view keep the plane // to VP & shorter edge ⊥ to HP

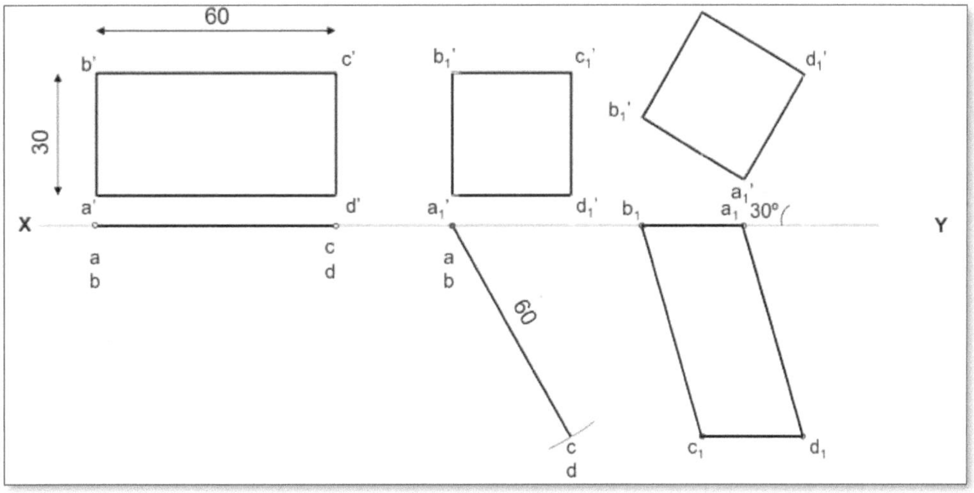

Figure 3.49

Problem 22: Draw isometric view of a hexagonal prism having a base with 30 mm side and a 70mm long axis resting on its base on the HP. With an edge of the base parallel to the VP when:

(a) using Box Methods

(b) using Off-set Method

Solution:

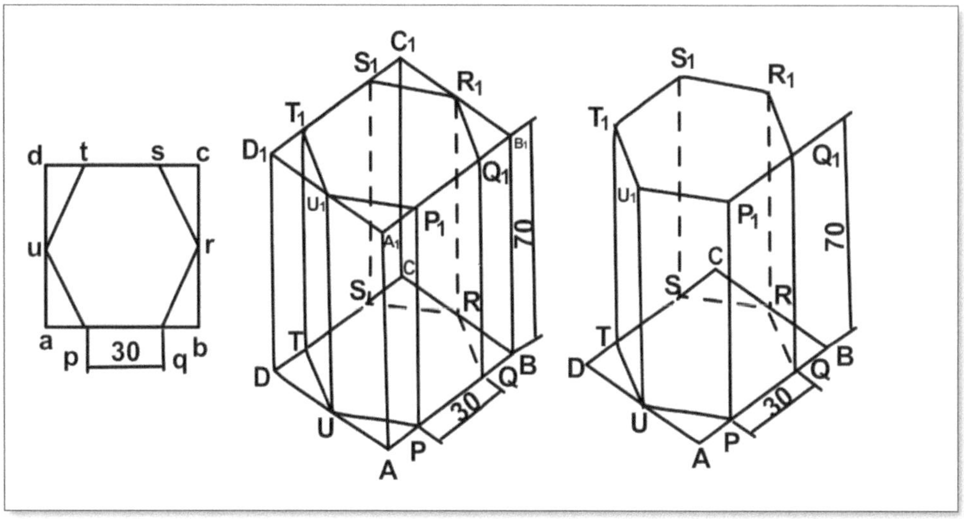

Figure 3.50

3.8.2 PYRAMID

A polyhedron formed by a plane surface as its base and a number of triangles as its side faces, all meeting at a point, called vertex or apex.

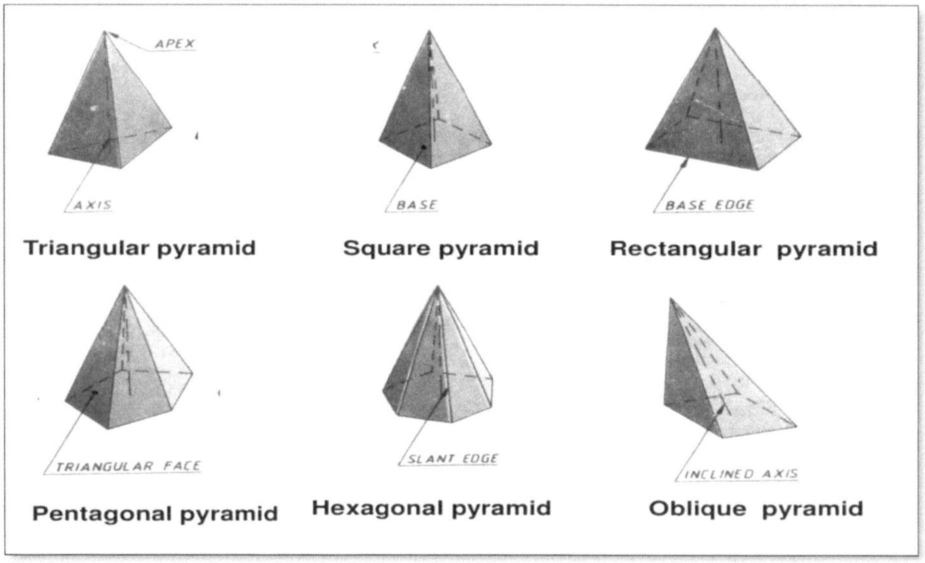

Figure 3.51

Pyramid placed with their axis perpendicular to VP in three different positions.

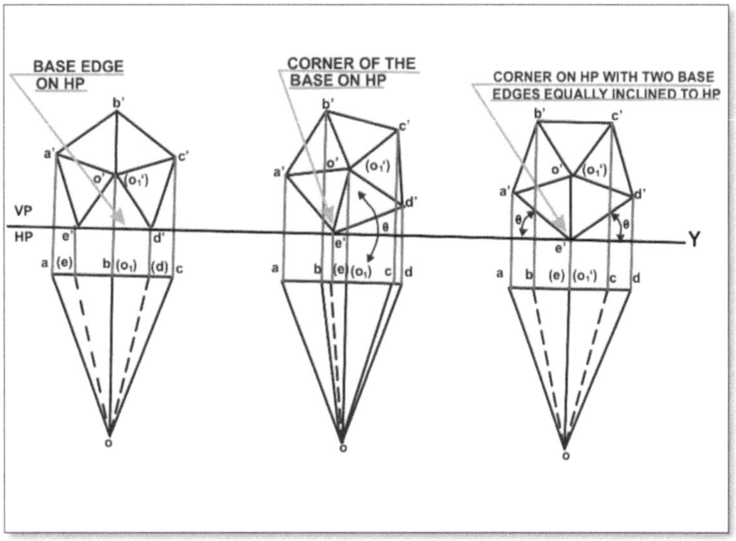

Figure 3.52

Problem 23: The isometric view of a hexagonal pyramid of side of base 30 mm and height 75 mm, when it is resting on HP such that an edge of the base is parallel to VP
Solution:

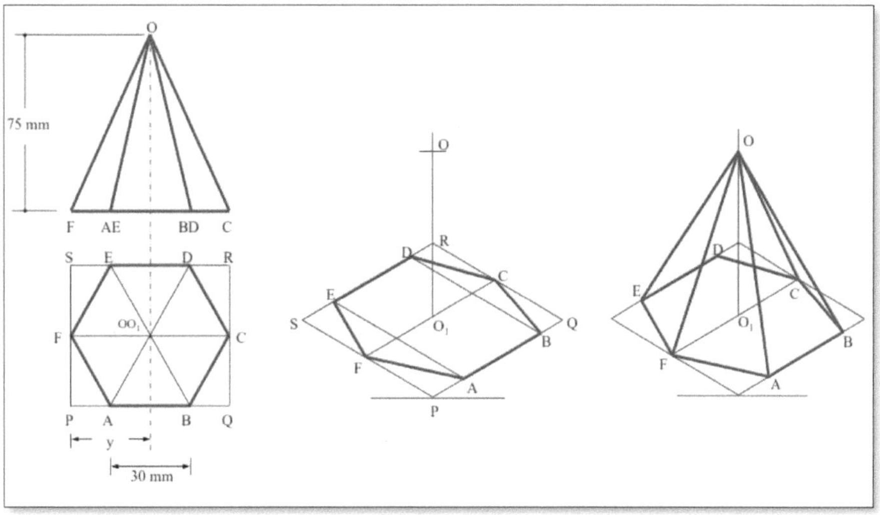

Figure 3.53

Problem 24: Draw the Isometric view of the pentagonal pyramid, the projections of which are given in figure 3.54(i)
Solution:
(i) Enclose the base (in the top view) in an oblong.
(ii) Draw an offset oq (i.e. pq) on the line ab.
(iii) Draw the isometric view of the oblong and locate the corners of the base in it [figure 3.54(ii)].
(iv) Mark a point Q on the line AB such that AQ = aq. From Q, draw a line QP equal to qo and parallel to 2C. At P, draw a vertical OP equal to o'p'.
(v) Join O with the corners of the base, thus completing the isometric view of the pyramid.

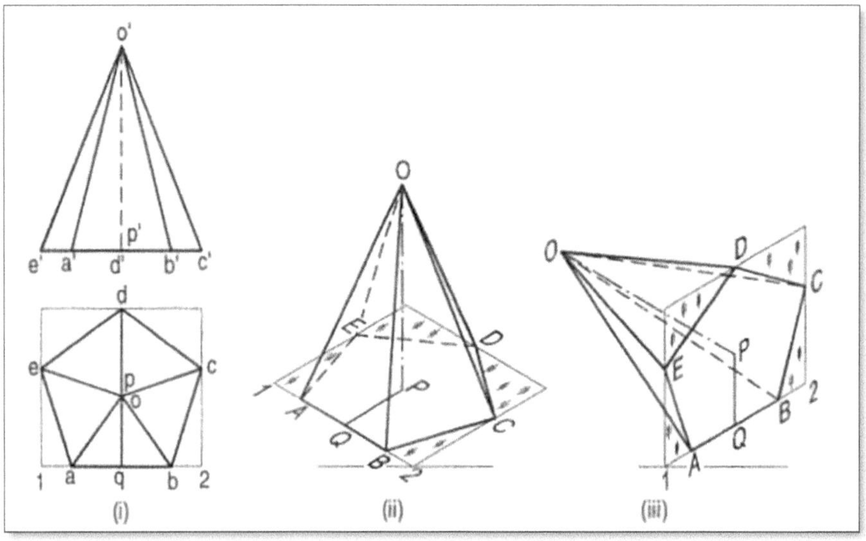

Figure 3.54

Problem 25: Draw the isometric view of the truncated triangular pyramid shown in figure 3.55(i).
Solution:
(i) Draw the perpendiculars d x', e' y' and f z' the front view and the offsets dq, er and fc in the top view.
(ii) Draw the isometric view of the whole pyramid [figure 3.55(ii)].
(iii) Transfer the offsets and the verticals to this view and obtain points D, E and F on the lines OA, OB and OC respectively.
(iv) Draw lines DE, EF and FD and complete the required isometric view.

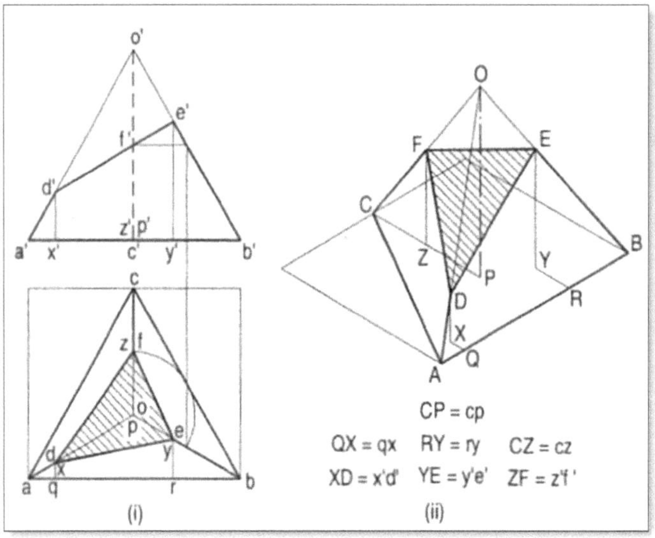

Figure 3.55

Problem 26: Draw an isometric view of a pentagonal pyramid having a base, with a 30 mm side and 50 mm long axis:
(a) when the its axis is vertical?
(b) when the its axis is horizontal?
Solution:

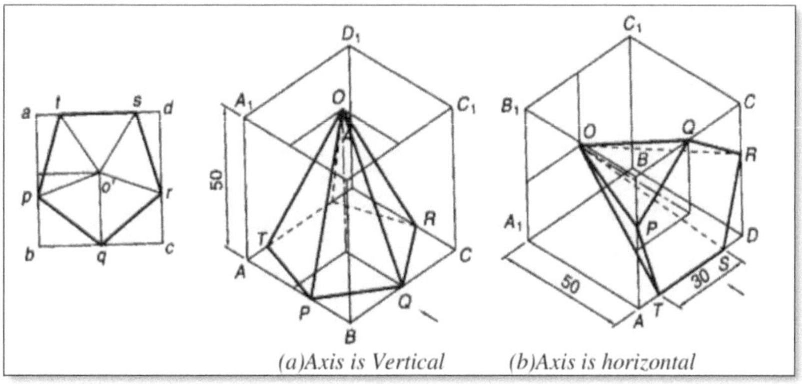

Figure 3.56

3.8.3 CONE

It revolved about one of its sides, the hypotenuse of the right triangle generates a cone.

Problem 27: Draw the isometric view of a cone, base 40 mm diameter and axis 55 mm long:

(i) when its axis is vertical and

(ii) when its axis is horizontal.

Solution:

(i) Draw the ellipse for the base [figure 3.57(a)]. Determine the position of the apex by the offset method.

(ii) Draw tangents to the ellipse from the apex. Erase the part of the ellipse between the tangents and complete the view as shown.

(iii) See figure 3.57 (b) which is self-explanatory.

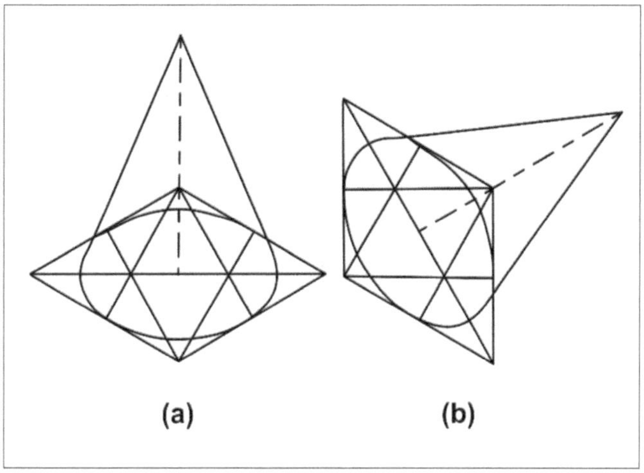

Figure 3.57

Problem 28: Draw an isometric view of Cone with a base diameter is 50 mm side and 70 mm long axis:

(a) When the base is on the HP

(b) When the base is on the VP?

Solution:

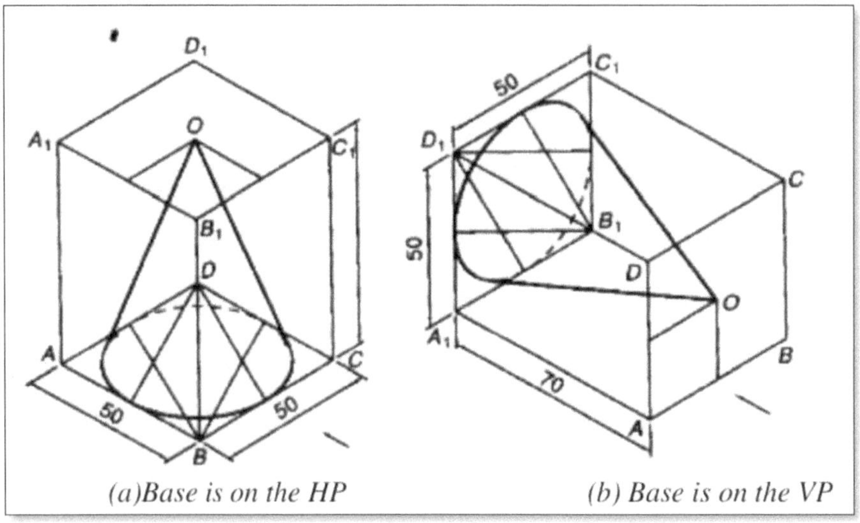

Figure 3.58

Problem 29: Draw the isometric view of the frustum of a cone shown in figure 3.59(i).
Solution:
(i) Draw the ellipse for the base [figure 3.59(ii)]. Draw the axis.
(ii) Around the top end of the axis, draw the ellipse for the top.
(iii) Draw common tangents, erase the unwanted part of the ellipse and complete the view as shown.

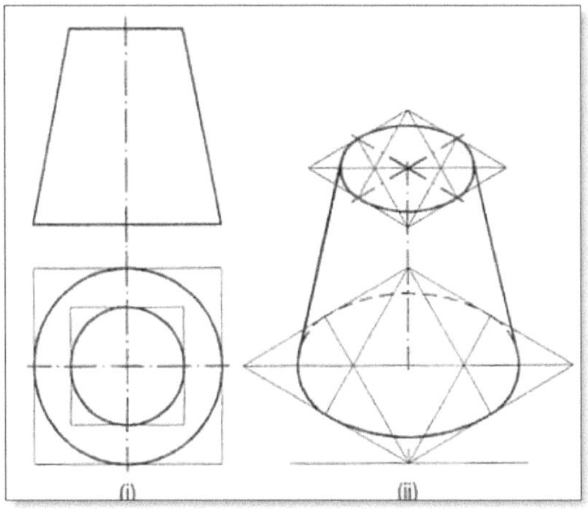

Figure 3.59

3.8.4 CYLINDER

When a rectangle is revolved about one of its sides, the other parallel side generates a cylinder.

Problem 30: Draw the isometric view of the cylinder shown in figure 3.60 The axis of the cylinder is vertical, hence its ends are horizontal. Enclose the cylinder in a square prism.
Draw in
Method I: Draw the isometric view of the prism.
Method II: Draw the rhombus for the upper end of the prism.

Solution:
Method I:
Draw the isometric view of the prism [figure 3.60(ii)]. In the two rhombuses, draw the ellipses by the four-centre method.

Draw two common tangents to the two ellipses. Erase the inner half of the lower ellipse and complete the required view.

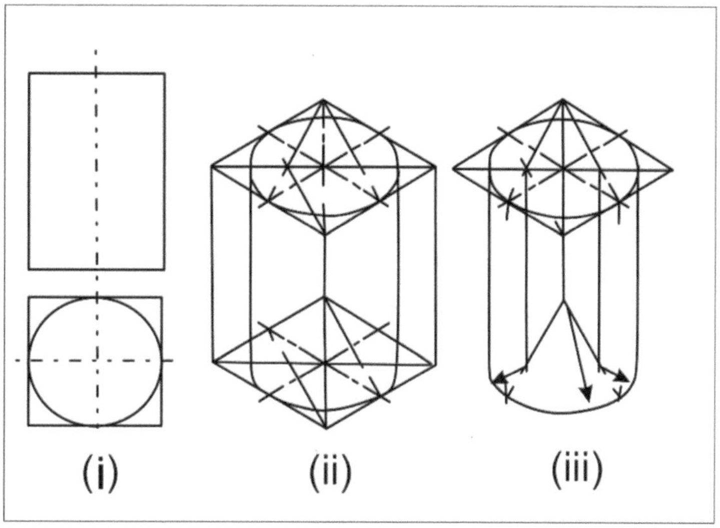

Figure 3.60

Method II:
Draw the rhombus for the upper end of the prism [figure 3.60(iii)] and in it, draw the ellipse by the four-centre method. From the centres for the arcs, draw vertical lines of length equal to the length of the axis, thus determining the centres for the lower ellipse.
Draw the arcs for the half ellipse. Draw common tangents, thus completing the required view.

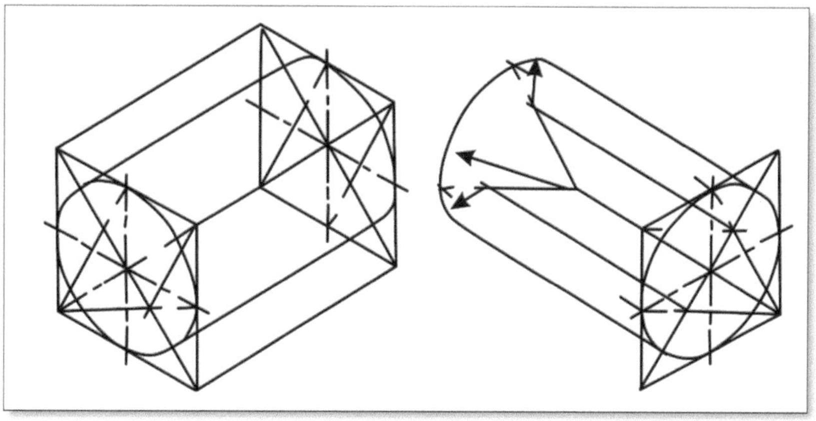

Figure 3.61

When the axis of the cylinder is horizontal, its isometric view is drawn by method I as shown in figure 3.61(i).

Figure 3.61(ii) shows the view drawn by method II, but the axis is shown sloping in the other direction.
Figure 3.62 respectively show the isometric views (drawn by method II) of a half-cylindrical disc with its axis in vertical and horizontal positions.

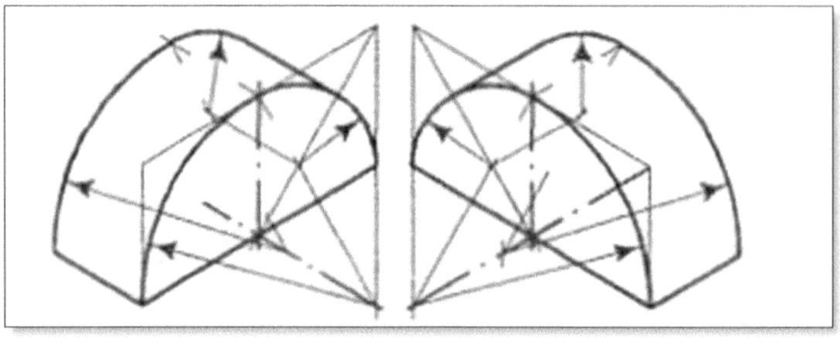

Figure 3.62

Problem 31: Draw an isometric view of a cylinder, with a 50mm base diameter and a 70mm long axis when:
(a) The base is on the HP.
(b) When one of the generators is on the HP?
Solution:

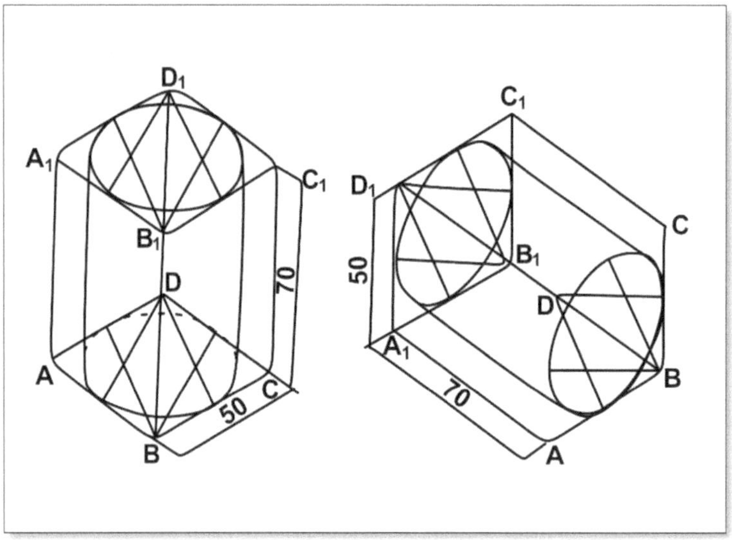

Figure 3.63

3.8.5 SPHERE

An orthographic view of a sphere viewed from any direction is a circle with a diameter equal to that of the sphere. Therefore, the isometric projection of a sphere will also be a circle of the same diameter, as explained below.
The figure shows the front and top views of a sphere on the ground. Figure 3.64(i). C is the centre, D is the diameter, and P is the point of contact with the ground.
Assume a vertical section through the centre of the sphere. Its shape will be a circle of diameter D. The isometric projection of this circle is shown in figure 3.64(ii) by ellipses 1 and 2, drawn in two different vertical positions around the same centre C.
The length of the major axis in each case is equal to D. The distance of the point P from the centre C is equal to the isometric radius of the sphere.
Again, we assume a horizontal section through the centre of the sphere. An isometric projection of this circle is represented by an ellipse 3 drawn horizontally around the same centre C.
Also, in this case the distance from the centre C to the outermost point on the ellipse is equal to OSD.
From this we can see that in the isometric projection, the distance from the centre of every point on the surface of a sphere is equal to the radius of the sphere.

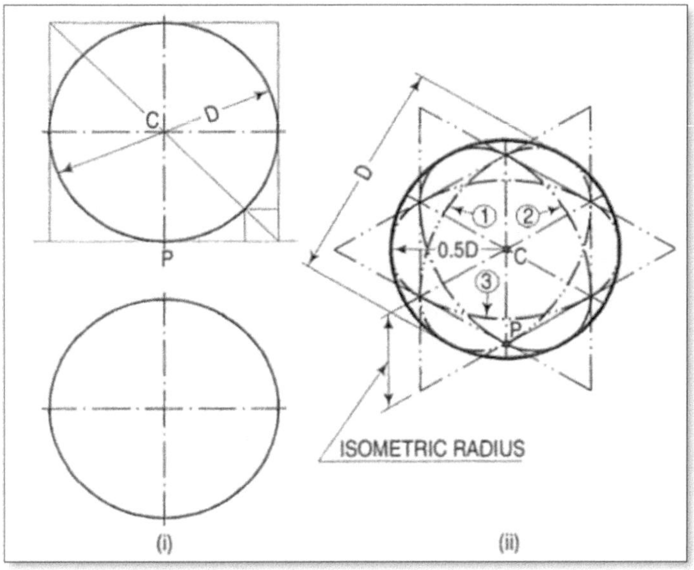

Figure 3.64

Therefore, the isometric projection of a sphere is a circle whose diameter is equal to the true diameter of the sphere. Also, in this case the distance from the centre C to the outermost point on the ellipse is equal to O.SD.
From this we can see that in the isometric projection, the distance from the centre of every point on the surface of a sphere is equal to the radius of the sphere.
Also, the distance from the centre of the sphere to the point of contact with the ground is the equiangular radius of the sphere, or HCP.
Therefore, it is very important to note that when drawing an isometric projection of a solid associated with a sphere or spherical part, you should always use an isometric scale. Hence, an orthographic view of a sphere viewed from any direction is a circle with a diameter equal to that of the sphere.

Problem 32: A Sphere with a 60 mm diameter is resting centrally on the Top of the Square Block with a 70 mm side 20 mm thickness. Draw an isometric view of the arrangement.
Solution: The isometric view of the arrangement is self explanatory.

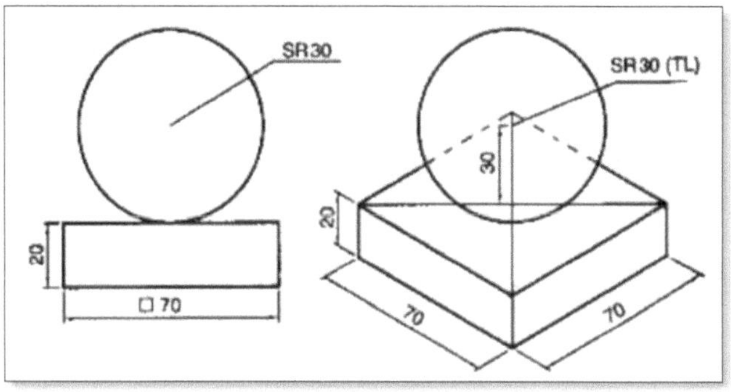

Figure 3.65

Problem 33: Draw the isometric projection of a sphere resting centrally on the top of a square prism, the front view of which is shown in figure 3.66

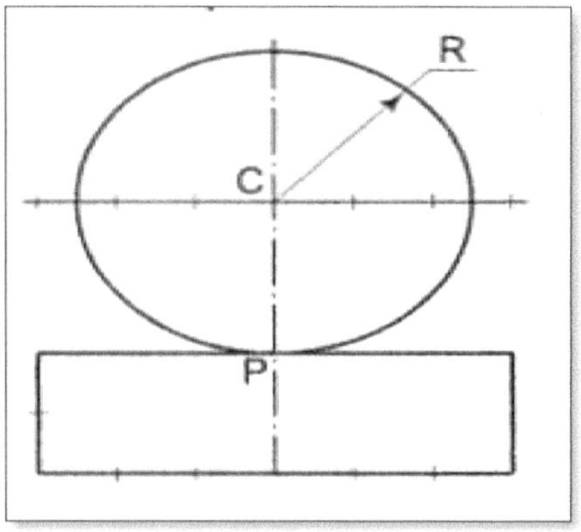

Figure 3.66

Solution:
(i) Draw the isometric projection (using isometric scale) of the square prism and locate the centre P of its top surface [figure 3.67].
(ii) Draw a vertical at P and mark a point C on it, such that PC = the isometric radius of the sphere.
(iii) With C as centre and radius equal to the radius of the sphere, draw a circle which will be the isometric projection of the sphere.

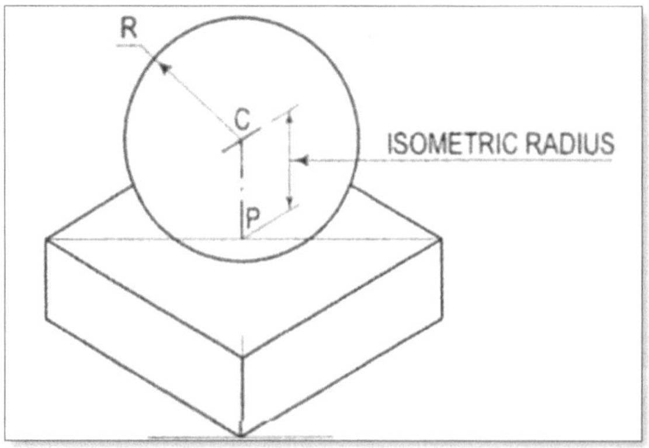

Figure 3.67

3.9 PROBLEMS RELATED TO OBJECTS CONTAINING LINES, CIRCLES AND ARCS

Solutions to the following typical problems are mostly self-explanatory. Construction line remains it is left as is for orientation purposes. Hidden lines were showing dotted lines in some views. This is to make the structure clearer. All dimensions apply unless otherwise stated Specified in millimetres. On an isometric drawing, circles appear as ellipses and arcs as elliptical arcs. It is obtained by using four-centre method explained below.

Circles and arcs pose a particular problem when drawing in perspective. When the perspective is altered, they change in terms of both length (as lines do) and shape – they become ellipses.

A straight line is the shortest route to join any two given points. It is a one-dimensional object having only length (l). The projection of straight lines are obtained by joining the top and front views of the respective end points of the line. The actual length of the straight line is known as true length (TL).

Problem 34: Draw the isometric view of a square with 40mm side.

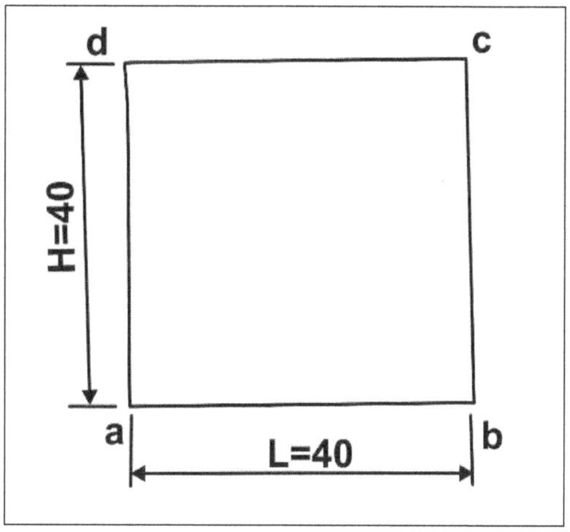

Figure 3.68

Solution: The isometric view of a square with 40mm side is self explanatory.

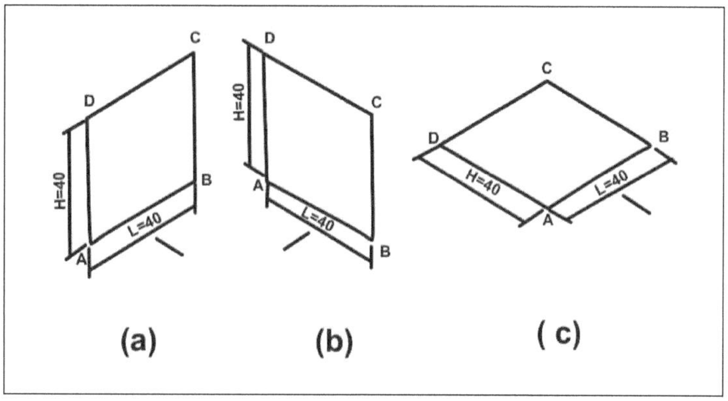

Figure 3.69

Problem 35: Draw the isometric view of a Hexagon with 40mm side such that its surface is Parallel to the HP and a side Parallel to the VP?

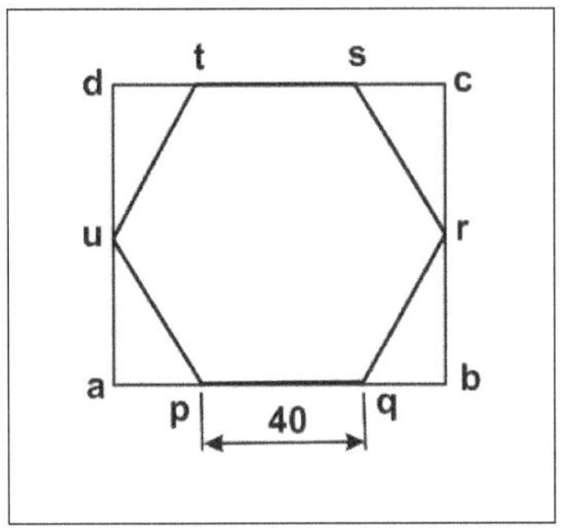

Figure 3.70

Solution: The isometric view of a Hexagon with 40mm side such that its surface is Parallel to the HP and a side Parallel to the VP is self explanatory as shown in the figure 3.71

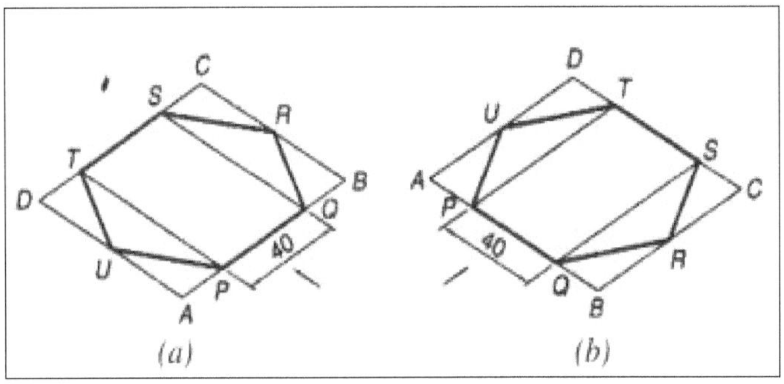

Figure 3.71

Problem 36: Draw the isometric view of a Circle with a 60mm Diameter on all three principle. Planes Using Coordinate methods.

Solution:

Construction Procedure:

1. Draw a circle with 60mm Diameter and enclose it in a square abcd.
2. Mark midpoints of the sides 1, 2, 3 and 4, where the square touches the circle tangentially
3. Draw the Diagonals of the square which cut in the circle at points 5, 6, 7 and 8 as shown in fig (1).
4. Draw a Rhombus ABCD to represent Isometric view of a square abcd.
5. Mark points 1, 2, 3 and 4 on it as the midpoint of the sides.
6. Mark points 5, 6, 7 and 8 on it, such that they are at a distance equal to Ax from the side of the square join points to obtain isometric view as shown in figures(2)(3)(4).

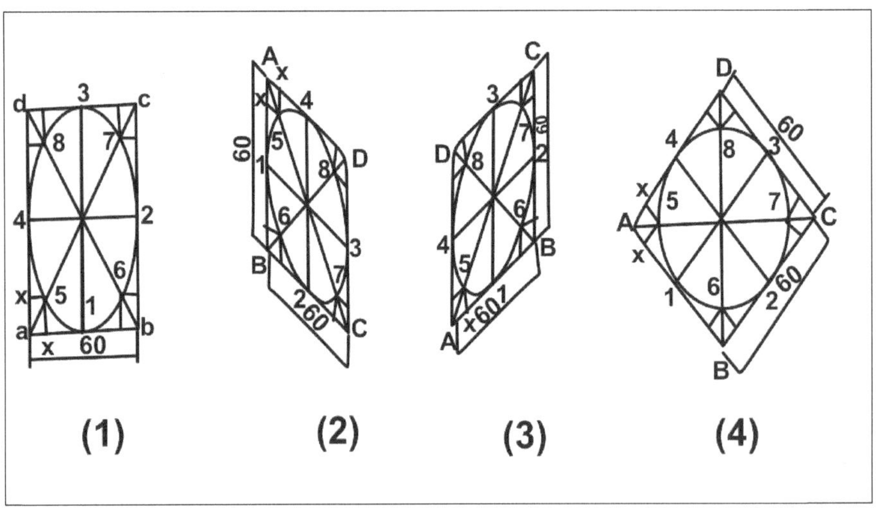

Figure 3.72

Problem 37: Draw the isometric view of a Circle lamina with a 60 mm Diameter on all three principal Planes using for centre methods.

Solution: Construction:

1. Draw a Rhombus ABCD of 60 mm side to represent isometric view of a square
2. Mark 1, 2, 3 and 4 as a mid-points of the sides AB, BC, CD and DA respectively join (the ends of the minor diagonals) B to meet points 3 & 4 and D to meet points 1 & 2. Let B4 and D1 intersect at point E and B3 and D2 intersect at a point F. then B,E,D and F are the Four centres for drawing the ellipse.
3. With centre B and radius B3 draw Arc 3-4. With centre D and Radius D1 draw Arc 1-2. With centre E and radius E1 draw Arc 1-4. With centre F and radius F2 draw Arc 2-3. 4.

These Arcs join in the form of an Ellipse which represents the required isometric as shown in figure (a) (b) (c).

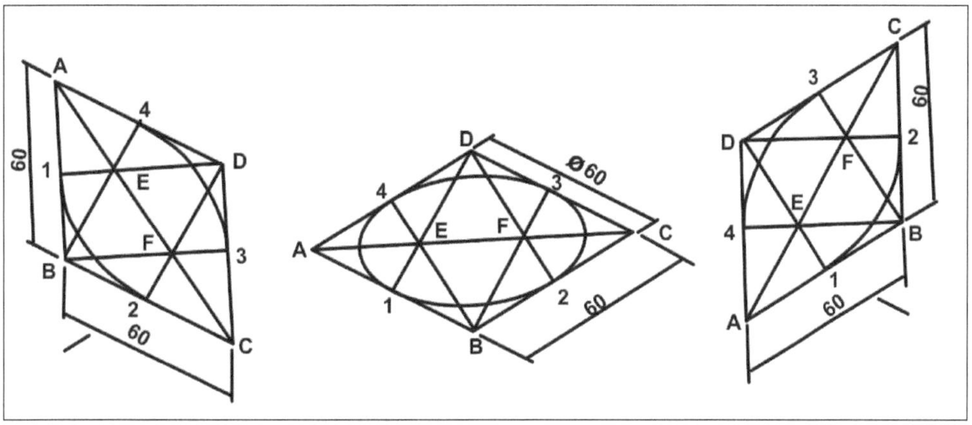

Figure 3.73

3.10 CONVERSION OF ORTHOGRAPHIC VIEWS INTO ISOMETRIC VIEWS/PROJECTION

Orthographic Projections - Orthographic Projections is a technical drawing in which different views of an object are projected on different reference planes.

Different Reference planes are:

– Horizontal Plane (HP)
– Vertical Plane (VP)
– Side or Profile Plane (PP)

Different views are:

– Front View (FV) – Projected on VP
– Top View (TV) – Projected on HP
– Side View (SV) – Projected on PP

Isometric projection - The **Isometric projection** of an object on a vertical plane of projection by placing the object in such a way that its three perpendicular edges make equal inclinations with the plane of projection.

DIFFERENCE BETWEEN ISOMETRIC VIEW AND ORTHOGRAPHIC VIEW

The main difference between isometric view and orthographic view is that an isometric view presents a three-dimensional object in a two-dimensional plane, while an orthographic view depicts the same three-dimensional object from multiple different angles at the same time.

Isometric views are often used for artistic or technical representation, as they can represent complex objects in one single 2D image.

Orthographic views, on the other hand, are mostly used in manufacturing designs as they provide detailed measurements of each side of the object to the manufactured.

Problem 38: Draw the isometric view of given orthographic drawing.

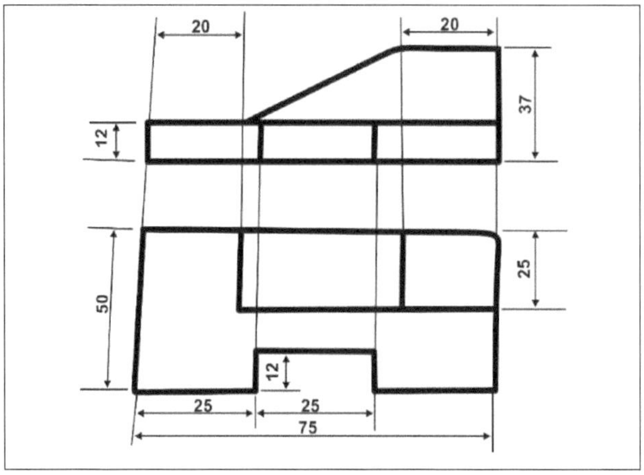

Figure 3.74

Solution: The conversion orthographic views into isometric views/projection is given below.

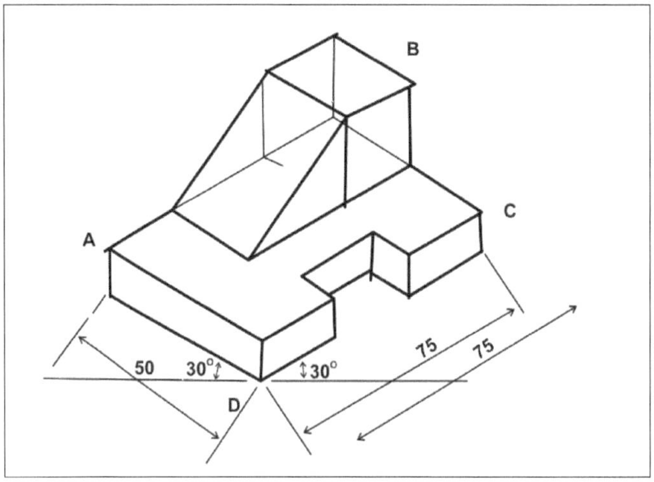

Figure 3.75

Problem 39: Draw the isometric view of given orthographic drawing.

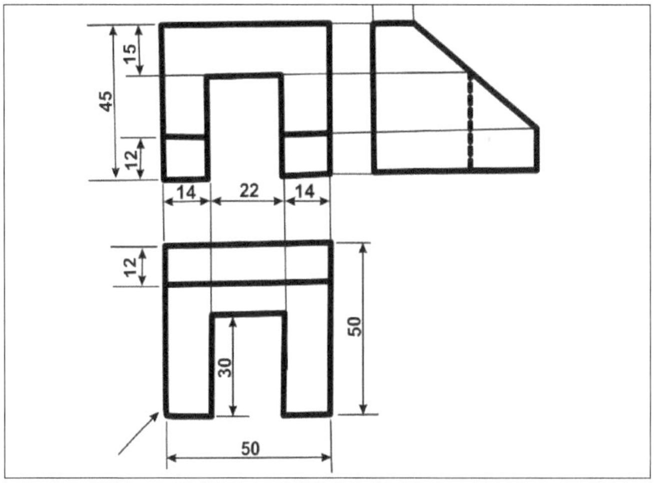

Figure 3.76

Solution: The conversion orthographic views into isometric views/projection is given below.

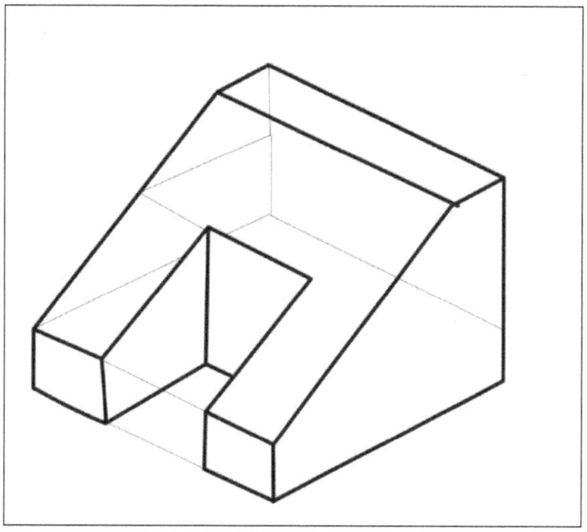

Figure 3.77

Problem 40: Draw the isometric view of given orthographic drawing whose F.V. and T.V. is given below. (2D→3D).

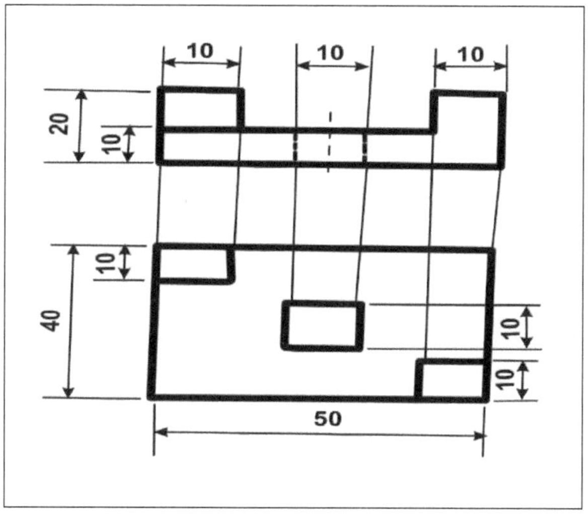

Figure 3.78

Solution: The conversion orthographic views into isometric views/projection is given below.

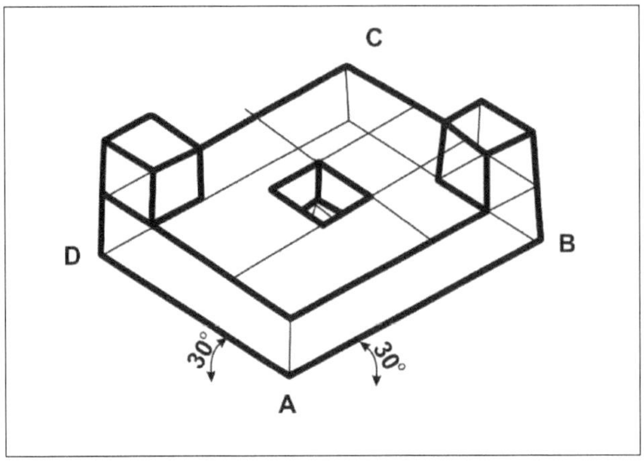

Figure 3.79

Problem 41: Draw the isometric view of given orthographic drawing.

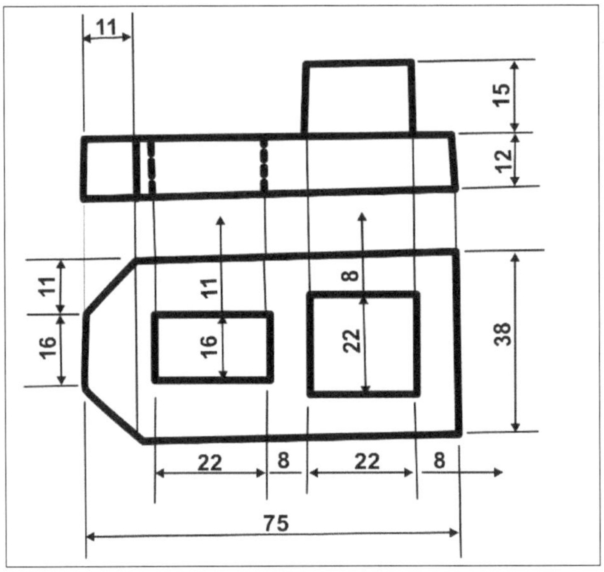

Figure 3.80

Solution: The conversion orthographic views into isometric views/projection is given below.

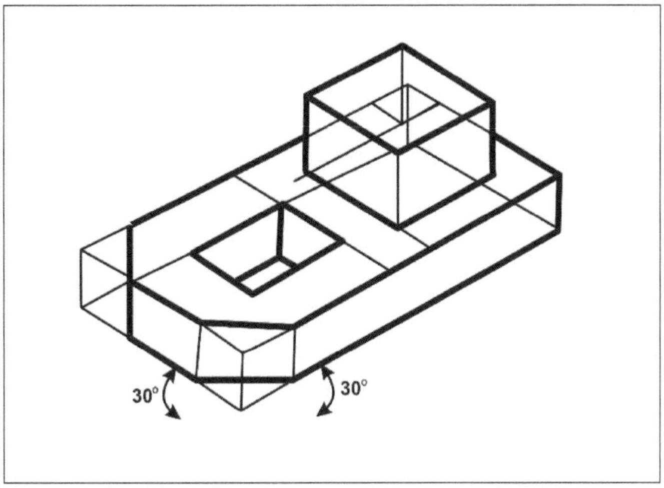

Figure 3.81

Assignment

1. An Axonometric Plane is inclined to ___ of the Principal Planes.
a) One
b) Two
c) Three
d) None
2. If Axonometric Plane is equally inclined to all the three principal plane then that plane is known as
a) diametric plane
b) isometric plane
c) trimetric plane
d) orthographic plane
3. If Axonometric plane is having different inclination with all the principal planes, then that plane is known as
a) orthographic plane
b) isometric plane
c) diametric plane
d) trimetric plane
4. In isometric projection, direction of sight is ____ to the plane of projection. i.e., Axonometric plane.
a) parallel
b) perpendicular
c) inclined
d) any of the above
5. Line in space ___ to principal planes are known as isometric lines.
a) perpendicular
b) inclined at 30°
c) inclined at 45°
d) parallel
6. Three lines ___ apart with one of them vertical, are known as Isometric Axes.
a) 30°
b) 60°
c) 90°
d) 120°
7. Isometric plane is
a) parallel to one of the principal planes and perpendicular to other two principal planes
b) parallel to two of the principal planes and perpendicular to other one principal plane
c) inclined to one of the principal planes and perpendicular to other two principal planes
d) parallel to one of the principal planes and inclined to other two principal planes
8. In isometric projection, the square faces of the cube are seen as

a) squares
b) rectangles
c) rhombuses
d) any of the above
9. In preparing isometric scale, true or actual scale is drawn at ___ to the horizontal.
a) 15°
b) 30°
c) 45°
d) 60°
10. Dimension of isometric projection is reduced by ___ approximately to the actual dimension.
a) 51%
b) 61.5%
c) 71.5%
d) 81.5%
11. Isometric projection is reduced volume wise by ____ to the actual size.
a) 34.13%
b) 44.13%
c) 54.13%
d) 64.13%
12. While drawing isometric view/drawing, scale is reduced by
a) 61.5%
b) 71.5%
c) 81.5%
d) no size reduction is required
13. Following is (are) the method(s) to construct isometric projection
a) conventional method
b) exact method
c) both (a) and (b)
d) none of the above
14. Isometric view of square is a
a) square
b) rectangle
c) rhombus
d) any of the above
15. Isometric view of a rectangle is a
a) rhombus
b) square
c) rectangle
d) parallelogram
16. 4-centre method is used to draw isometric view of a
a) triangle
b) square
c) circle
d) all of the above
17. Iso-metric view of a circle is a(n)
a) circle
b) ellipse
c) parabola
d) line
18. Box method is used to draw isometric view of
a) Pyramid
b) Prism
c) Cylinder
d) All of the above
19. Isometric projection of a sphere is a(n) ___ .
a) ellipse

b) semi-circle
c) circle
d) arc
20. In isometric projection, Extension lines and dimension lines are drawn ___ to the isometric axis.
a) parallel
b) perpendicular
c) inclined
d) any of the above
21. If isometric projection of an object is drawn with true lengths the shape would be same and size is how much larger than actual isometric projection?
a) 25%
b) 29.5%
c) 22.5%
d) 33.3%
22. If an isometric projection is drawn with true measurements but not with isometric scale then the drawings are called ____________
a) Isometric projection
b) Isometric view
c) Isometric perception
d) Orthographic view
23. If an isometric projection is drawn with true measurements but not with isometric scale then the drawings are called ____________
a) Isometric projection
b) Isometric view
c) Isometric perception
d) Orthographic view
24. Identify the front view of the below isometric view.

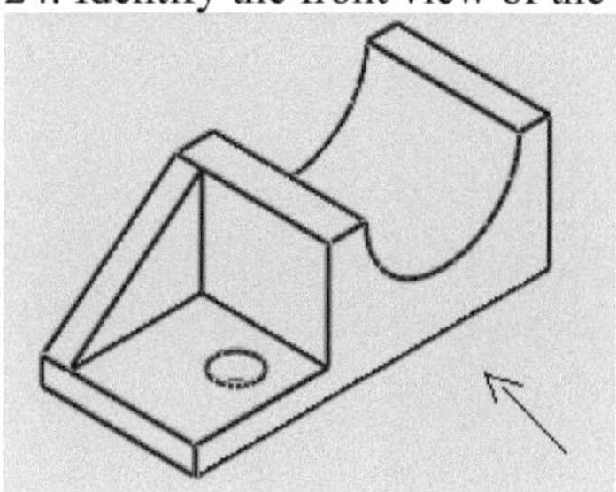

a) 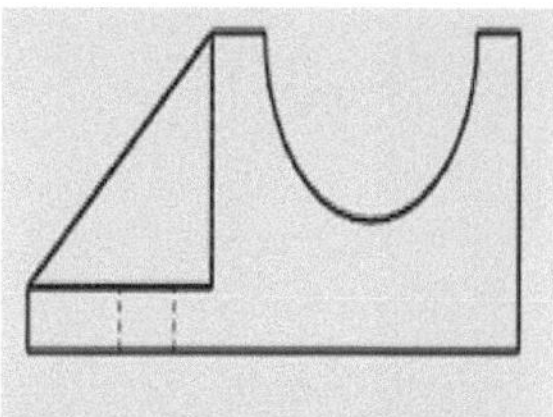

b) 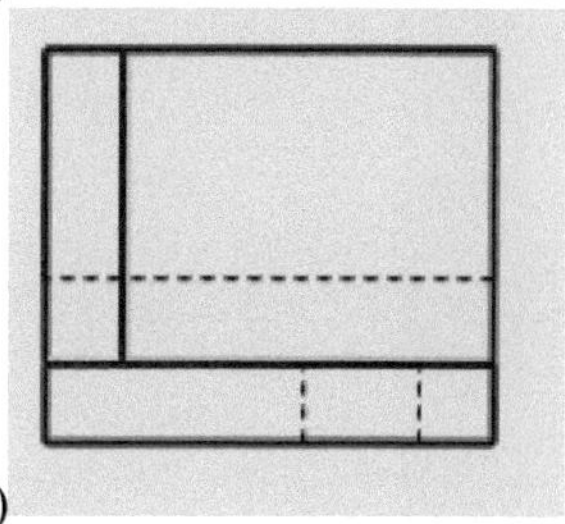

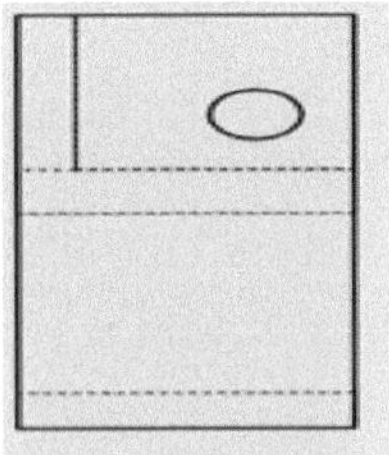
c)

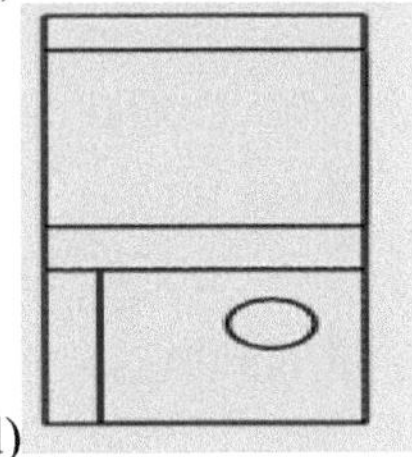
d)

25. Identify the top view of below isometric view.

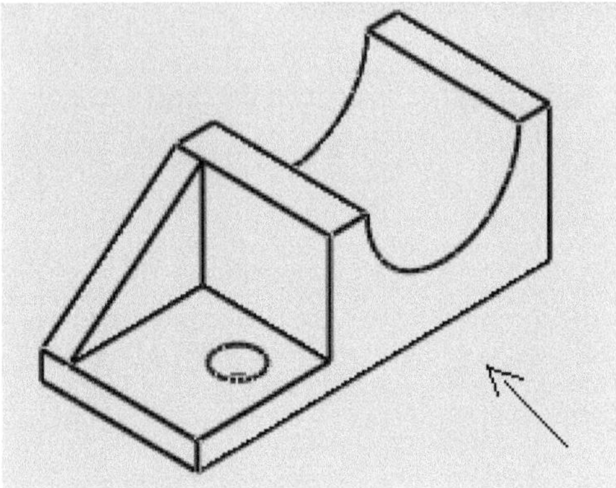

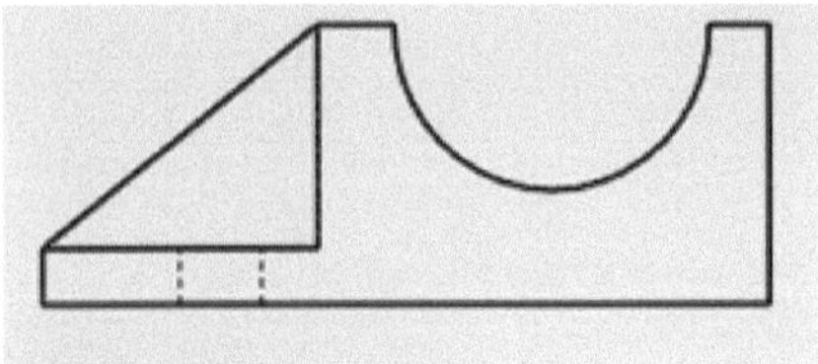
a)

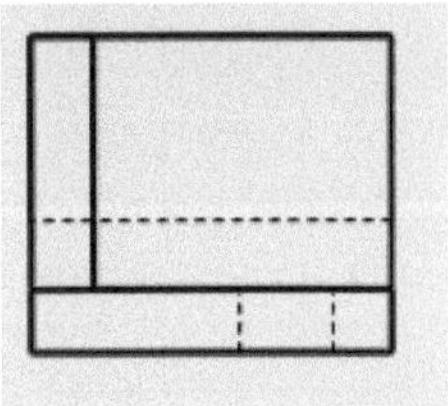
b)

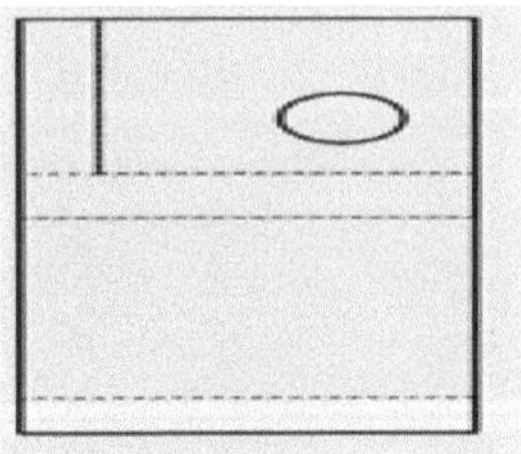
c)

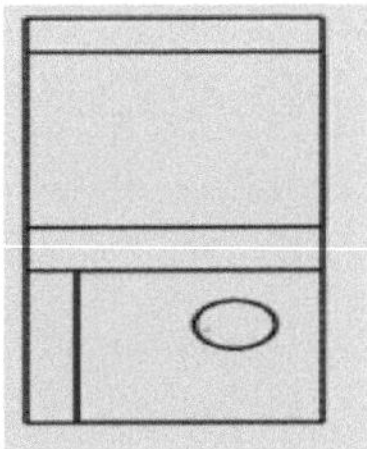

d)

26. Identify the side view of the below isometric view.

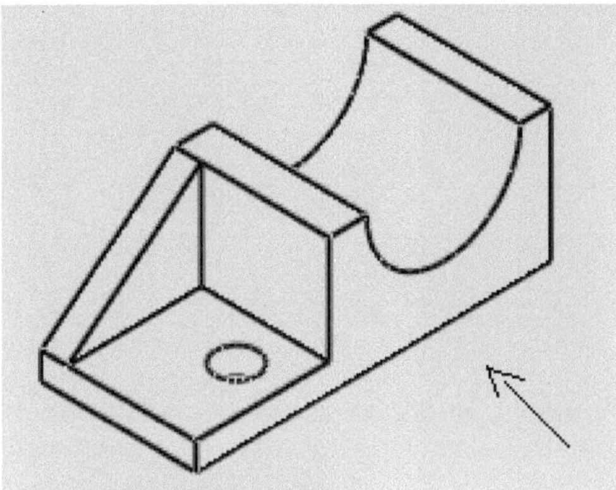

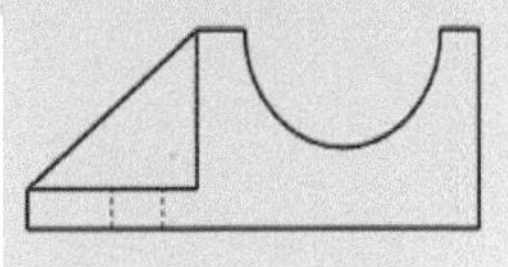

a)

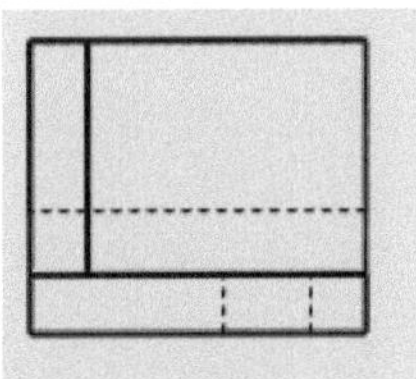

b)

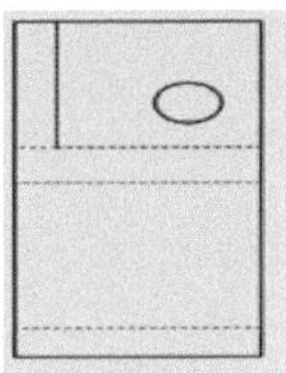

c)

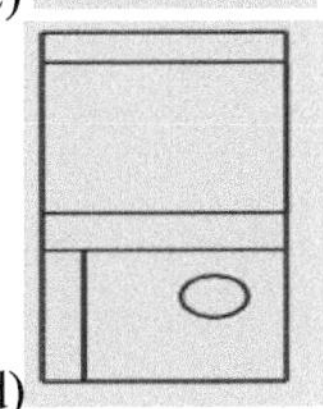

d)

27. Identify the side view of the below isometric view.

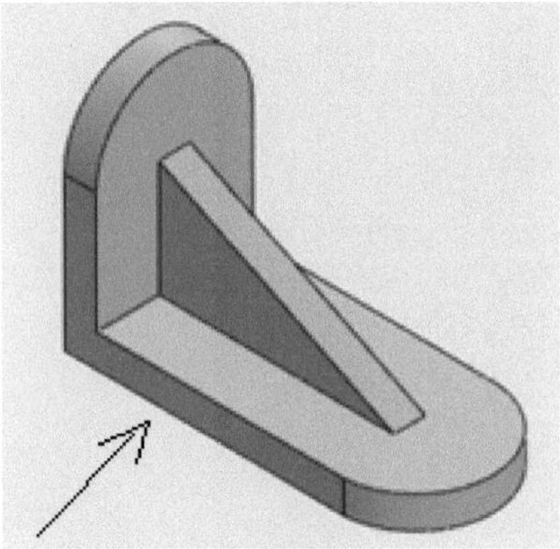

a) 

b) 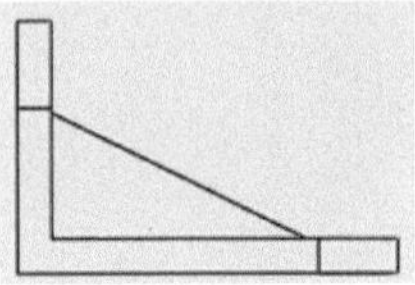

c) 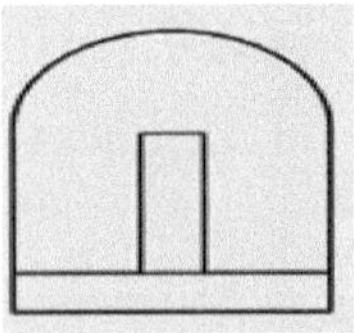

d) 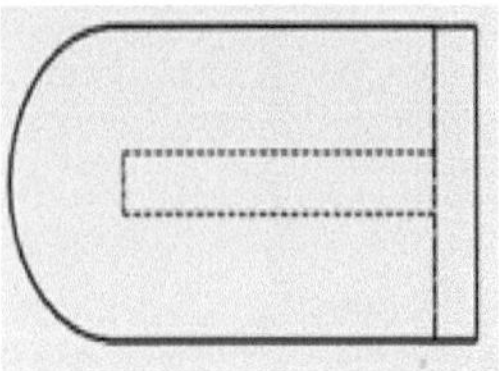

28. Identify the top view of the below isometric view.

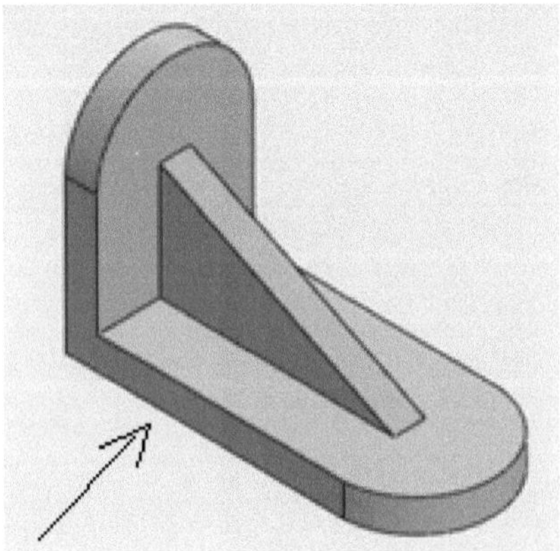

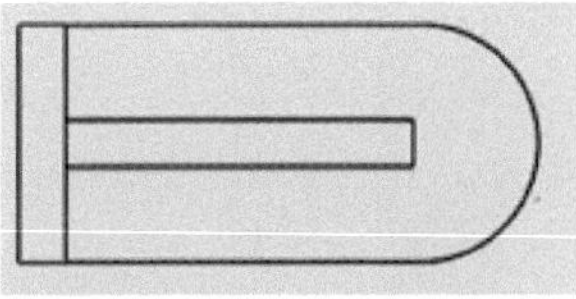
a)

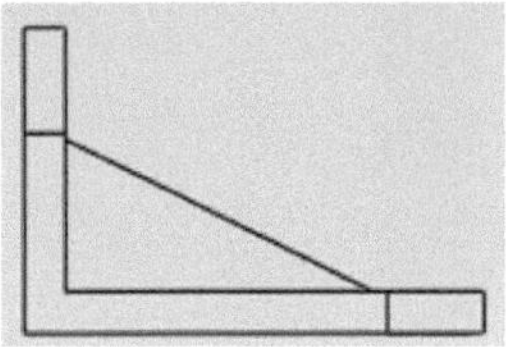
b)

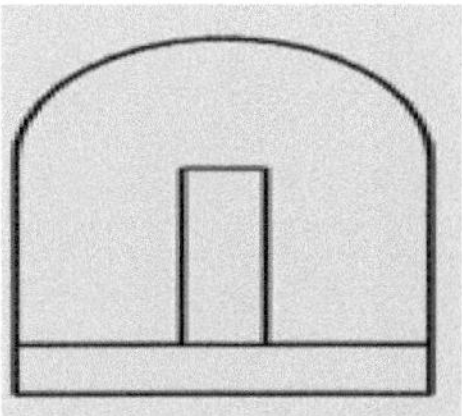
c)

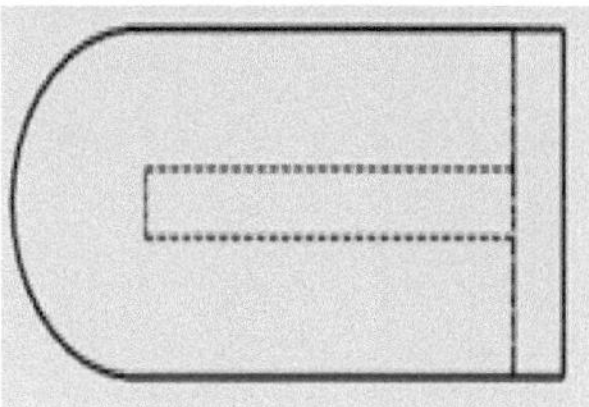
d)

29. Identify the front view of the below isometric view.

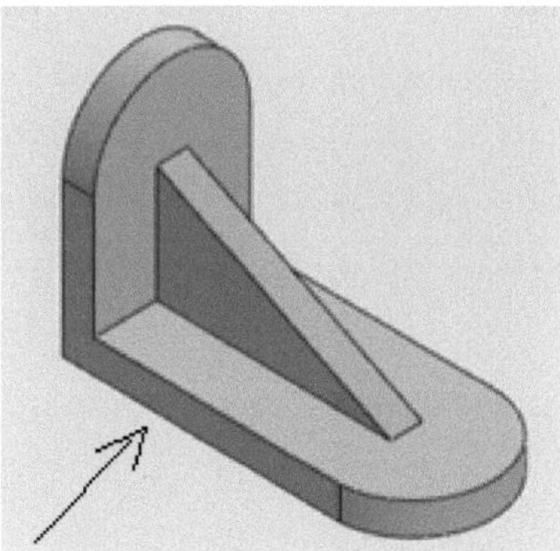

a)

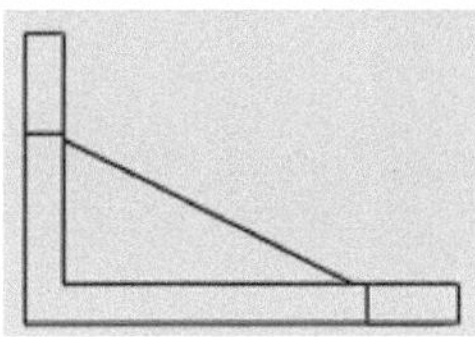
b)

c) 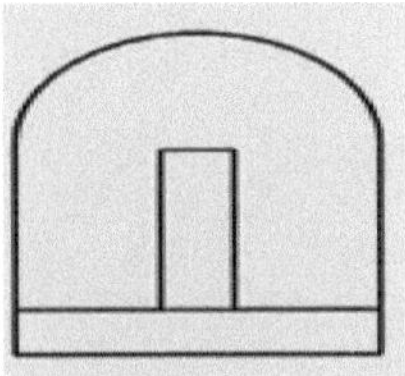

d) 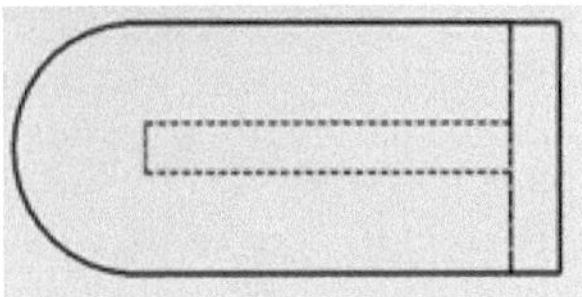

30. Identify the front view of the below isometric view.

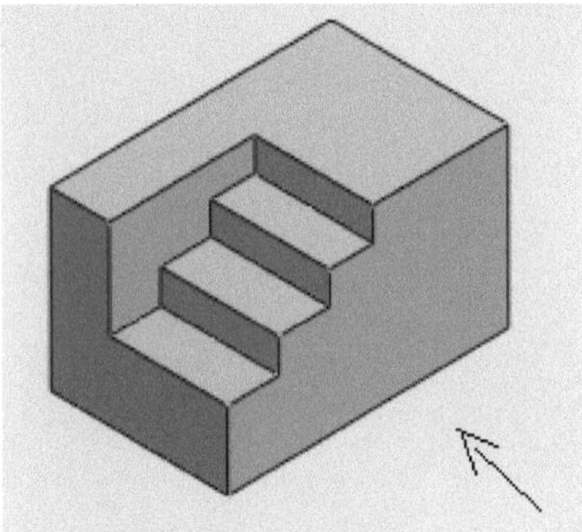

a) 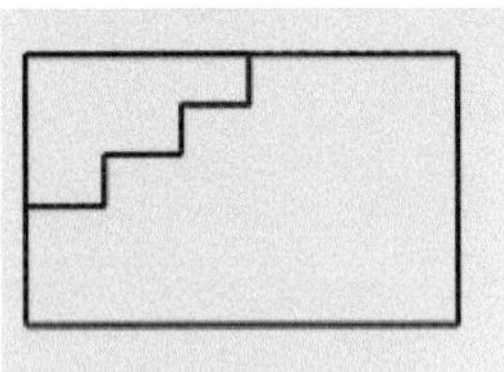

b) 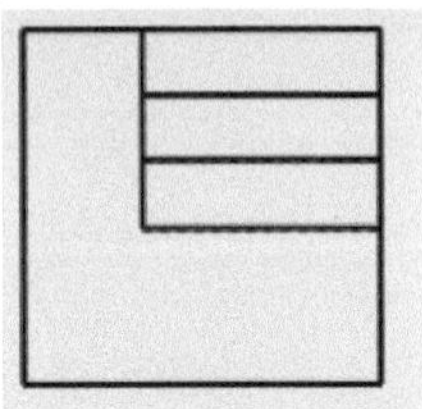

c) 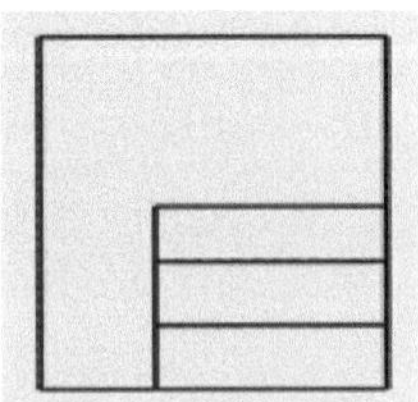

d) 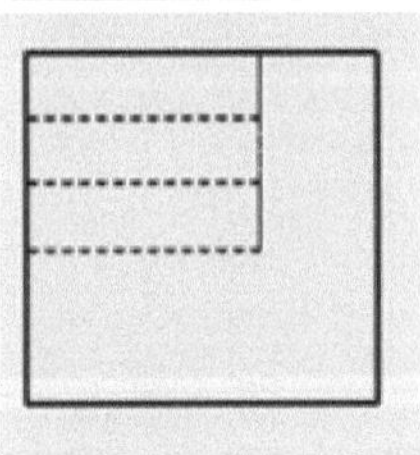

31. Identify the top view of the below isometric view.

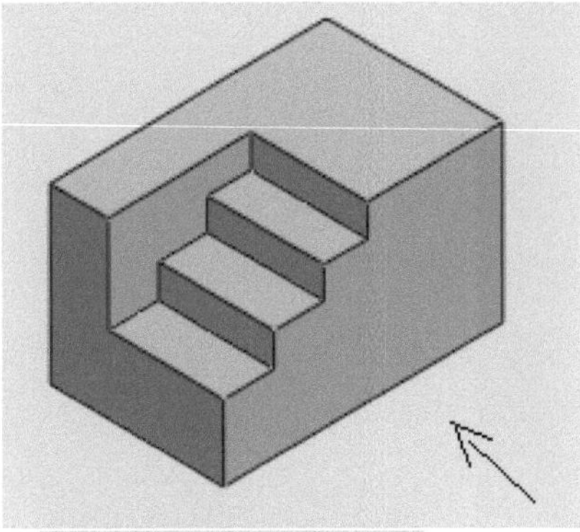

a) 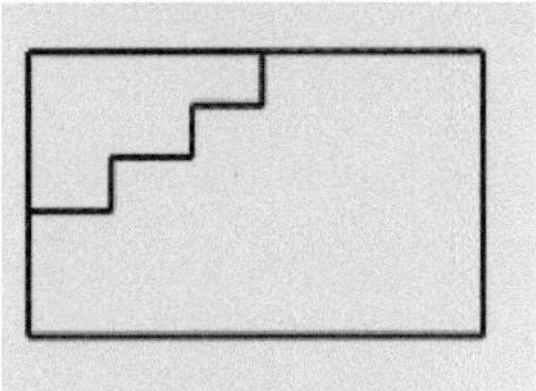

b) 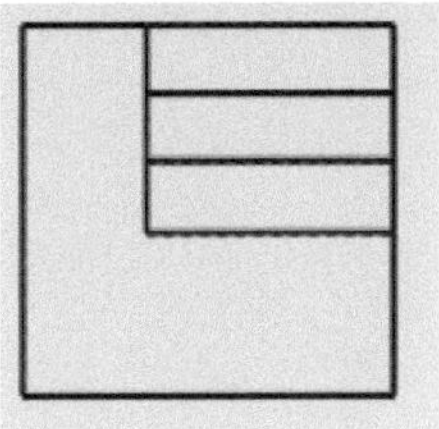

c) 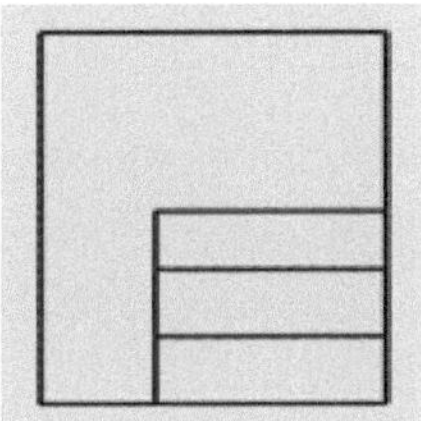

d) 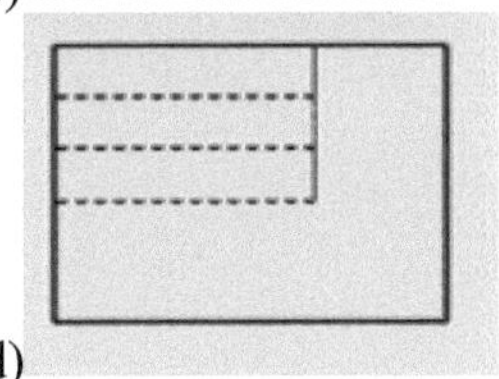

32. Identify the side view of the below isometric view.

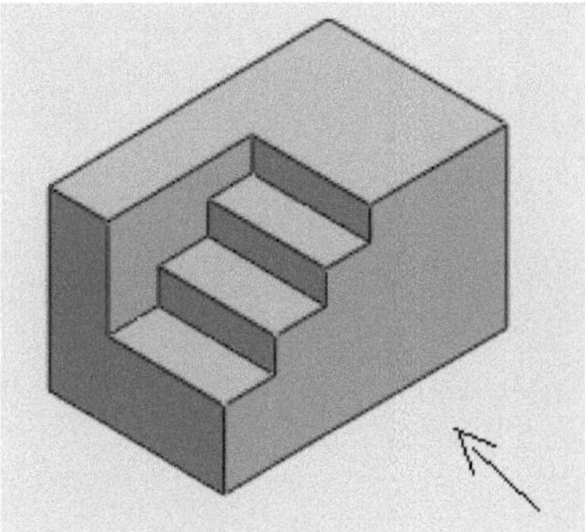

a) 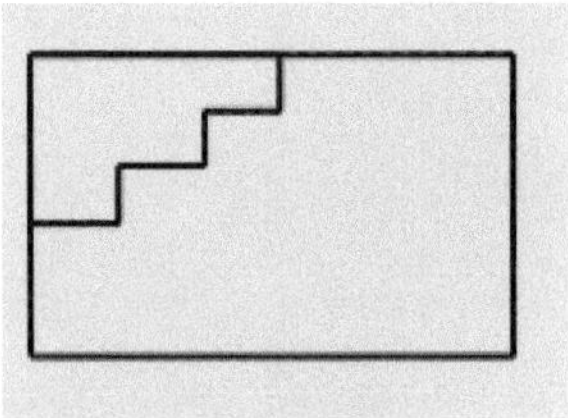

b) 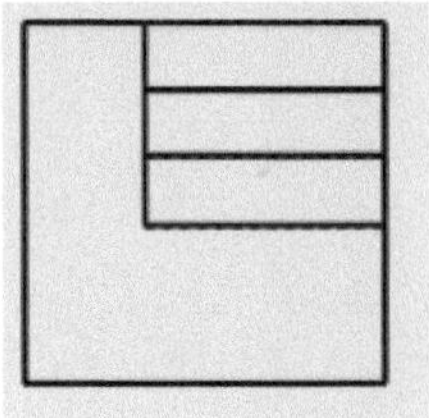

c) 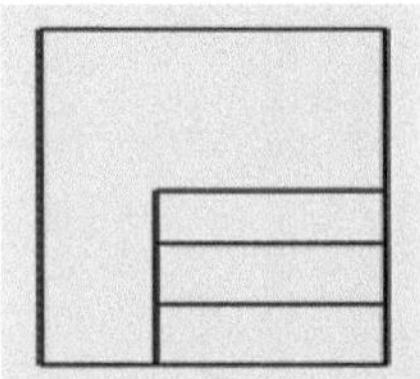

d) 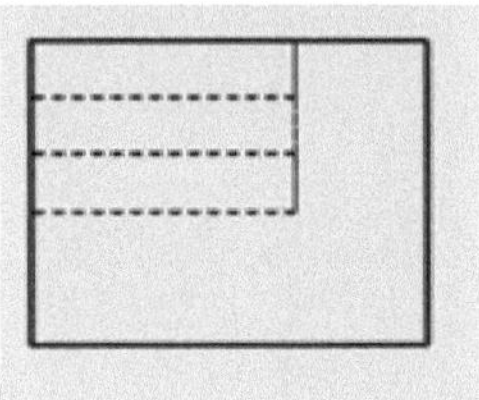

33. An orthographic projection map is a map projection of __________
a) Sphere
b) Earth
c) Cartography
d) Top view
34. Taking 'A' as the FRONT VIEW. Which view will letter 'D' represent in the following figure?

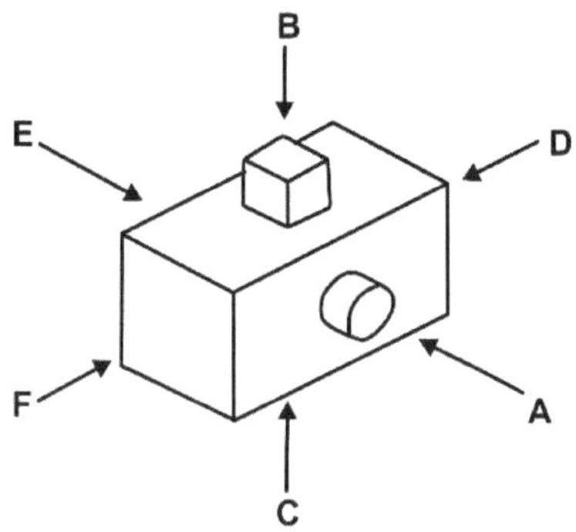

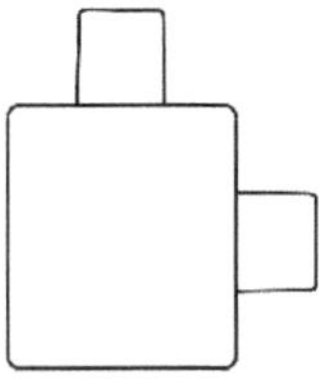

a)

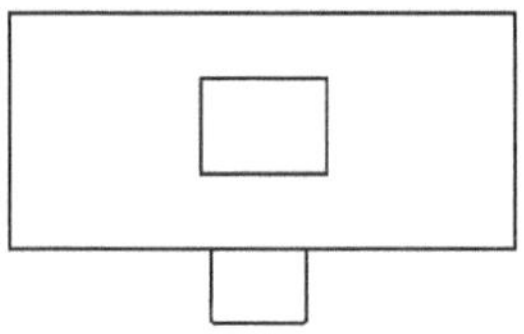

b)

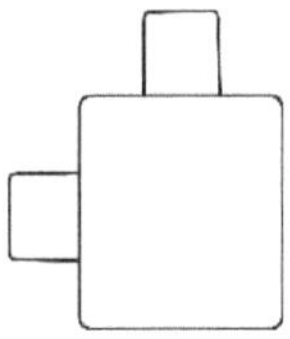

c)

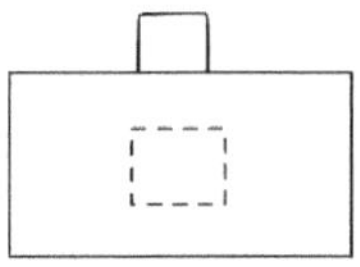

d)

35. Which view will letter ‘C’ represent in the following figure?

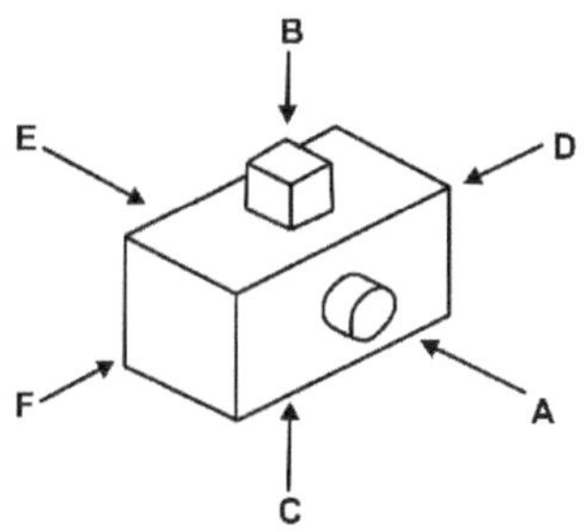

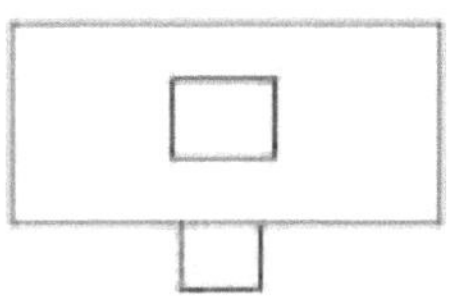

a)

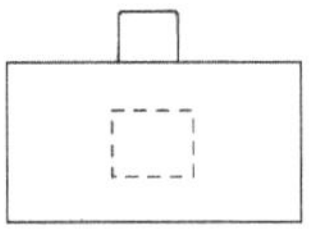

b)

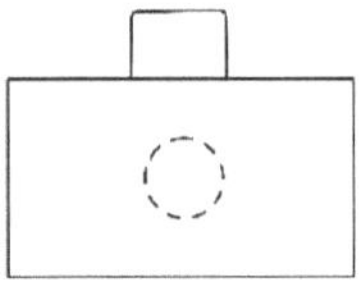

c)

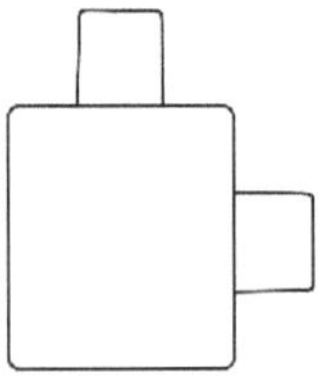

d)

36. Which view will letter ‘E’ represent in the following figure?

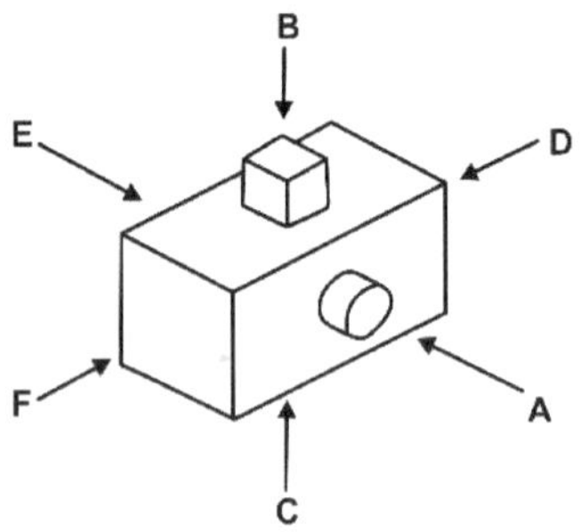

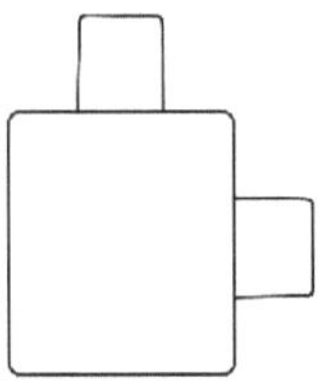

a)

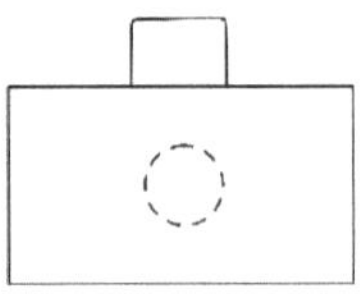

b)

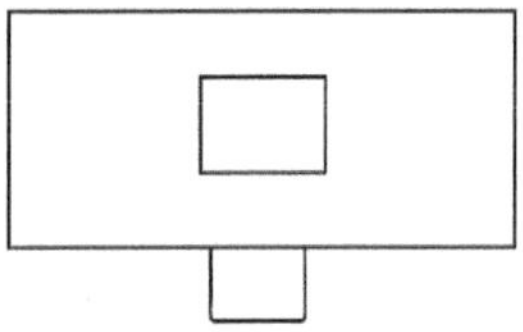

c)

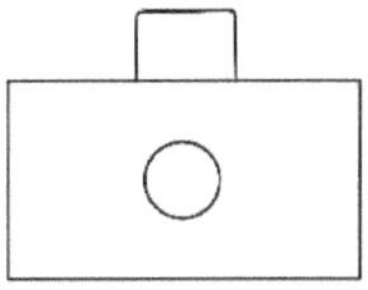

d)

37. Which view will letter ‘B’ represent in the following figure?

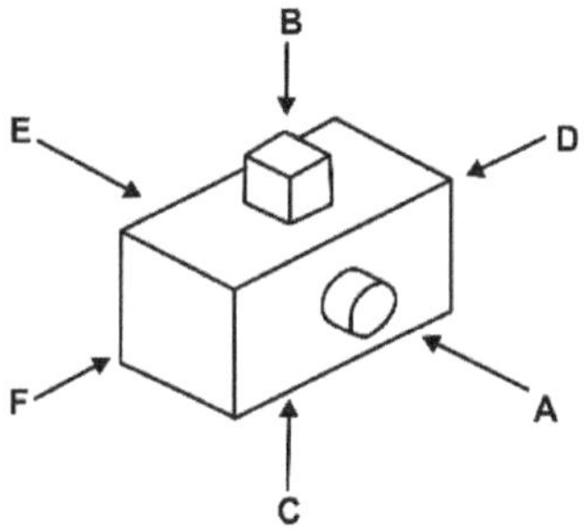

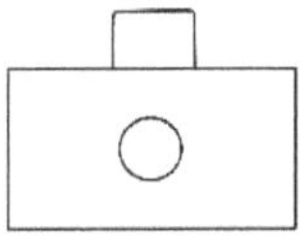

a)

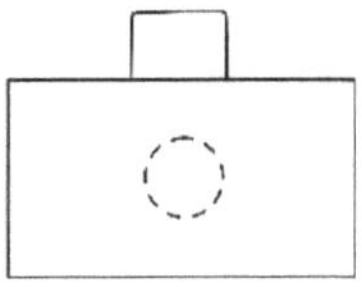

b)

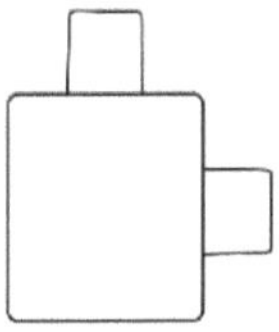

c)

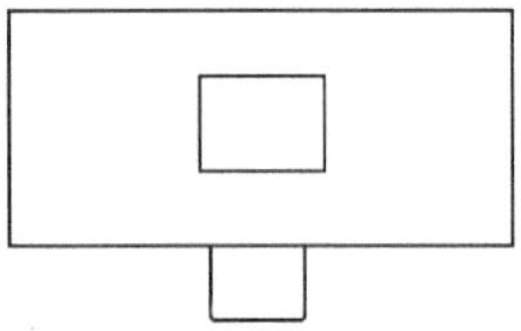

d)

38. 'A', 'B' view will be represented by which figure?

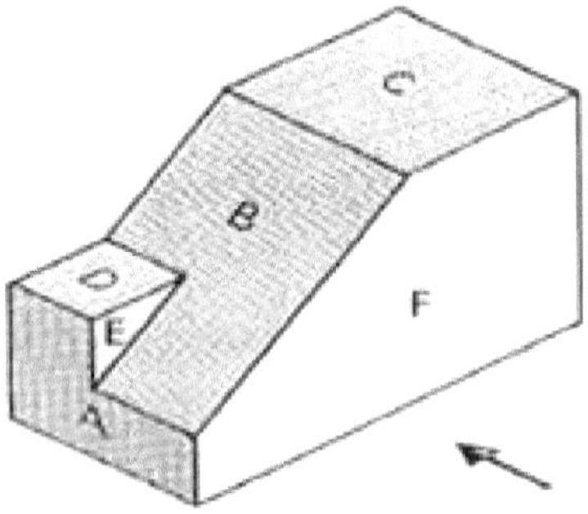

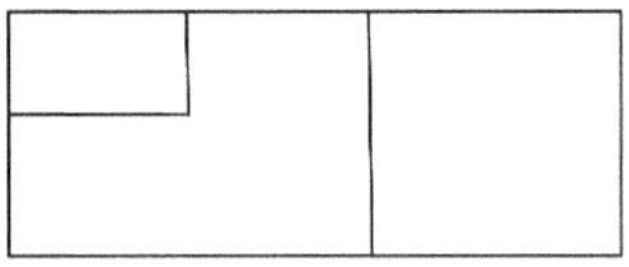

a)

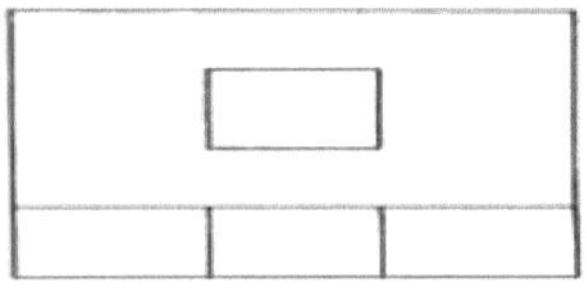

b)

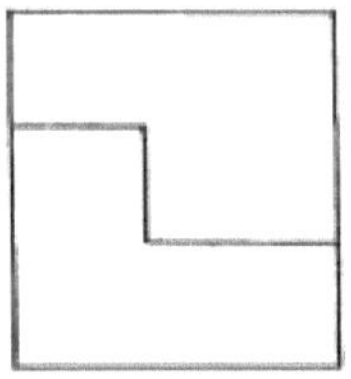

c)

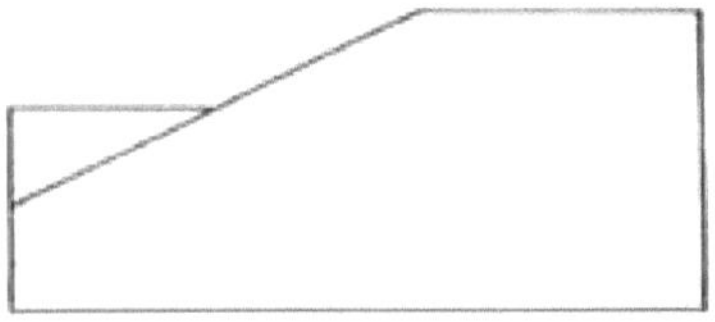

d)

39. 'F', 'E' view will be represented by which figure?

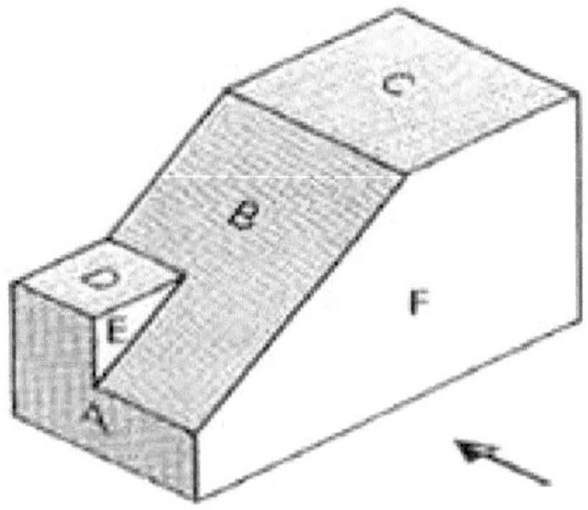

a) 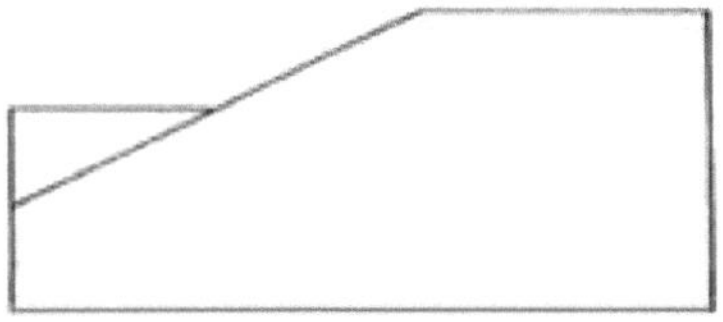

b) 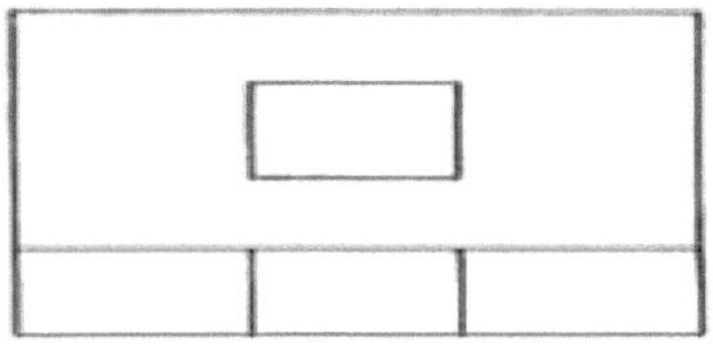

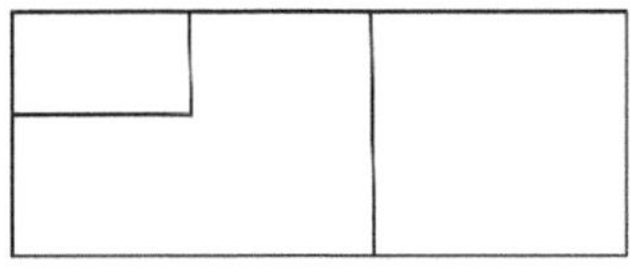

c) 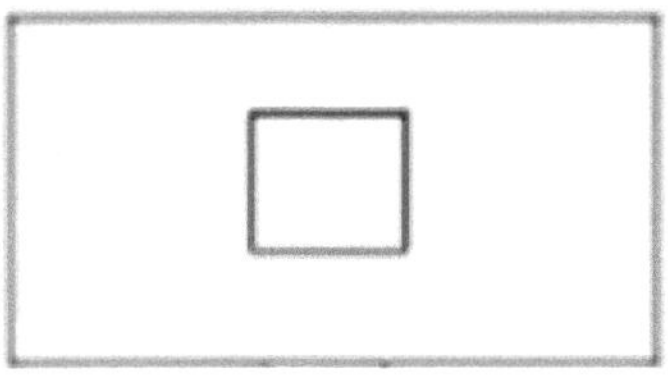

d)

40. The given figure below will represent which type of three views?

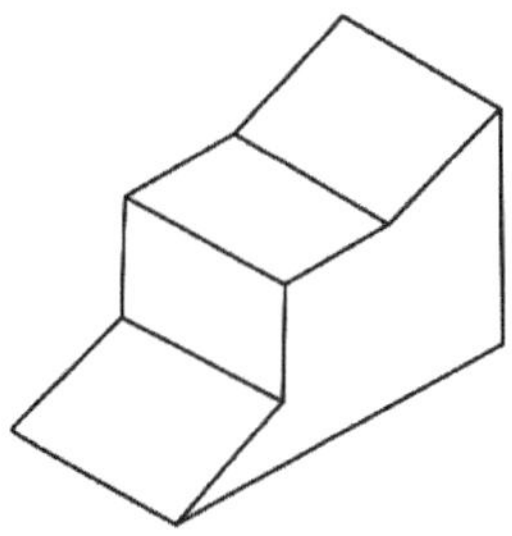

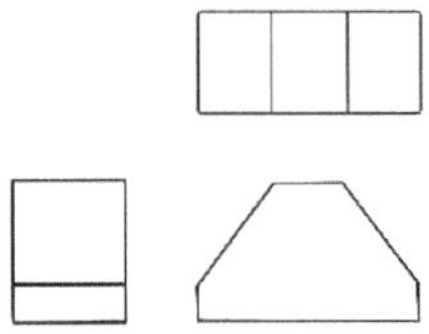

a)

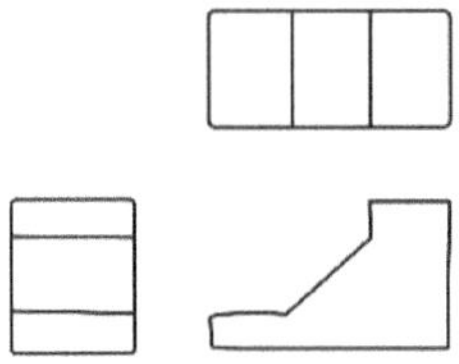

b)

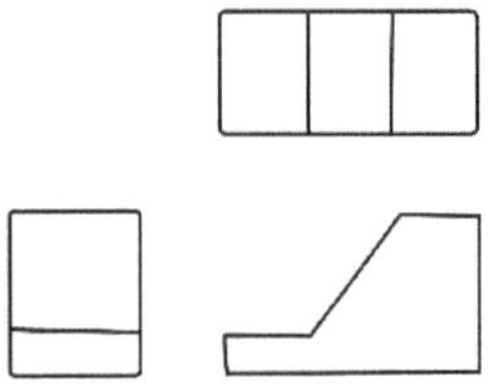

c)

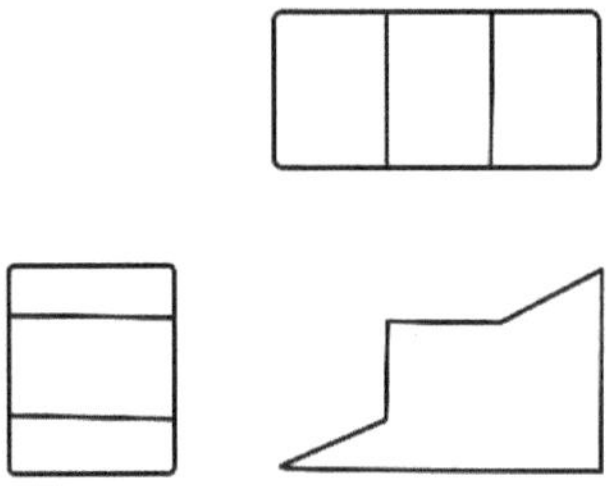

d)

41. The given figure below will represent which type of three views?

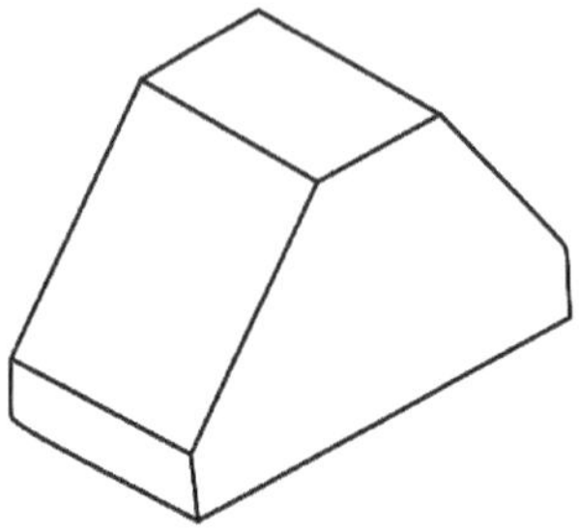

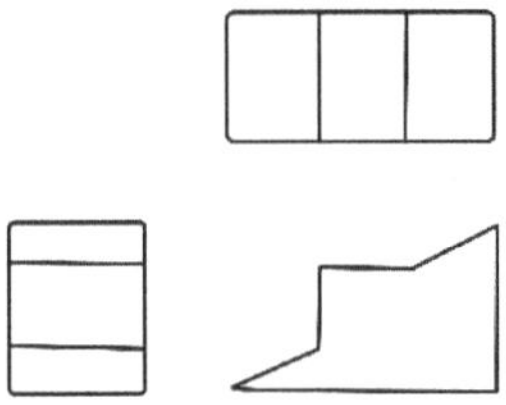

a)

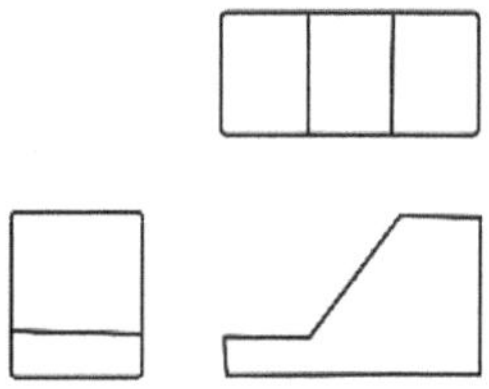

b)

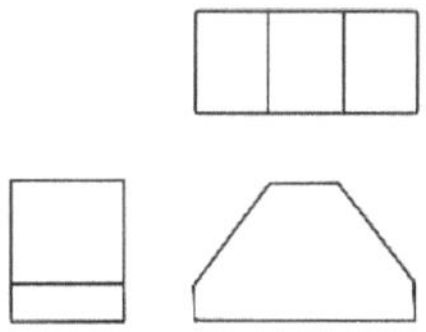

c)

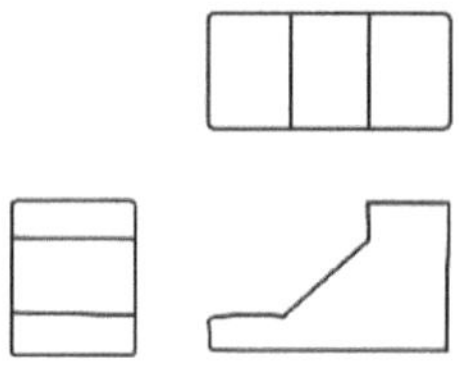

d)

Answers

1. Answer: b
2. Answer: b
3. Answer: d
4. Answer: a
5. Answer: c
6. Answer: b
7. Answer: b
8. Answer: c
9. Answer: c
10. Answer: d
11. Answer: b
12. Answer: c
13. Answer: c
14. Answer: c
15. Answer: d
16. Answer: a
17. Answer: a
18. Answer: d
19. Answer: a
20. Answer: b
21. Answer: c
22. Answer: a
23. Answer: a
24. Answer: a
25. Answer: d
26. Answer: b
27. Answer: c
28. Answer: a
29. Answer: b
30. Answer: a
31. Answer: c
32. Answer: b
33. Answer: a

34. Answer: c
35. Answer: b
36. Answer: b
37. Answer: d
38. Answer: c
39. Answer: a
40. Answer: d
41. Answer: b

CHAPTER 4: FREE HAND SKETCHES OF ENGINEERING ELEMENTS

4.1 FREE HAND SKETCHES OF MACHINE ELEMENTS

Free hand sketching in engineering graphics are very critical part of any engineering project. It may be used at any stage of a project's life cycle, whether it is the planning, execution or even end phase of the project, it is a way to convey critical information regarding certain aspects of the project.

Free Hand Sketching is used to transfer the idea of an engineer's abstract world to the real world in 2D and 3D. It is used for daily communication in the engineering world as drawing is the international language of engineering. Free hand drawing is also used to quickly present an idea in a graphical form to non-technical peoples.

It improves the communication between all the members of a team and also between the customer. Free Hand Sketching is such a drawing which is drawn without measuring instruments. This drawing is drawn with the help of pencil and eraser only. Such drawing is drawn before every type of actual drawing because it takes less time. After doing such drawing, it is pondered over, and necessary alterations are made in it if needed.

WHERE AND WHEN ARE FREE HAND SKETCHES USED?

- Free hand sketches are used during the beginning and early stages of a design when the designer illustrates certain configurations or fits to the draftsman
- Modifications or alterations can be indicated with a free hand sketch during the drafting or even manufacturing phase. Often times modifications are made and the modification is captured with a free hand sketch that is later incorporated into the drawing
- Often times a draftsman would visit the client to measure and make free hand sketches of equipment or components that must be drawn
- A sketch is made of what is required on a drawing
- The component may only be available for a short time thus, a free hand sketch needs to be made to use for the creation of a drawing

REQUIREMENTS OF A FREE HAND SKETCH

- Free hand sketches should be made proportional. As in the example above, I used each block as a 5mm representation. This helped to keep the sketch of the different parts proportional
- Details should be indicated correctly inaccurate information will translate to mistakes on the final drawing
- The free hand sketch should have as much detail as possible
- It should include all features that may have an impact on the design of the part/section
- It should be readable by someone else. Sometimes it may happen that someone else will be creating the drawings, so one should make sure that someone else can work from the sketch
- Neat and tidy sketches make creating the drawings easier
- Complete with all the relevant dimensions. Missing dimensions will cause difficulties when the actual drawings are made.
- Correct and complete. Remember, the sketch is the main information where the drawing is created from, correctness and completeness is vital.

HOW TO DRAW A FREE HAND SKETCH

PAPER TO BE USED

If possible, use a page with blocks on (as shown above). It will make it easier to draw straight lines, draw in proportion and just the overall neatness of the sketch

If you do not have lined or blocked paper, blank paper can also be used. When using blank paper, draw the lines very faint first and then draw dark over it. It will help to keep sketches neat and with less mistakes

CIRCLES, RADII & ELLIPSES

Circles, radii and ellipses are often difficult to draw. Instead of just jumping in and drawing a circle, make small marks on the paper to mark reference points of the circle. Then, just "connect the dots" and it can make a circle or radius look much neater.

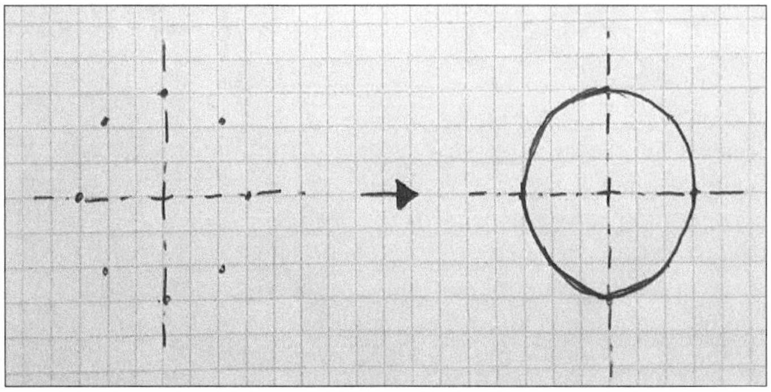

Figure 4.1

IMPORTANCE OF FREE HAND SKETCHING

Free hand drawing has much importance in practical works because a draftsman has to check the drawings of every object from the engineer.

At the checking of complete drawing, much time is wasted on the alterations, if needed.

Since less time is needed for free hand drawing, therefore, initially such drawing should be depended upon.

PRINCIPLES OF FREE HAND SKETCHING

Following principles should be followed for free hand sketching:

1. The object should be seen thoroughly and pondered over its objectives and concepts.
2. More detailed aspect is considered for selection of a view.
3. Space is specified on a drawing sheet according to the sizes of views.
4. Firstly, the dim lines are drawn so that extra lines may be erased easily.
5. Scale and ruler are not used in drawing, although, ratio and proportion of different parts of an object are considered.
6. Firstly, horizontal lines are drawn then vertical lines are drawn, and arcs and circles are drawn lastly.
7. After completing the diagram, a clear scale is written in some suitable manner.

SKETCHING OF LINES IN FREE HAND DRAWING

1. SKETCHING OF STRAIGHT LINES

The line is the most important part of free hand drawing as sketching is done by different lines. Straight lines can be divided into different lines such as oblique, horizontal, and vertical lines. Drawing straight lines for the first time is a difficult task. Mostly result in curved and arcs.

This happens due to the forearm when remaining in the fixed position.

So, when drawing the straight lines, moving your forearm is important. Basically, when drawing horizontal straight lines. The distance between your hand and pencil tip should be 40 mm minimum, and your hand should rest on the edge of the paper. Straight lines are drawn with the thumb and index fingers.

The most important part of sketching straight lines is that your movement should be in a controlled manner. You can also draw oblique lines by just changing the position of your arm.

The easiest way to draw long straight lines is by sketching short lines and then connecting it with one another. By doing this, there will be little to no chances of errors.

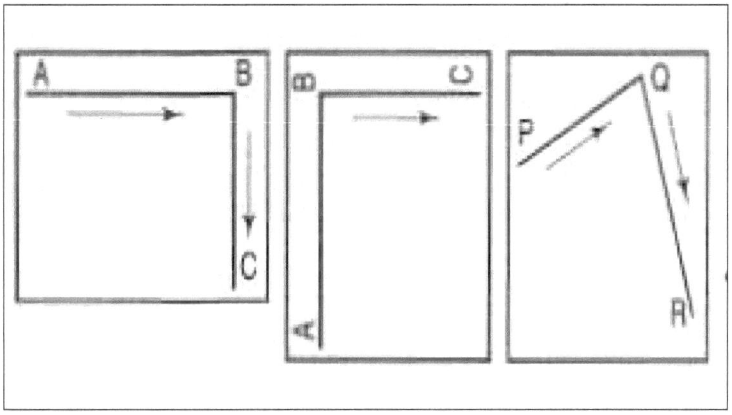

Figure 4.2

(I) SKETCHING OF VERTICAL LINES
A vertical line is drawn from the top to the bottom of the drawing sheet. It takes a lot of practice to draw vertical lines.
However, you can draw vertical lines using the edge of a paper as a guide and your hand as a scale. (As shown in below figure 4.3).

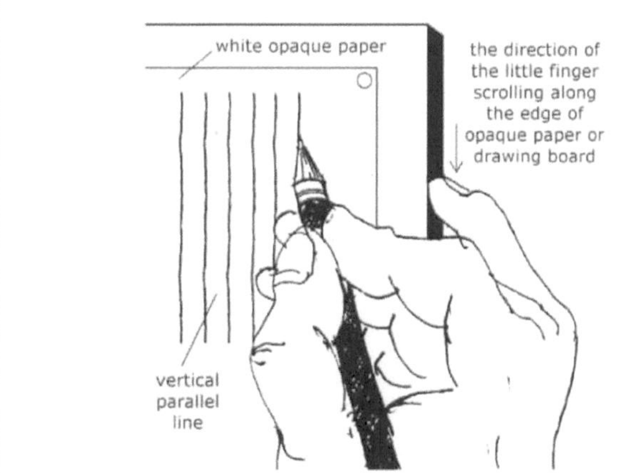

Holding the pencil when drawing *vertical straight lines* [4]

Figure 4.3

(II) SKETCHING OF HORIZONTAL LINES
When sketching horizontal lines, your hand and forearm should turn at the elbow, and the edge of your hand should slide horizontally on the paper. Horizontal lines are sketched with the motion of the wrist and the fore-arm. They are sketched from left to right. To sketch a horizontal line, mark the end points. Hold the pencil at about 30 mm distance from the lead point. Swing it from left to right and backwards, between the two points and without touching the surface of the paper, till the correct direction is achieved. Then begin to draw the line (with the wrist-motion) with short and light strokes. Shift the hand after each stroke. Keep your eyes on the point at which the line is to end. Finish finally with a dark and firm line. Take proper care to maintain straightness and correct direction of the line.

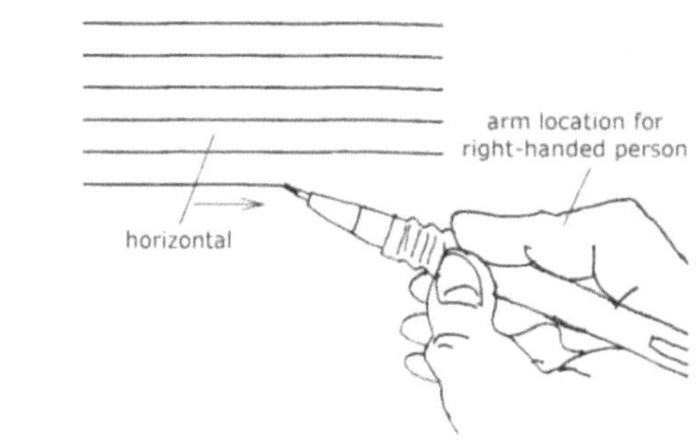

Freehand sketching of *horizontal parallel lines*

Figure 4.3

When drawing a horizontal line, you should first try drawing a line in the air to see how the line goes. Then apply a little pressure on the pencil tip and draw a horizontal line lightly.

(III) SKETCHING OF OBLIQUE LINES

The oblique lines are drawn from the one corner of the page to the other diagonally. Usually, the right-handed people draw sloping or oblique lines from the bottom left to the top right of the paper.

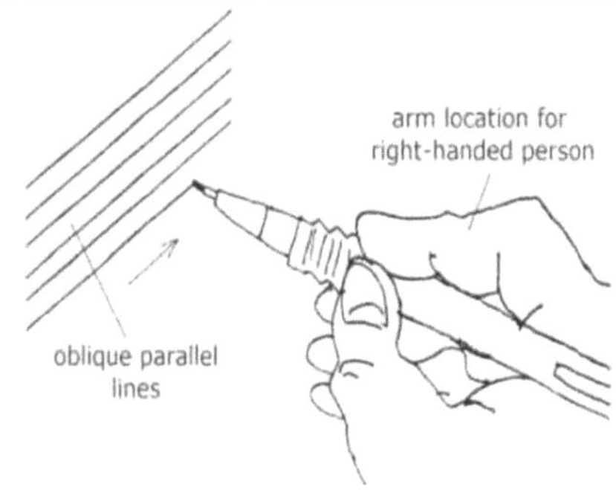

Figure 4.4

However, the sketching can be made much easier by just rotating the paper into a position where you are going to draw an oblique line.

2. SKETCHING OF RECTANGLES

A rectangle is a common shape in technical drawings. To draw rectangles via free hand sketching different techniques are used.

They can be drawn in several ways, but the easiest way to draw a rectangle is by drawing vertical or horizontal lines and rotating the paper at 90 degrees angle.

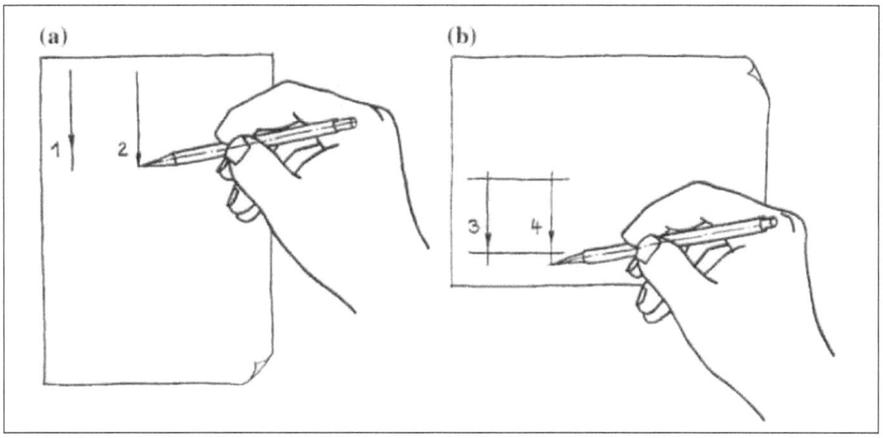

Figure 4.5

3. SKETCHING OF CURVED LINES

Curved lines in the drawing can be arcs, circles, etc. To draw a circle, first of all, you should draw a centre line and mark it with radii. Shape the radii points into a box, inside in which you can draw a circle.
First of all, sketch the top-left part of the circle (rotate the pencil in an anti-clockwise direction) then draw the bottom-right part of the circle (rotate the pencil in a clockwise direction).

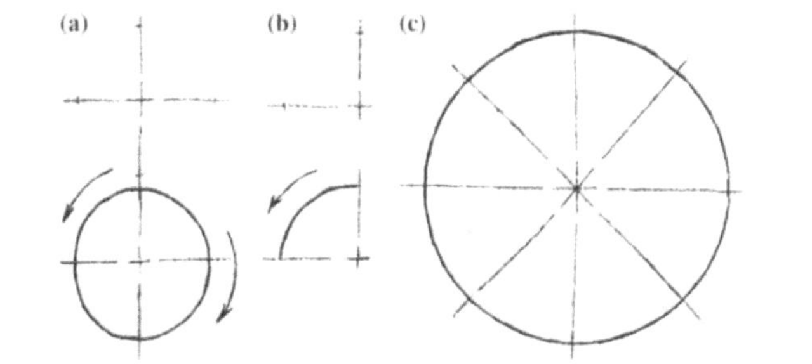

Sketching a circle and an arc [19]. Principles: two arcs for small circle (a), sketching arc into clockwise direction (b), larger circle with eight arcs (45 degrees) (c)

Figure 4.6

FREE-HAND SKETCHING INSTRUMENTS

Following instruments are needed for the drawing of free hand sketching.
1. Soft Lead Pencil
2. Eraser
3. Sharpener
4. Drawing Sheet
5. Graph Sheet

4.1.1 SCREW THREADS

Fasteners are used to connect various parts of constructions, machineries, or other parts. The fasteners used for this purpose are just temporary; so, by simply unscrewing or removing the fastener, the sections can be readily separated without breaking. For this purpose, fasteners such as keys, cotters, pins, studs, nuts, and bolts are utilised. The screw thread(s) on a bolt, stud, set screw, cap screw, wood screw, etc. serve as the functional component of temporary fastening. Cutting a helical groove on a circular surface, either on the inside or outside, produces screw threads. A screw is the name for a part's threaded surface.

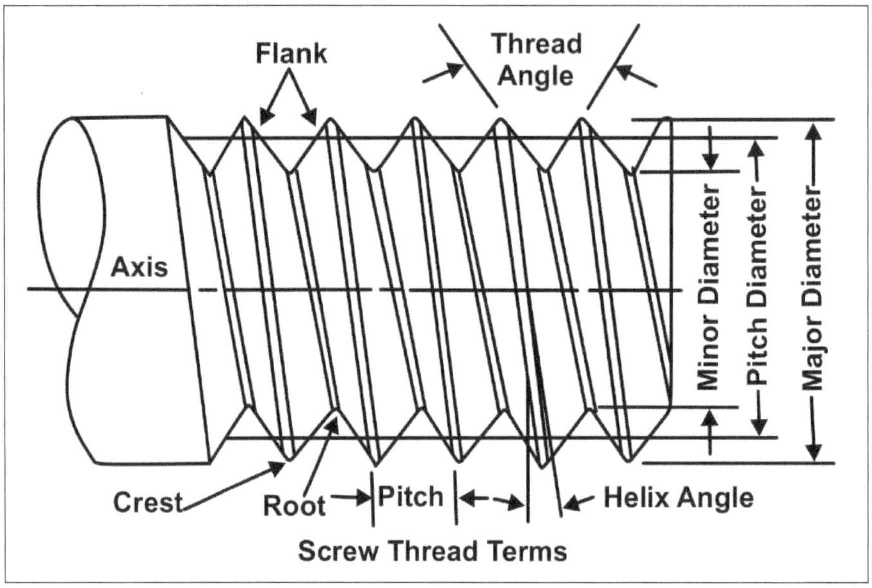

Figure 4.7

PITCH: It is the distance from top of one thread to the top of the adjacent thread along the axis of the rod.
ROOT: It is the surface of a thread which connects adjacent flanks at the bottom of the groove.
FLANK: The flanks are the parts of the surface between root and crest. These are inclined lines starting from crest towards roots forming a solid surface with inclined side surfaces.
CREST: It is the part of the surface of the thread which connects adjacent flanks at the top of ridge.
LEAD: It is the distance moved by a nut or bolt in the axial direction in one complete revolution. In single start threads, the lead is equal to the pitch. In multi start threads, lead is equal to the number of starts multiplied by pitch.
ANGLE OF THREAD: It is the angle between the flanks measured in an axial plane.
MAJOR DIAMETER: The diameter of the imaginary cylinder at crest or on which threads are cut is known as major diameter.
MINOR DIAMETER: The diameter of the imaginary cylinder at root is known as minor diameter.
PITCH DIAMETER: The mean of major and minor diameter is known as pitch diameter.
EXTERNAL THREAD: A thread on the outside surface of a member such as bolt, stud, screw, pipe, cylinder, shaft etc.
INTERNAL THREAD: A thread on the inside surface of a member such as threads in the hole of the nut, pipe, cylinder etc.
RIGHT HAND THREAD: The thread on the outer surface of bolt rod which advanced into the nut when turned in a clockwise direction or in which thread slopes up towards the left when screw is placed horizontally.

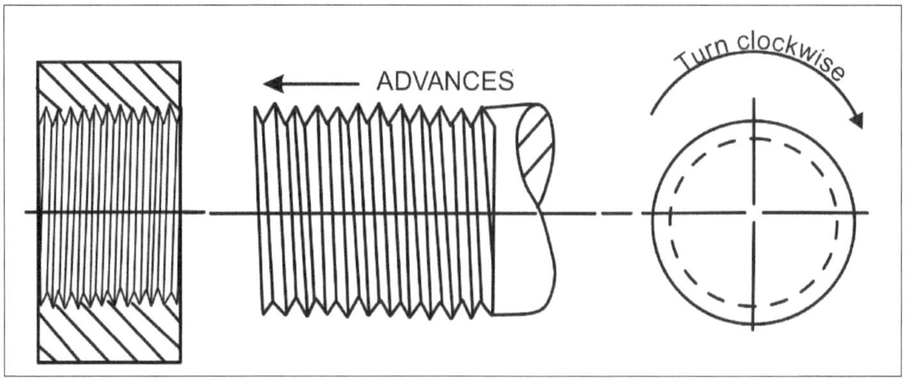

Figure 4.8

LEFT HAND THREAD: The thread on the outer surface of bolt rod which advanced into the nut when turned in an anticlockwise direction or in which the direction of the thread slopes up towards the right when screw is placed horizontally.

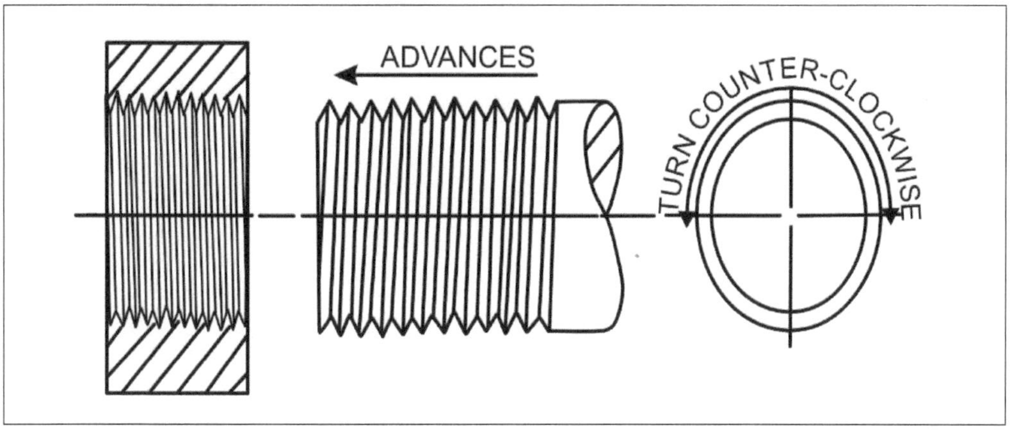

Figure 4.9

SINGLE START THREAD: In a single start thread, the pitch is equal to the lead, when only one helix or one starting point is seen forming the thread, run on a cylinder throughout its length. If a nut is screwed on the threaded bolt and held firmly at one point, the rod is rotated by 360°, the nut slides along the axis equal to pitch of the thread.

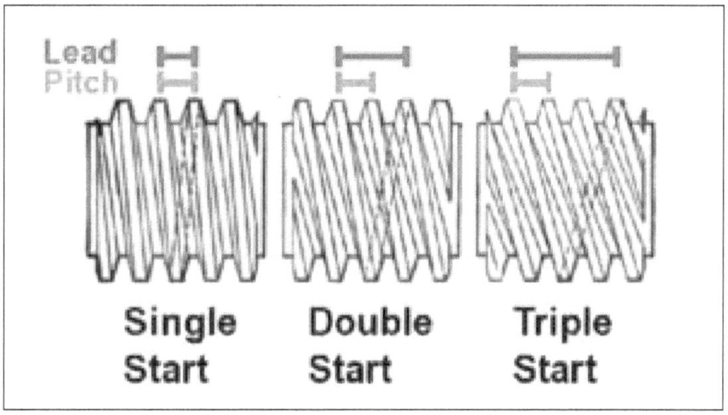

Figure 4.10

DOUBLE START THREAD: In double start thread, the threads in the inner surface of the hole of nut should also be double start thread. If a nut is screwed on the double start threaded bolt or rod and the rod is rotated by 360°, the nut will lead or shift twice the pitch of the thread.

MULTI-START THREAD: In multiple start threads, two or more threads having the same pitch run parallel to one another. Multiple start threads are used wherever, quick motion in minimum rotations is required and application of great force is not allowed such as between fountain pen and its cap, bottles, toothpaste caps, valves, water tap spindles etc. to impart quick movement in opening and closing.

FORMS OF SCREW THREADS:

B.S.W. (British Standard Whitworth):

BSW thread is used as a standard thread in British for general purposes. It is the modified form of V thread with angle of threads **55°** and corners equally rounded off at crest and root for smooth movement of threads. This type of thread is used on bolts, nuts and other screwed fastenings.

$d = D$

$d1 = D1 = d - 1.28 \times P$

$d2 = 0.640327 \times P$

R = 0.137329 X P

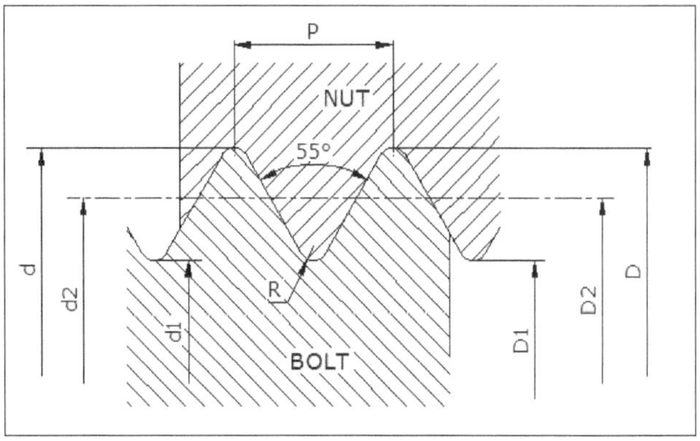

Figure 4.11

B.A. (British Association) Thread Section:
BA thread is recommended by British standard and institute for use in preference to BSW thread for all small screws and small instruments works. The angle between flank is **47.5°**.

It is used where accuracy is main consideration such as aircrafts mathematical instruments.

A set of small screw threads known as British Association screw threads, or BA screw threads, is practically extinct. The largest is 0BA, with a diameter of 6 mm. They served as models and small instruments. They stand out because, starting with 0BA at 6.0 mm in diameter and 1.0 mm in pitch, they were likely the most "scientific" screw design ever created. They then progressed in a geometric series, with each greater number being 0.9 times the pitch of the previous one.

They ruined this by rounding metric values to two significant figures. After that, convert to inches and round to the nearest thousandth. This was nearly a century ahead of the global metrication trend. The British Association first put out the design in 1884, using the Swiss Thury thread as a model for the thread angle and depth. The Association accepted it in 1903.

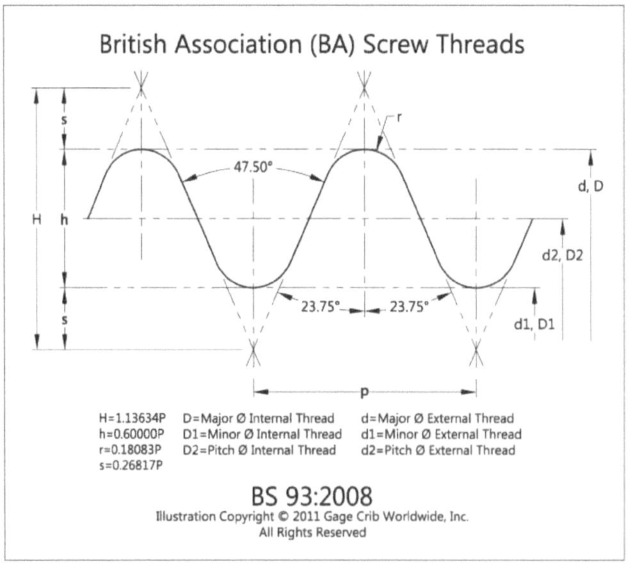

Figure 4.12

Seller Thread Section or American National Thread Section:
This thread is used as a standard thread in America for general purposes. The angle between two flanks of the thread is **60°**. It is used on bolts, nuts, machine screws, etc.

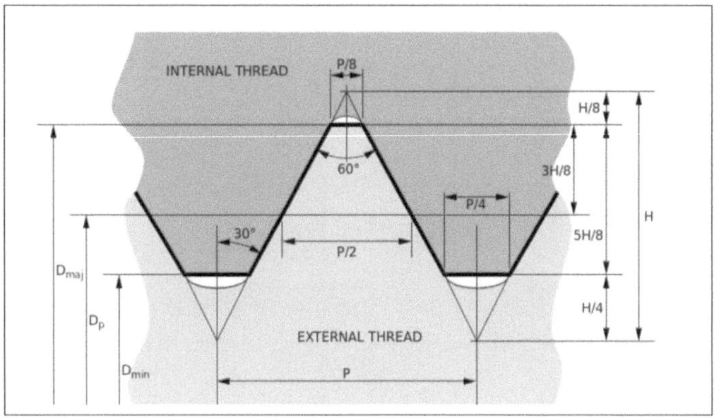

Figure 4.13

Metric Thread:

This thread is used as a standard thread in India for general purposes. The metric thread is a symmetrical V thread with an angle of **60°** between flanks. The thread on the bolt is rounded off at the root and flat at the crest and the thread on the nut is rounded off at the root but the crest is flat.

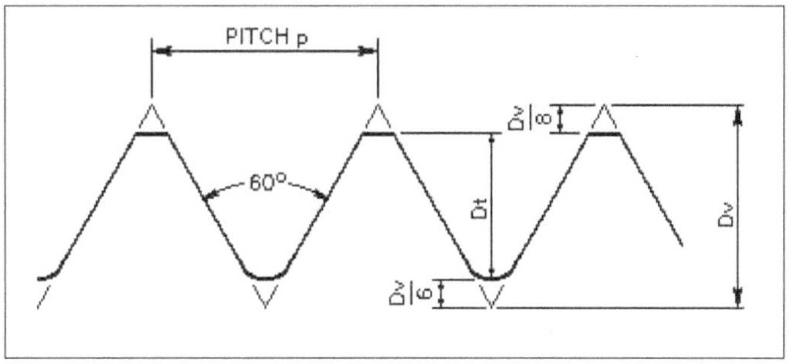

Figure 4.14

Square Thread Section:

The square thread form is a common screw thread profile, used in high load applications such as leadscrews and jackscrews. It gets its name from the square cross-section of the thread.

This thread is of square shape and it is used for power transmission. The depth and thickness are equal to half the pitch. It is used in screw jack, vice screws, large valve spindles because they offer at least frictional resistance to motion.

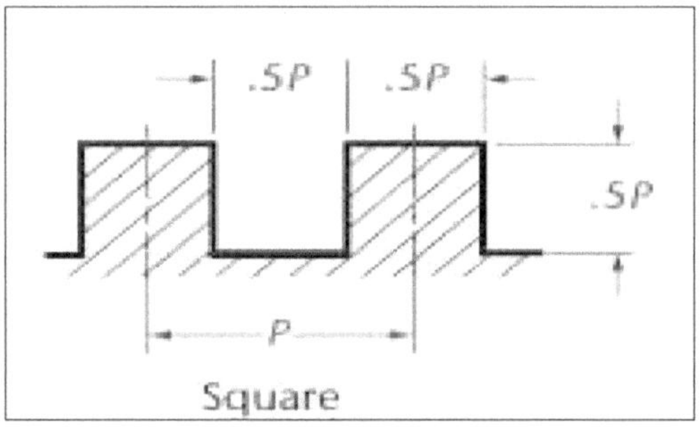

Figure 4.15

Knuckle Thread Section:

Knuckle threads are generally used in some type of electric bulbs, bottle tops, fire hydrant and railway couples etc. It is the modified form of square thread. It is semi-circular shaped thread. It is not easily damaged when subjected to rough usage.

Another variation of a square thread is a knuckle thread. The square thread's pointed corners have been softened, resulting in semicircular crests and roots. The roots and the crests cross each other at a right angle. Since the profile is challenging to machine, these threads are typically made via rolling or casting.

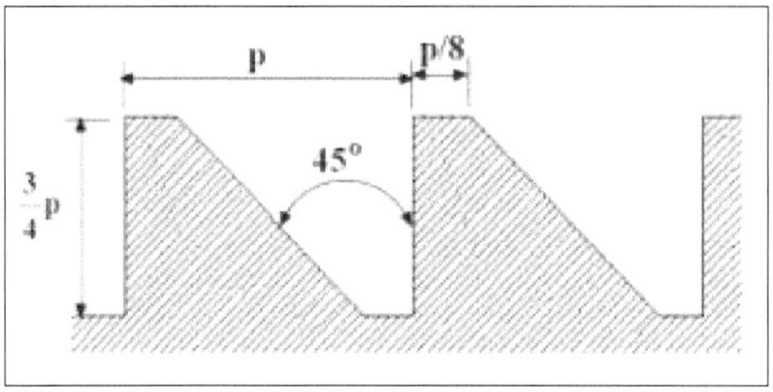

Figure 4.16

Buttress Thread:

In the buttress thread section, one side of the thread is perpendicular to axis whereas the other side is kept at **45°** inclined to axis. The buttress threads are used for the transmission of power in one direction only such as in spindles of bench vices, propellers etc.

Buttress thread forms, also known as sawtooth thread forms or breech-lock thread forms. are screw thread profiles with an asymmetric shape, having one square face and the other slanted.

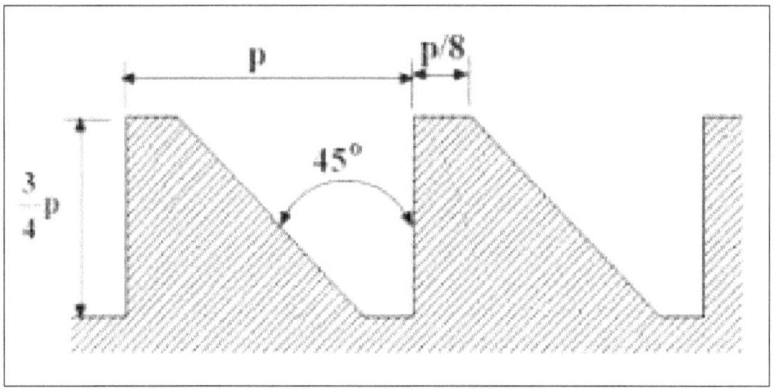

Figure 4.17

Acme Thread Section:

It is the modification of square thread. The shape of the thread can be cut very easily which gives more strength to the threads because of its broader base and narrow top. The angle between thread flanks is **29°**.

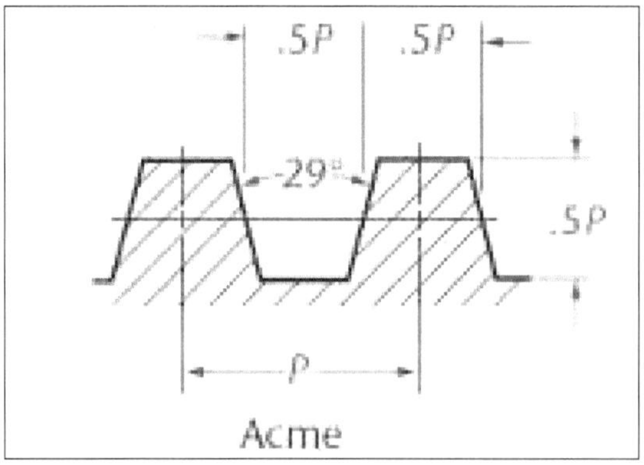

Figure 4.18

METHODS OF REPRESENTING SCREW THREADS

There are three methods of representing screw threads on a drawing.

- Detailed
- Schematic
- Simplified

Screw thread representation present in this chapter is in accordance with the **ASME Y14.6-2001** standard.

1. DETAILED REPRESENTATION

A detailed representation is a close approximation of the appearance of an actual screw thread.

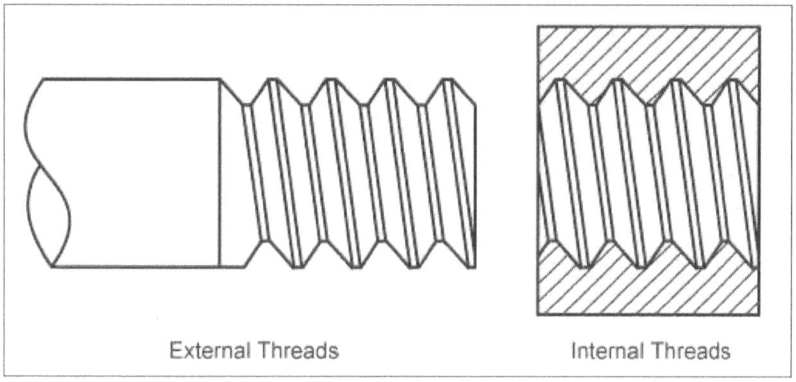

Figure 4.19

2. SCHEMATIC REPRESENTATION

The schematic representation uses staggered lines to represent the thread roots and crests.

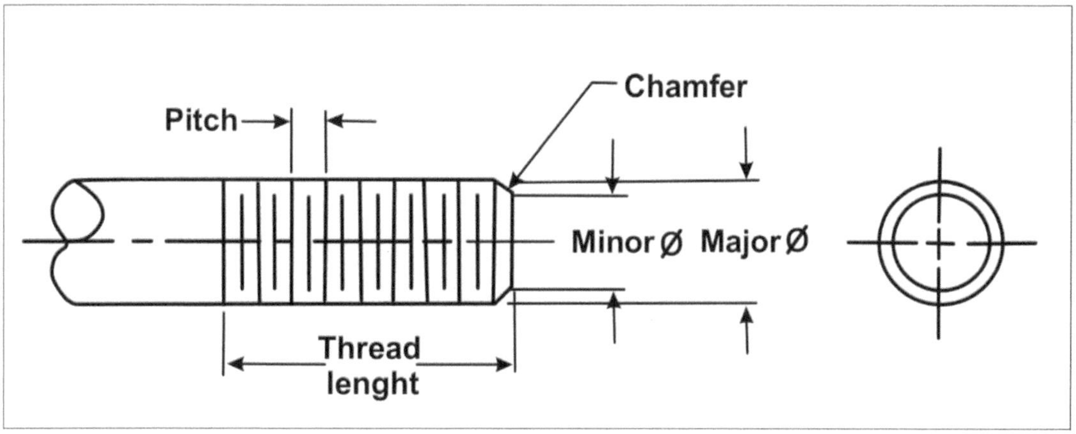

Figure 4.20

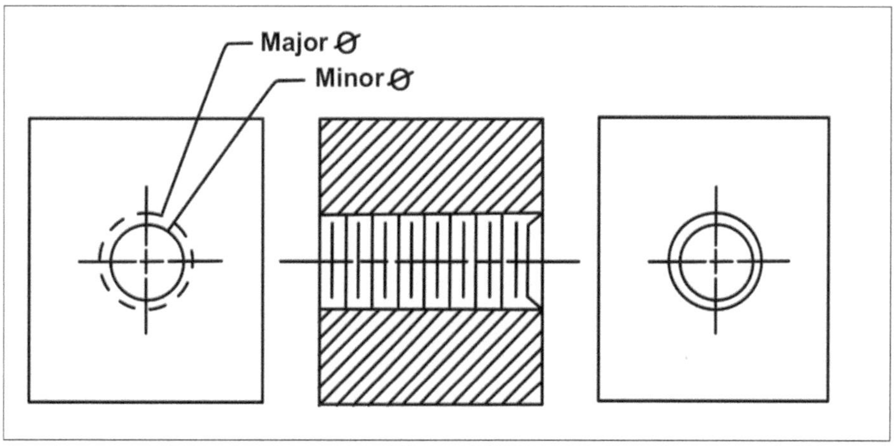

Figure 4.21

3. SIMPLIFIED REPRESENTATION

The simplified representation uses visible and hidden lines to represent the major and minor diameters.

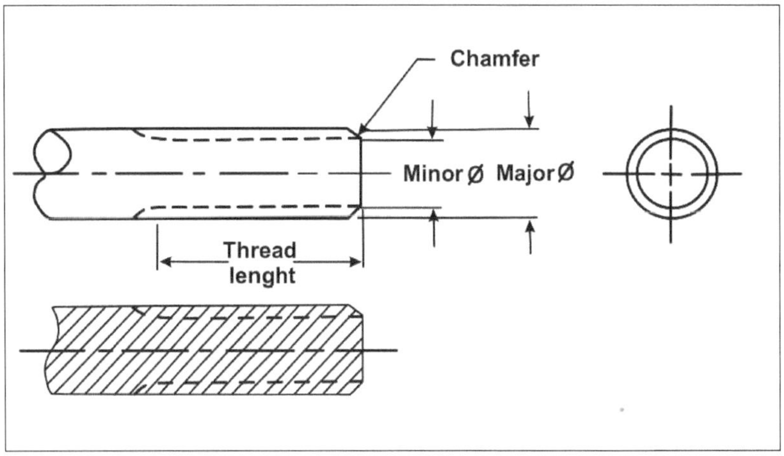

Figure 4.22

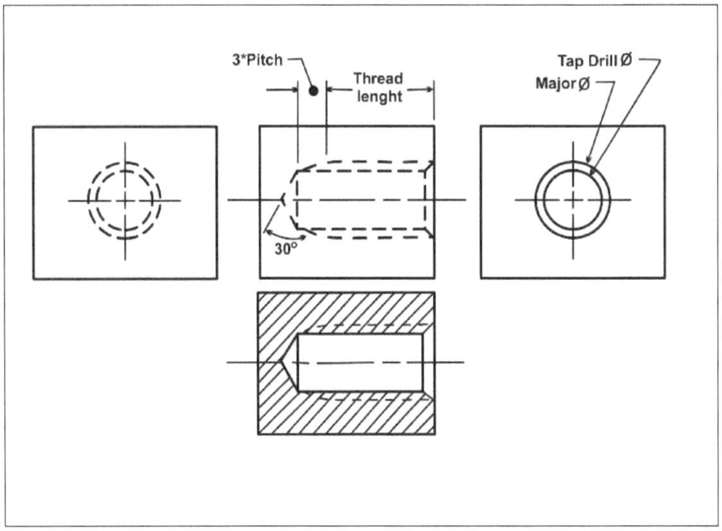

Figure 4.23

DRAWING SCREW THREADS

Thread tables in the appendix of the Engineering Graphics Essentials book can be used to look up value for the following

- Pitch
- Minor diameter
- Tap drill diameter

If screw thread tables are not available, the minor diameter can be approximated as 75% of the major diameter.

UNIFIED NATIONAL THREAD NOTE COMPONENTS

After drawing a thread, we need to identify the size and thread form in a thread note.

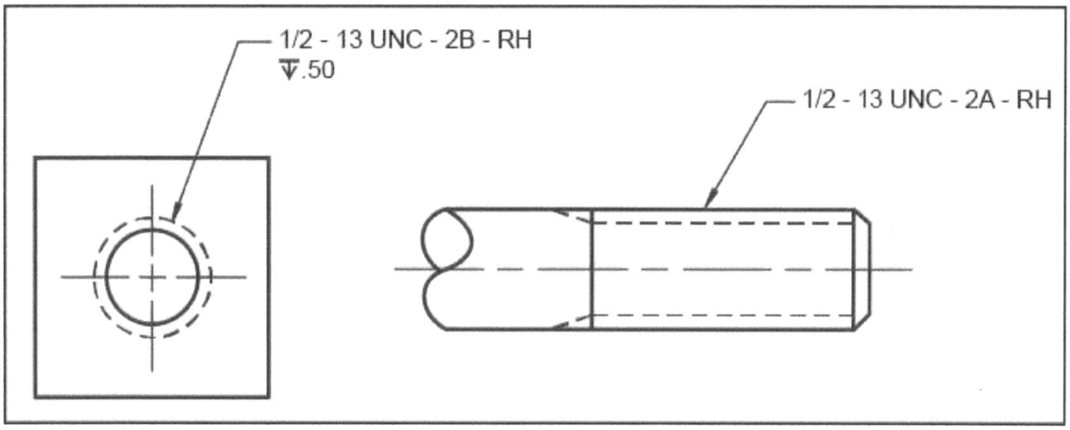

Figure 4.24

The following items are included in a thread note.

Major Diameter: The largest diameter.
Threads per inch: Number of threads per inch for a particular diameter. This is equal to one over the pitch (1/P).
Thread Form and Series: The shape of the thread cut.
UNC = Unified National coarse. For general use.
UNF = Unified National fine. Used when high degree of tightness is required.
UNEF = Unified National extra fine. Used when length of engagement is limited (Example: Sheet metal).
Thread Class: Closeness of fit between the two mating threaded parts. A thread class of 2 is assumed if none is stated.
1 = Generous tolerance. For rapid assembly and disassembly.
2 = Normal production
3 = High accuracy
External or Internal Threads:
A = External threads
B = Internal threads
Right handed or Left handed Thread:
RH = Right handed (right handed threads are assumed if not stated.)
LH = Left handed
Depth of thread: The thread depth is given at the end of the thread note and indicates the thread depth for internal threads. This is not the tap drill depth.

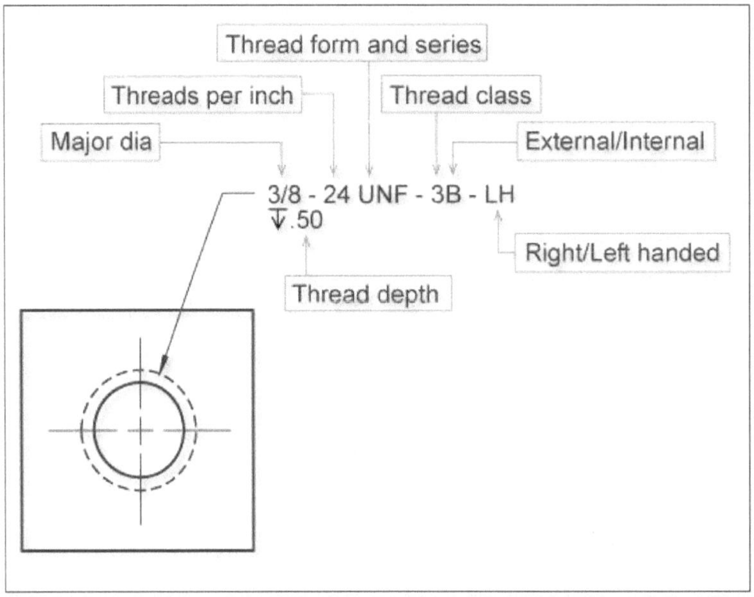

Figure 4.25

4.1.2 BOLTS

The bolt is a straight rod having an integral head on one side and threaded on the other end. It is passed through clearance holes in two or more aligned parts together. A nut is screwed on the threaded end of the bolt to tighten the parts together. A bolt and nut form one screw fastener as a unit.

A Bolt is a cylindrical item with a hexagonal (now and again square) head on one give up and threads (fully or partly) on the alternative stop. A Nut is the counterpart of the Bolt and its miles hexagonal (sometimes square) item with threads at the inside. The pitch of the threads at the bolt and nut ought to be the same for securing them properly.

HEXAGONAL HEADED BOLT:

The bolt with a hexagonal head on one side and threaded on the other end is called bolt. The bolt may be threaded throughout its length or partly threaded.

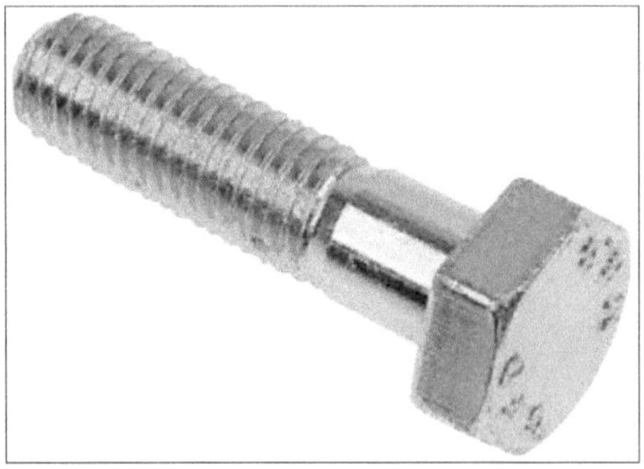

Figure 4.26

Ratio and proportions of hexagonal headed bolt:

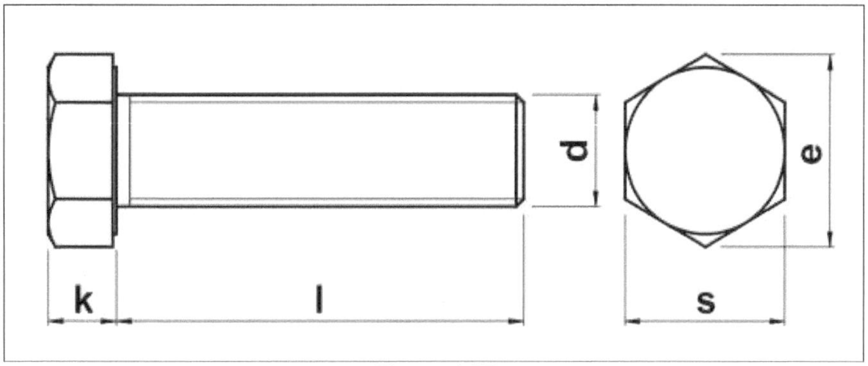

Figure 4.27

Width across flats, W = 1.5 D + 3 mm
Thickness of head, T = 0.8 D
Angle of chamfer = 30° to the base of head
Radius of chamfer, R = 1.4 D
Length of threaded portion, L1 = 2D + 5 mm up to 80 mm diameter
or L1 = 2D + 10 mm from 81 to 200 mm diameter
or L1 = 2D + 20 mm above 200 mm diameter.

SQUARE HEADED BOLT:

It is another common form of bolt. The contact area of the head with the part to be tightened is comparatively small as compared to hexagonal nut. It is generally used where the head of the bolt is to accommodate in a recess which is itself of square form. The head rests in the square recess with little clearance. When the nut is tightened or screwed off, the square recess prevents the head from turning.

Figure 4.28

Ratio and proportions of square headed bolt:

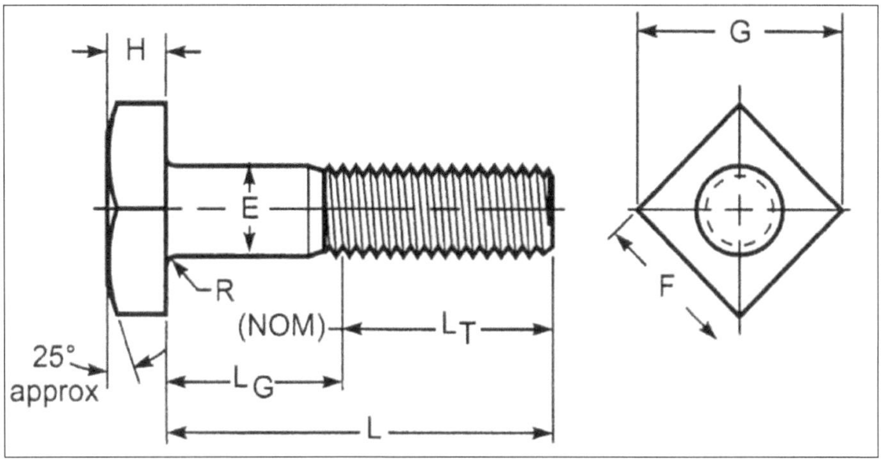

Figure 4.29

Width across flats, W = 1.5 D + 3 mm
Thickness of head, T = D
Angle of chamfer = 25° to the base of head
Radius of chamfer, R = 2 D
Length of threaded portion, L1 = 2D + 5 mm up to 80 mm diameter
or L1 = 2D + 10 mm from 81 to 200 mm diameter
or L1 = 2D + 20 mm above 200 mm diameter.

BOLT MATERIALS
The material used in the manufacturing of the bolts will determine its strength, applications, corrosion resistance, etc. There are several materials of Bolts and some of the popular ones are:

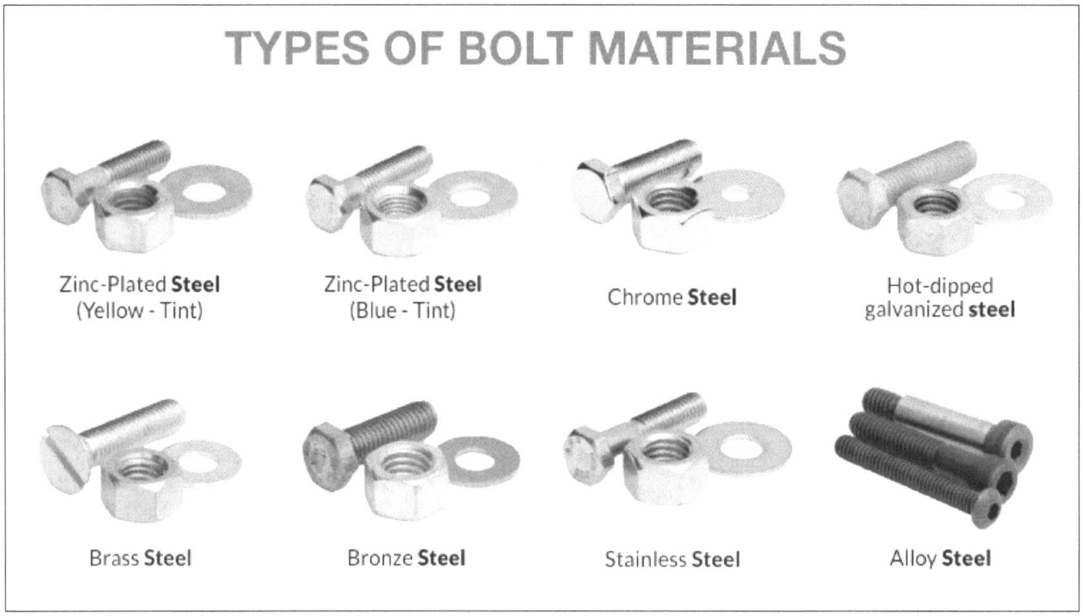

Figure 4.30

Zinc-Plated Steel: It is one of the popular and commonly used bolt materials for general purpose applications, usually indoor. Depending on the type of manufacturing, the colours of the bolt may be bluish or yellowish tint. It is a low carbon steel with a Zinc coating that provides moderate corrosion resistance. They are very cheap to manufacture as well.

Stainless Steel: One of the toughest bolt materials with very good corrosion resistance. As a result, stainless steel bolts are often used in industries, automobiles, ships (and other marine equipment) and outdoor objects in general. They are slightly expensive than the Zinc plated counterparts.
Chrome / Nickel Plated Steel: Although stainless steel already consists of chromium, additional Chrome and Nickel plating of steel further increases the corrosion resistance. The other benefit of this plating is the surface of the bolt looks very shiny, smooth and polished.
Hot-dipped Galvanized Steel: In this type, the steel is coated with even thicker layer of Zinc for increased corrosion resistance and thus making them suitable for outdoor use. Even the nuts and washers must be of the galvanized type for proper coupling.
Brass / Bronze: These types of bolts have a natural resistance to corrosion as they are alloys of copper. They are generally expensive than other bolt materials and as a result they are often used in arts and decorative pieces.
Alloy Steel: The Alloy Steel contains additional elements other than iron and carbon such as manganese, nickel, chromium etc. The result is a bolt with very good strength, corrosion resistance and wear resistance. It is often black in colour due to additional black oxide coating.

COMMON TYPES OF BOLTS

As Bolts and Nuts are one of the commonly used fasteners, there are different types of bolts for different types of applications. Some of the more common and popular types of bolts are:

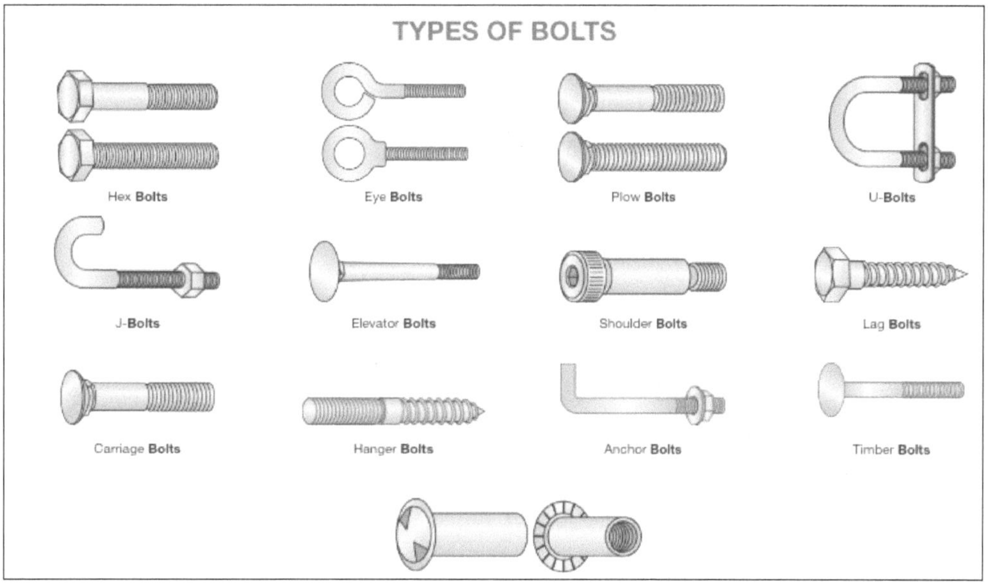

Figure 4.31

Hex Bolt
Lag Bolt
J-Bolt
U-Bolt
T-Bolt
Anchor Bolt
Sex Bolt
Carriage Bolt
Flange Bolt
Eye Bolt
Elevator Bolt
Plow Bolt
Hanger Bolt
Timber Bolt
Shoulder Bolt
Stud Bolt
Toggle Bolt

BOLT DIMENSIONS

Before looking at the Bolt Size Chart, let us first try to understand different proportions and dimensions associated with a bolt. The following image shows a typical hex bolt with all the important dimensions marked on it.

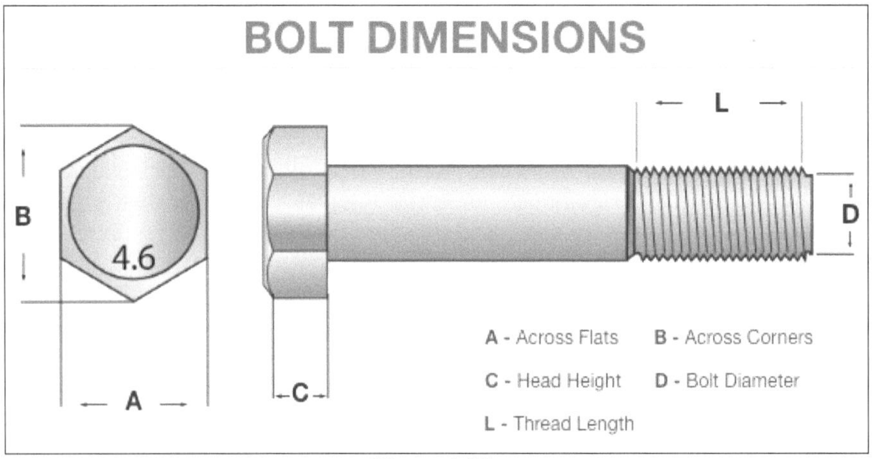

Figure 4.32

A bolt can be divided into two parts: Bolt Head and Bolt Body:

BOLT HEAD

The head of the bolt is usually hexagonal shape (sometimes square or circular) and is used to tighten or loosen the bolt. The size of the head is dependent on the shank's nominal diameter.
Bolt head can have three important dimensions. They are:
Size across the Flats
Size across corners
Height of the Head

BOLT BODY

The Bolt Body consists of the main shank with full or partial threads. Bolt Body can have important dimensions.
They are:
Bolt Diameter
Body Length
Thread Length
Thread Pitch

The diameter of the Bolt is the actual size of the bolt and it is measured across the bolt shank or the threaded area. The body length of the bolt is measured from the tip of the bolt to the point where the shank meets the head. If the bolt is partially threaded, then the thread length can also be measured.
Thread Pitch is an important dimension associated with a Bolt. In imperial system, the thread pitch of the bolt is determined by the number of threads in one inch. In metric system, the distance between two adjacent threads in millimetres is considered as the Thread Pitch.
The thread pitch can be further categorized into coarse (or standard), fine and super fine (or extra fine) threads. Coarse Threaded bolts have lower thread count than the other types are very common in general purpose applications. For precision applications and situation where you require more holding force, fine threaded bolts are used.

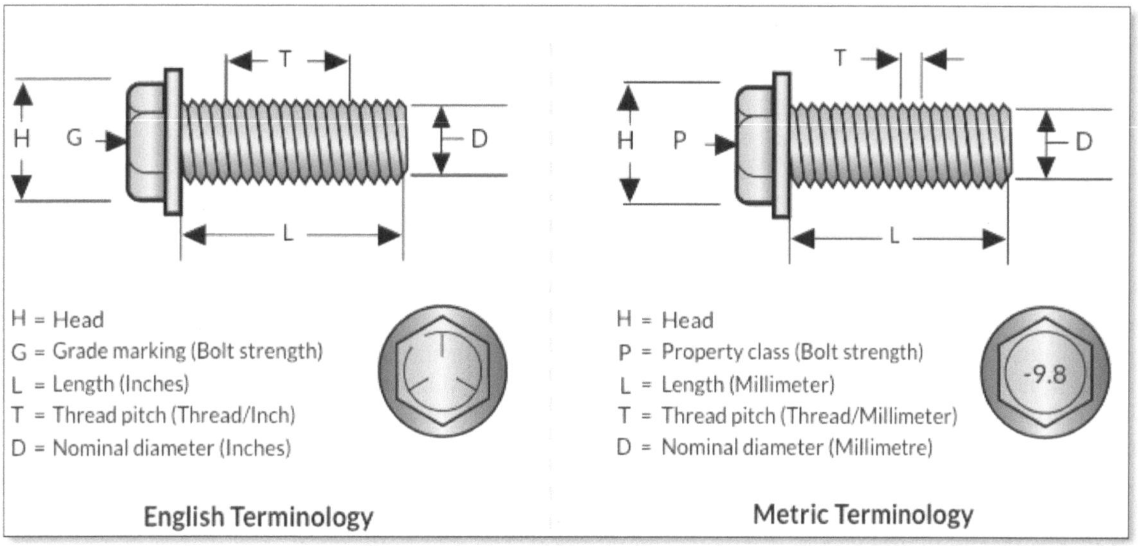

Figure 4.33

GRADE OR CLASS

Another important specification associated with bolts is its grade (imperial) or class (metric), which determines the tensile strength of the bolt. The tensile strength of a bolt is the maximum stress / stretch it can handle before it breaks. In USCS imperial system, the bolt's tensile strength is represented using 'grade marks' on the top of the head. The 'grade marks' are nothing but radial lines and more the number of lines on the head of the bolt, greater is the strength of the bolt.

In metric system, the strength of the bolt is determined using its class number, which is also printed on the head of the bolt. Higher the class number, greater is the strength of the bolt.

If the number is followed by another number (for example, 10.9), then the second number represents the yield strength of the bolt, which determines the stress it can handle before it cannot return to its original shape without damage.

Consider a Class 10.9 Bolt. Here, this bolt has a tensile strength of 1,000MPa (145,000PSi) and a yield strength of 900MPa (130,000PSi).

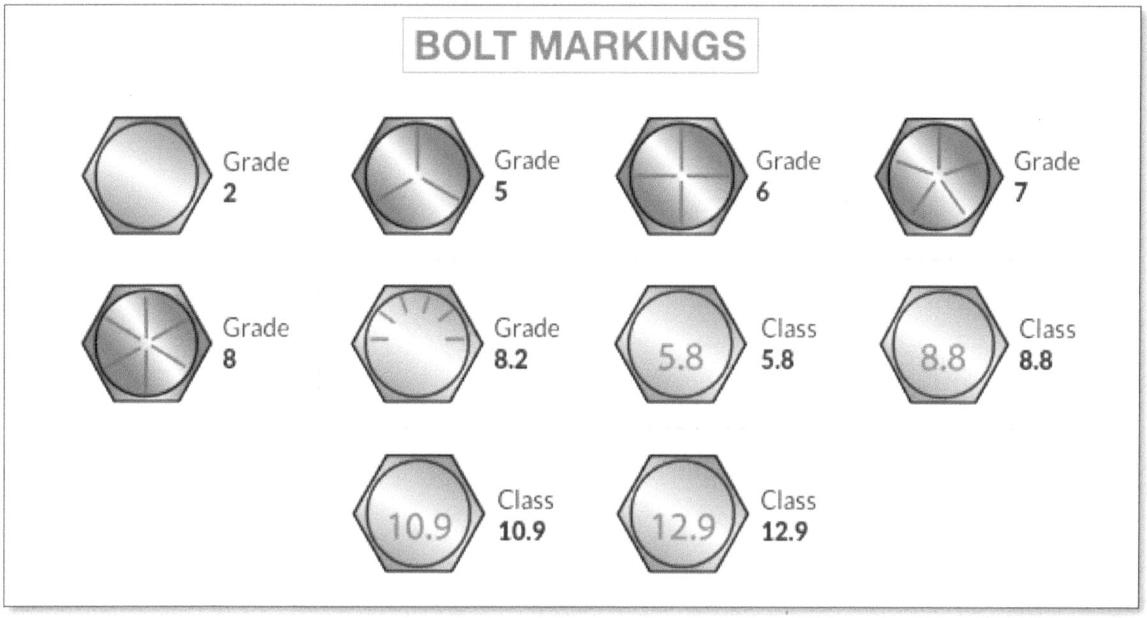

Figure 4.34

METRIC BOLT SIZE CHART

In metric system, the bolts (and nuts) are usually referred using the letter 'M' followed by additional dimensions. For example, a common metric specification of a bolt is "M6-1.0 x 20". Here,

'M' indicates the bolt uses metric threads

'6' indicates the nominal diameter of the bolt (at the shank or the threads)

'1.0' indicates the thread pitch of the bolt in millimetre

'20' indicates length of the bolt in millimetre.

4.1.3 NUTS

Nuts are generally in the form of hexagonal or square prisms. Other forms are also used to suit specific requirements.

Hexagonal nuts are generally used in most of the applications. The spanner used for turning the nut can have better hold on a square nut than on a hexagonal nut, but the angle through which the spanner will have to be turned to get another hold is only 60° in case of a hexagonal shape, while it is 90° in case of a square shape. Hence, it is more convenient to screw-on a hexagonal nut than a square nut.

An octagonal nut would require only turn, but there would be greater tendency for the spanner to slip. Hence, this shape is seldom used. Nuts of forms other than the above two are usually provided with special facilities for screwing them on or off the bolts.

HEXAGONAL NUT:

It is general purpose nut used mostly on fasteners. I t has six surfaces having hexagonal shape, that is why it is called hexagonal nut. The nut are rounded off or chamfered at an angle of 30° but on one side only. Due to rounding or chamfering, an arc is formed on each vertical face and a circle is thus formed on the top surface of nut.

The upper corners of this nut are rounded-off or chamfered. The chamfering is generally conical. The angle of chamfer is 30° or 45° with the base of the nut. Due to chamfering, an arc is formed on each vertical face, and a circle is formed on the top surface of the nut.

The dimensions of the hexagonal nut cannot be expressed exactly in terms of the nominal diameter of the bolt. The standard proportions for nuts and bolt-heads may be obtained from tables. For elementary work, the following approximately standard dimensions may be adopted

If D = the nominal diameter of the bolt,

Thickness of the nut, T = D

Width across flats, W = 1.5D + 3 mm

Angle of chamfer = 30°

Radius of chamfer arc, R = 1.4D (approx.)

Very often, and especially when a nut is shown in one view only, the following rough rule dimensions are used.

Thickness of the nut, T = D

Distance across diagonally opposite corners = 2D

Angle of chamfer = 30°

Radius of chamfer arc, R = 1.5D(approx.).

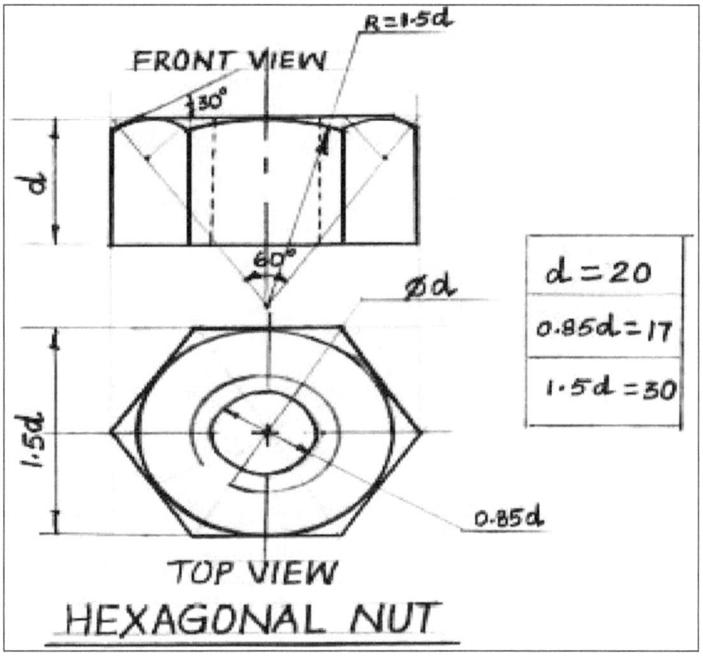

Figure 4.35

SQUARE NUT:
The nut having a square shape is known as square nut. This type of nut is usually used in conjunction with the square headed bolt. The corners of the square nut are also chamfered in the same way as for hexagonal nut. The upper corners of the square nut are also chamfered in the same manner as those of the hexagonal nut. The widths across flats of a square nut and a hexagonal nut, for the same size of bolt, are also equal.
Dimensions of a square nut:
If D = the nominal diameter of the bolt,
Thickness of the nut, T = D
Width across flats, W = 1.5D + 3mm
Angle of chamfer = 30°
Radius of chamfer arc, R = 2D (approx.).

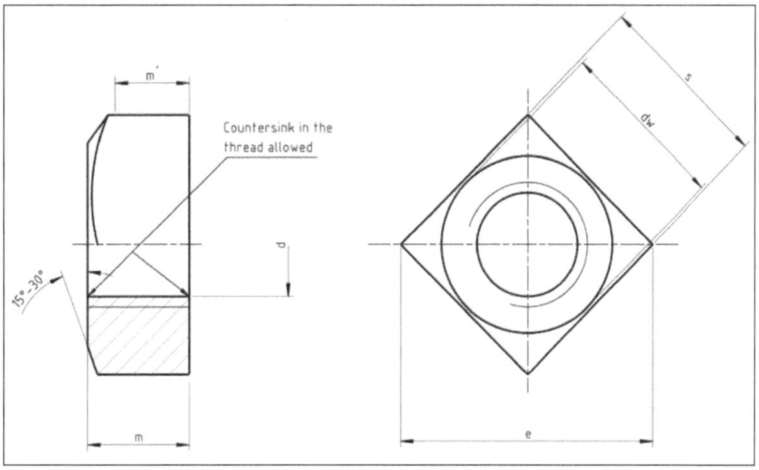

Figure 4.36

FLANGED NUT:

The hexagonal nut provided with a flange of circular washer at the base of the nut. Because of increased bottom surface of the nut due to flanged surface, the bearing surface of the nut is increased and therefore used in such places where bolt of large size diameters are required to pass through the required hole.

Figure 4.37

PROPORTIONS OF THE FLANGED NUT:

D is the nominal diameter of the bolt
Width across flats, W = 1.5 D + 3 mm
Thickness of nut, T = 0.8 D to D
Thickness of flange, t = 0.25 D
Angle of chamfer = 30° to base of the nut
Diameter of flange = 2.2D

CAP NUT:

The cap nut is a hexagonal nut provided with an integral cylindrical cap at the top to protect the end of the bolt from corrosion. The nut can be seen in smoke boxes of locomotives on the main pipes, connections, etc. It also prevents leakage through threads.

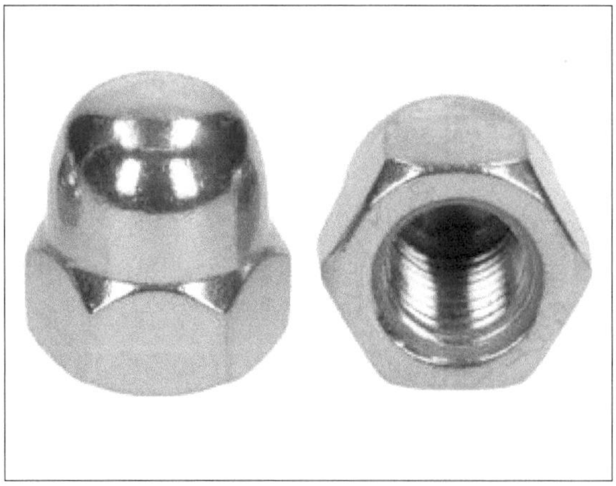

Figure 4.38

It consists of a hexagonal prism having a circular collar at the top with a central threaded hole, six slots are cut in the collar which are radially opposite. This nut is used to provide as a hold on nut which is subjected to sudden and variable vibration such as on piston rod, scooter wheel etc. A pin is fixed across the two slots which passes through the bolt after tightening the nut in position. Thus, the nut is prevented from becoming loose.

DOME NUT:

The top of the dome nut looks like a dome on the hexagonal nut. Dome nut serves the same purpose as the cap nut.

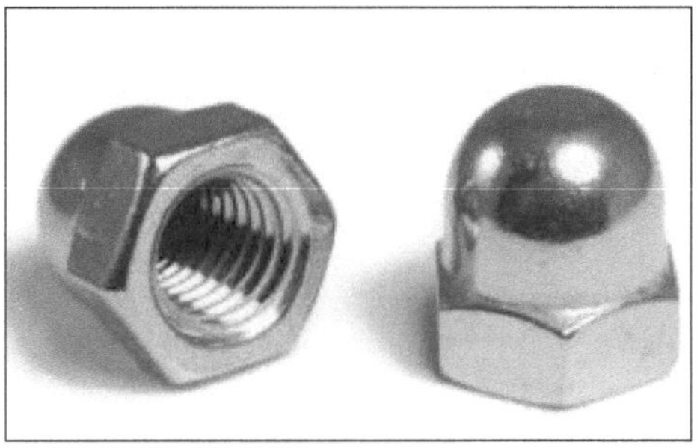

Figure 4.39

WING NUT:

It is a conical nut having two wings attached with its slant surface. It has a threaded hole at the centre. These nuts are designed to operate by thumb and finger and are frequently used for securing adjustable fittings as the wind screen of a motor car. No spanner is required to tighten or open the nut.

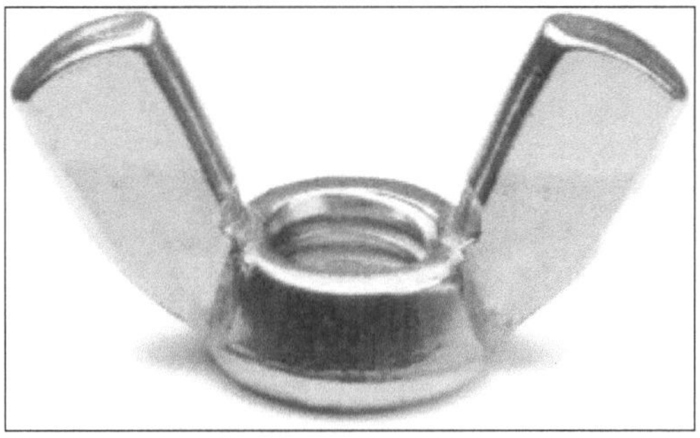

Figure 4.40

CAPSTAN NUT:

The circular nut provided with various holes on the outer circular surface of the nut. These holes are generally six or eight in number for turning the nut with a tommy bar.

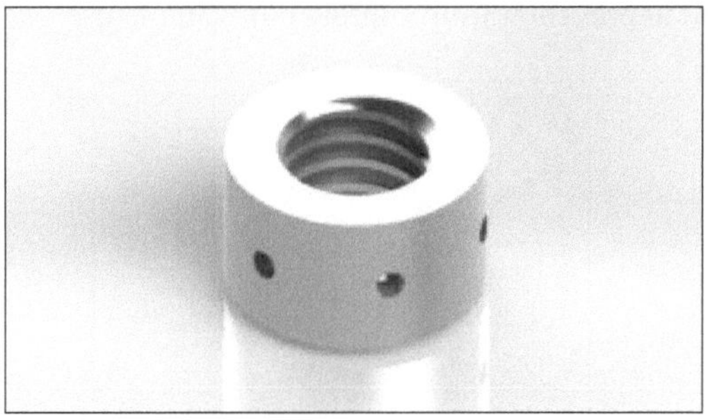

Figure 4.41

CASTLE NUT:

It consists of a hexagonal prism having a circular collar at the top with a central threaded hole, six slots are cut in the collar which are radially opposite. This nut is used to provide as a hold on nut which is subjected to sudden and variable vibration such as on piston rod, scooter wheel etc. A pin is fixed across the two slots which passes through the bolt after tightening the nut in position. Thus, the nut is prevented from becoming loose.

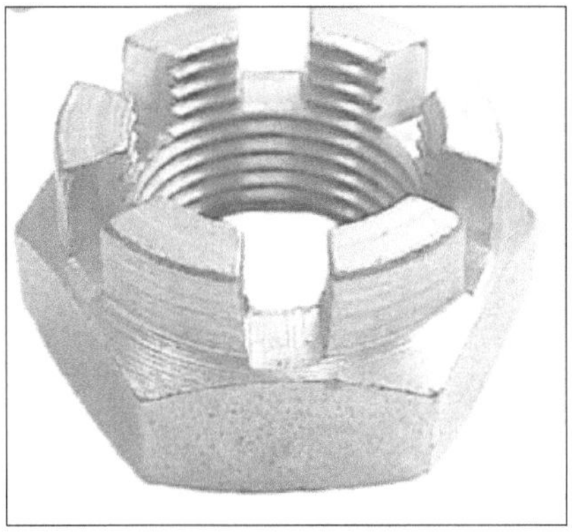

Figure 4.42

METHODS OF NUT LOCKING
#1 LOCKING OF NUTS

When a nut is tightened over the bolt, a small working clearance exists between the threads of a nut and bolt, even in the highest class of fits.

The types of nuts and bolts used in the moving parts of machinery will be subjected to continuous vibration. In addition, sometimes, they are required to carry varying axial loads also, for example, as the crossheads and connecting rods of an engine.

Under such conditions, because of the working clearance existing between the mating threads, the nut develops a tendency to work loose or unscrew by itself.

This consequently may result in a serious breakdown. To prevent this, some means of locking the nut from unscrewing is necessary. The various locking devices for nuts are classified into positive and friction types. However, the tendency of unscrewing is reduced if the threads of the nut and bolt are of a good fit.

The positive locking methods use a split pin, a screw, a lock plate, or a tab washer to lock the nut. In the friction method, an additional nut called a lock nut or a spring washer is used to lock the nut.

#2 SPLIT PIN LOCKING

The simplest method of locking a nut using a split pin is shown in the figure. After the nut has been finally tightened, a small hole is drilled through the blot close to the top face of the nut. The split pin is then inserted in the hole, and the split ends of the pin are opened to prevent it from coming out while in use.

Figure 4.43

#3 SET SCREW LOCKING USING GROOVED NUT

A hexagonal nut with a cylindrical grooved collar at the lower end is called Penn to ring or grooved nut. The end of the bolt hole is counterbored to receive the cylindrical lower grooved portion of the nut, as shown in the figure. Locking of the nut is done by a set screw screwed through the nearest face of the workpiece.

Figure 4.44

The projecting dog end of the set screw enters the groove in the cylindrical portion of the nut and prevents the slackening of the nut. This locking method is possible if the bolt hole is close to the nearest vertical edge, as in the case of the marine engine connecting rods.
When the bolt hole is not close to the vertical edge of the workpiece, this nut is used in conjunction with a separate collar, as shown in the figure. The dowel pin screwed into the bearing surface prevents the rotation of the collar.

#4 LOCKING BY A LOCK PLATE

This type of locking is employed in heavy engineering work, as in the case of connecting rods, wheel shafts, etc. The empirical proportions of the lock plate are shown in the figure. Initially, the nut is tightened, and then the lock plate is inserted through the nut.
The plate is grooved so that the grooves in the plate receive the hexagonal corners of the nut at every 30° rotation. The plate is fixed to the bearing surface by a tap bolt screwed into it. The figure shows the assembly of the locking of a nut by the lock plate. Since the nut is held tight within the grooves of the lock plate, the nut is prevented from slackening.

#5 LOCKING BY A TAB WASHER

A washer provided with a rectangular projection is called a tab washer. This method of locking of nut or bolt head is suitable when the nut or bolt head is placed near the vertical edge of the workpiece.
After the nut is tightened, the tab and the projecting portion of the washer itself are bent to bear against the vertical surface of the workpiece and one of the lateral faces of the nut or the bolt head.

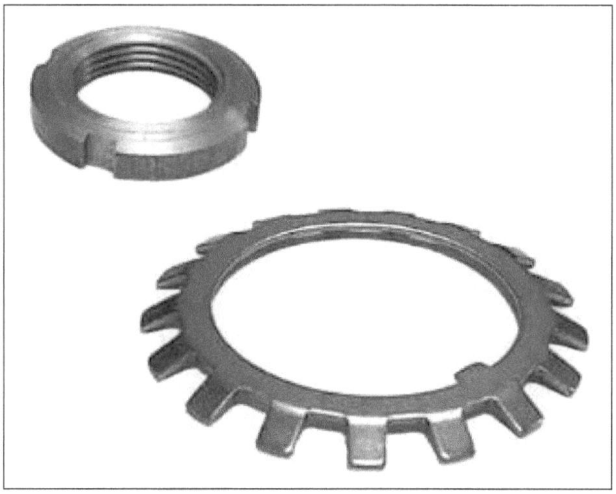

Figure 4.45

#6 LOCKING BY A LOCK NUT

In this type of locking, the friction between the mating threads of the bolt and nut locks the nut. When the nut is tightened over a bolt, the lower flanks of the bolt threads will be in contact with the upper flanks of the nut threads.

Figure 4.46

When another nut is screwed down hard over the lower nut, it tends to pull the bolt through the lower nut.

This virtually makes the upper nut take up all to load, While the lower nut to function simply as a washer since there will not be any contact between the threads of the bolt and the lower nut.

Now the upper nut is held by a spanner While the lower nut is pulled in the opposite direction. This cause the lower flanks of threads of the lower nut to come in contact with the upper flanks of the threads of the bolt, causing a wedging action between them.

This makes the lower nut act as a lock nut. Theoretically, the lower nut, which is a lock nut, may be the thin nut, as shown in Figure 4.46

But as it required a special thin spanner to turn it, in practice, it is more usual to place the taking nut below the lock nut as shown in Fig. such cases sometimes two nuts of equal thickness are used as shown in Figure 4.46

#7 LOCKING BY SPRING WASHER

In this method of locking a coiled spring, the washer is placed underneath the nut, when the nut is tightened, the spring force of the spring exerts an axial force on the underside of the nut and holds it by the friction grip. This prevents the slackening of the nut.

#8 SCREW PIN LOCKING

A nut may be locked by a screw pin, screwed in the bearing surface adjoining the nut touching one of the lateral faces of the nuts. This type of locking is employed when the nut is expected to remain without any adjustment for a long time.

4.1.4 Washer

It is one of the locking devices in which tightening of nut on spring washer results in exerting upward pressure on the nut in thrusting the nut threads against the bolt threads thus increasing the friction between them.

Its shape is like a round ring. Before tightening the nut and bolt, it is fitted on a stud or bolt. It is used to secure the tightness of a joint, screw, etc. and it becomes easy to open the nut-bolts and their threads are protected.

The washers are used to prevent a nut from loosening. These are used where there is vibration, or jolts occur. This puts pressure against the nut, preventing the latter (the nut) from turning and working loose.

Even if the fit is a bit larger than the diameter of the bolt, with the washers help nut bolt can be tightened easily.

The washer is a disc-shaped flat plate with a centre hole. They are used to give loads over a large area. It is mainly used for spacers, pre-loading screws, springs, and vibration reduction.

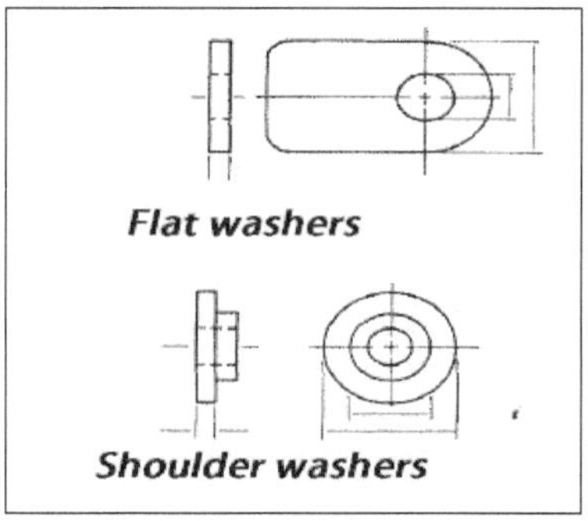

Figure 4.47

TYPES OF WASHERS

Following are the types of washers, as per their utility and functions:

1. Spring washer
    1. Belleville or conical washer
    2. Dome spring washer
    3. Wave spring washer
    4. Finger spring washer
    5. Crescent spring washer
2. Lock washer
    1. Split lock washer
    2. External tooth lock washer
    3. Internal tooth lock washer
3. Plain washer
    1. Torque washer
    2. Flat washer
    3. Fender washer
    4. Finishing or Countersunk washer
    5. Shoulder washer
    6. C-washer

1. SPRING WASHER

This washer is in the shape of a ring cut by spring steel. Its ends are slightly raised up and down due to spring tension. These washers are placed above or below the nut or bolt and tighten so much that both their ends face each other and become levelled.

There are several forms of spring washers. Each has its own advantages. Single pattern spring washer, standard pattern spring washer, double pattern spring washer, grip pattern spring washer, and girder pattern spring washer are the structures of spring washer.

#1 BELLEVILLE OR CONICAL WASHER

Figure 4.48

The Belleville washer consists of sides that support very heavy force with small deflections. It is also known as a conical spring washer.
These washers hold an arrangement under tension during thermal expansion and contraction requirements. Conical washer load-deflection properties depend on its height and thickness rate. It is used for projects including thermal expansion.

#2 DOME SPRING WASHER

Figure 4.49

The dome spring washers are similar to a Belleville washer but with rounded sides. Dome spring washers produce very high load capabilities with a comparatively small deflection range.
They possess ground curves to form a flatter load-bearing surface. It is used wherever you would use a crescent spring washer, but there is also a need to create a flatter surface.

#3 WAVE SPRING WASHER

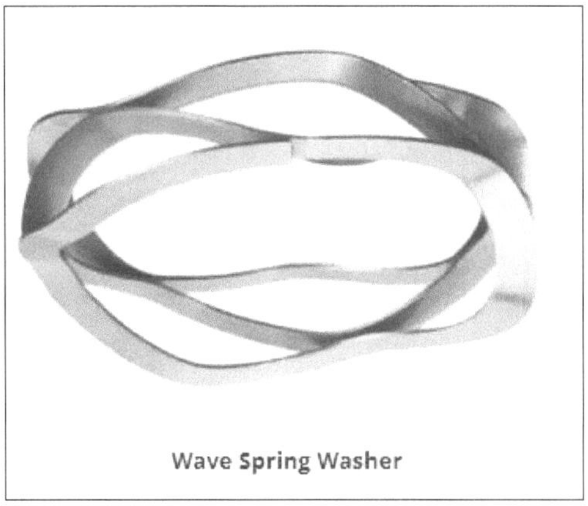

Figure 4.50

These types of washers rotate in two directions, forming a wave-like shape. Wave spring washers possess enough load capacity and deflection.
They serve as cushion springs or spacers on a shaft. Wave springs washers have average load capacity and deflection limits. It is most frequently used as a cushion or spacer.

#4 FINGER SPRING WASHER

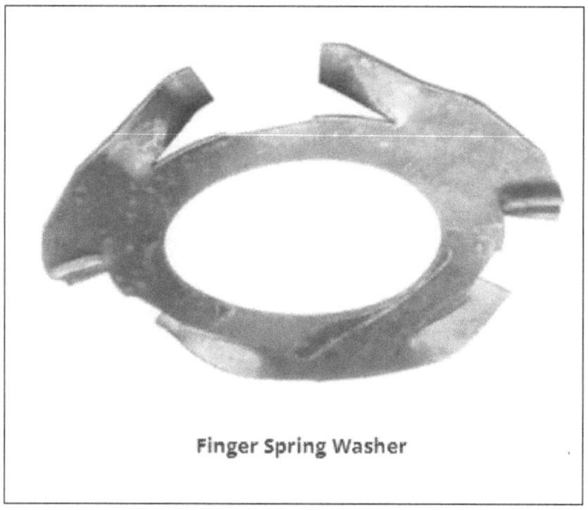

Figure 4.51

The finger spring washers consist of three curved flanges. Finger spring washers are used to reduce the damping and noise, extra wear, vibration, and skidding wear on revolving parts.
Finger spring washers are made of carbon steel. They combine the flexibility of a casing washer with the given load points of a wave washer.

#5 CRESCENT SPRING WASHER

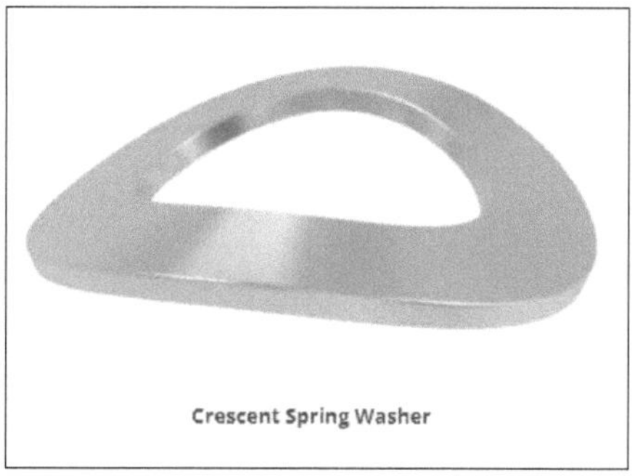

Figure 4.52

These washers look like flat washers that have been somewhat curved to give lighter pressure and maintain flexibility. It is also known as curved spring washers. They support almost small loads with a wide deflection range.
These types of washers provide uniform spring rates over the deflection range and have linear load-deflection properties. Curved spring washers must be used in flexible and load-cycling products. It is also employed to absorb movement.

2. LOCK WASHER

This form of lock washer is used to lock the nuts and bolts. These are many forms. Generally, six or more leaves are extended beyond the main diameter of the lock washer.
After fixing this washer its leaves are bent on the sides of the nut or bolt. Thus, this washer prevents the nut bolts from getting loose. This washer is generally used in cars, trucks, etc.

#1 SPLIT LOCK WASHER

Figure 4.53

Split lock washers are non-continuous rings that tilt each end slightly outside in opposite directions. When the splitting lock washers are secured, each end is dug into the mating surfaces. It is also sometimes known as helical washers. This washer develops preload on the screw while tightening. The developed preload gives protection from loosening during vibration and corrosion. It is used to lock a bolt into place.

#2 EXTERNAL TOOTH WASHER

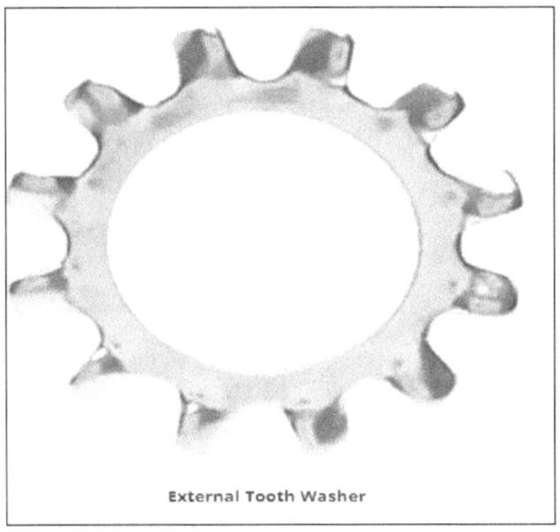

Figure 4.54

This type of lock washer has many teeth on its outer diameter that cut into a surface and maintain compression. The external tooth lock washer prevents a fastener head from loosening by using the strut action of the teeth.
They operate most suitable under large head screws. The external tooth lock washer provides maximum torsional resistance. It is used to lock deep-headed fasteners into place.

#3 INTERNAL TOOTH WASHER

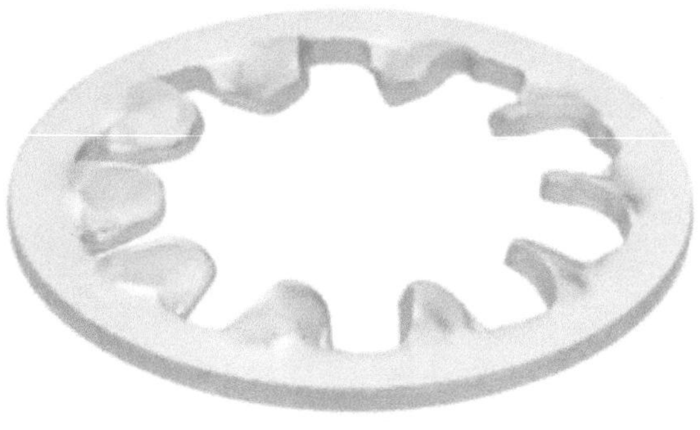

Internal Tooth Washer

Figure 4.55

This type of lock washer consists of many teeth on the inside diameter that cut into the nut or bolt. The internal tooth lock washer prevents a fastener head from loosening using the strut action of the teeth.

These washer teeth can also absorb shock and vibration. It is used to lock shallow-headed fasteners into place.

3. PLAIN WASHER

A plain washer is a flat disc. It has a hole in its centre through which a bolt, stud, or set-screw can pass. By using it, the surface below the bolts or nuts becomes plain. In this way, the pressure of the nut is uniform on the entire surface. This washer is of the following standard:

Diameter of the washer = 2D + 1/8 (D = Diameter of bolts)
The thickness of Washer = 1/8 D

The angle of Chamfer = 30 °
Diameter of inner hole = Diameter of bolt

#1 TORQUE WASHER

Torque Washer

Figure 4.56

The Torque washer consists of a square hole and outer forks that prevent the bolt from spinning with it when fastening a bolt. It is mainly used with a carriage bolt for woodworking projects.

#2 FLAT WASHER

The flat plain washers are thin, flat, and circular in shape with a hole in the centre. Flat plain washer application is intended to give the load over a relatively large area and give support to small head screws.

It is also known as Type A plain washers. A flat washer is what most people think of a normal washer. They are good for general use.

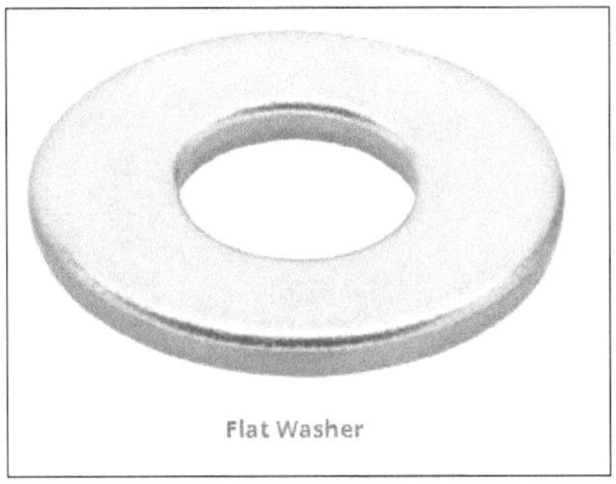

Figure 4.57

#3 FENDER WASHER

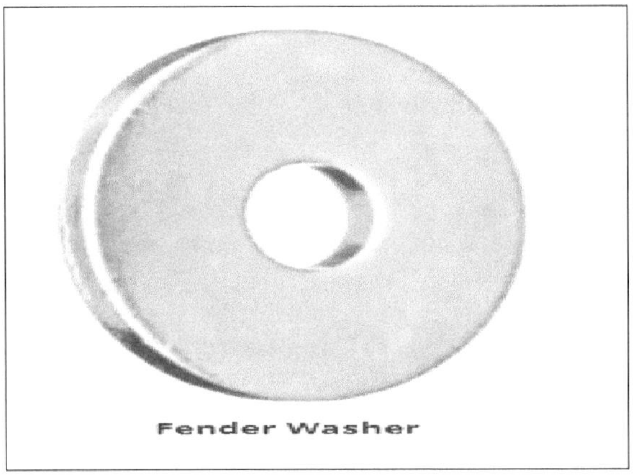

Figure 4.58

The fender washers are flat washers with a small hole and outside relatively large diameter. Hence this type of plain washer gives loads over a relatively larger area than flat washers.

Fender washers have applications for installing signboards, panels, drywall, automobiles, plumbing, sheet metal, and electrical applications. It is usually a thin metal. They have named it for their general use on car fenders.

#4 FINISHING OR COUNTERSUNK WASHER

Figure 4.59

The finishing washer has a top designed to hold the fastener. It is also known as a countersunk washer. The Countersunk washer has 90° or 120° counters at the top.

These washers served as flat washers for flat head counters screws. This gives a flush surface to the countersunk screw.

#5 SHOULDER WASHER

Shoulder washers consist of a shoulder type of structure. They are employed to protect screws, wires, or any element from the rest of the assembly. They are available in fibre-glass, phenolic, nylon, PCTFE, PTFE, and metals.

Figure 4.60

#6 C-WASHER

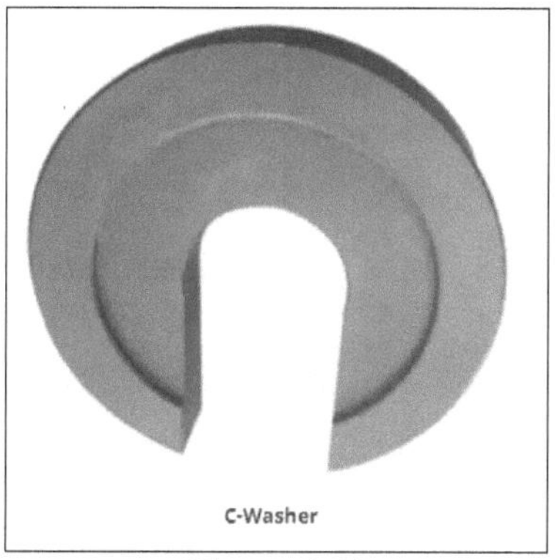

Figure 4.61

These look like a flat washer, however, it has a slot cut from the centre hole to form a "C" shape. A C-washer is made to slide in and out of place on the bolt or shaft.
The C-Washer used at any time may need to be adjusted or replaced, without removing the fastener.

4.1.5 COMBINATION OF BOLT, NUT AND WASHER FOR ASSEMBLING TWO PARTS TOGETHER

In common machineries used at home, we might have observed the assembly of bolt, nut and washer to connect two parts together. See Figure 4.62

Figure 4.62

In the earlier topics, we learnt how to draw the views of bolt, nut and washer separately. Here, we expect to understand the views of the assembly of bolt, nut and washer.

Problem 1: Draw to scale 1:1, the Front View, Top View and side view of a hexagonal headed bolt of diameter 25 mm with hexagonal nut and washer, keeping the axis parallel to V.P and H.P Refer Figure 4.63
Solution:
Steps Involved:
(i) Since the axis is parallel to both V.P and H.P, the side view reveals more information about the shape of the object. So, start with side view, where circles are seen.
(ii) Draw two circles of diameter d = 25 mm and 0.8d = 20 mm, in dotted lines to indicate the invisible feature from left side.
(iii) Draw the chamfering circle of diameter, 1.5d + 3 mm =40.5 mm
(iv) Circumscribe hexagon around the chamfering circle, using set-square and mini-drafter.
(v) Then draw a circle of diameter 2d + 3 mm = 53 mm for washer.
(vi) Project the side view to front view and top-view.
(vii) Both the views are completed as shown in the Figure 4.63S

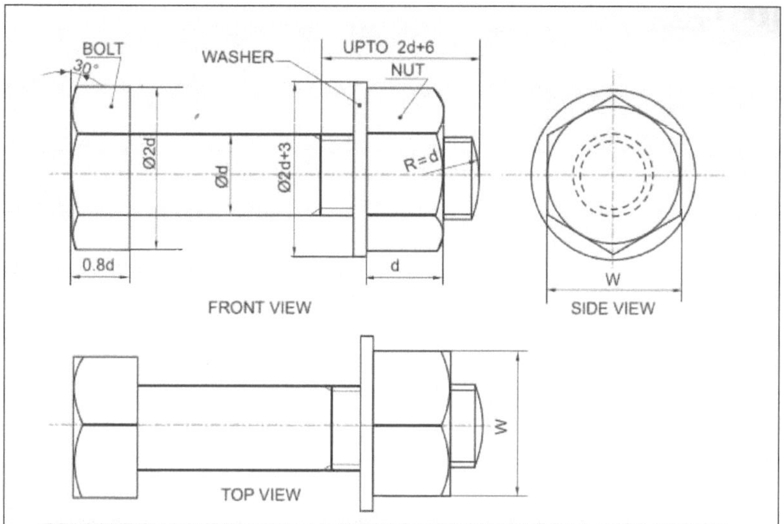

COMBINATION OF HEXAGONAL HEADED BOLT WITH HEXAGONAL NUT & WASHER

Figure 4.63

Problem 2: Draw to scale 1:1, the Front View and Side View of an assembly of hexagonal bolt of diameter 24 mm bolt length = 90 mm and a hexagonal nut, keeping the axis parallel to H.P and V.P

Solution:

Refer Fig 4.63 The steps involved are similar to the previous example

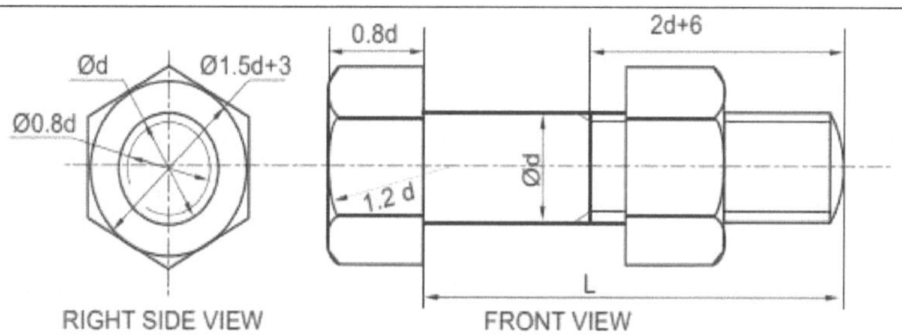

d	0.8d	1.2d	1.5d+3	2d+6	L
24	19.2	28.8	39	54	90

COMBINATION OF HEXAGONAL HEADED BOLT WITH HEXAGONAL NUT

Figure 4.64

Problem 3: Draw to scale 1:1, the Front View and Side View of an assembly of a square bolt of diameter 25 mm and a square nut, keeping the axis parallel to V.P and H.P. Take length of the bolt as 100 mm.

Solution: Refer Fig 2.31 The figure is self explanatory.

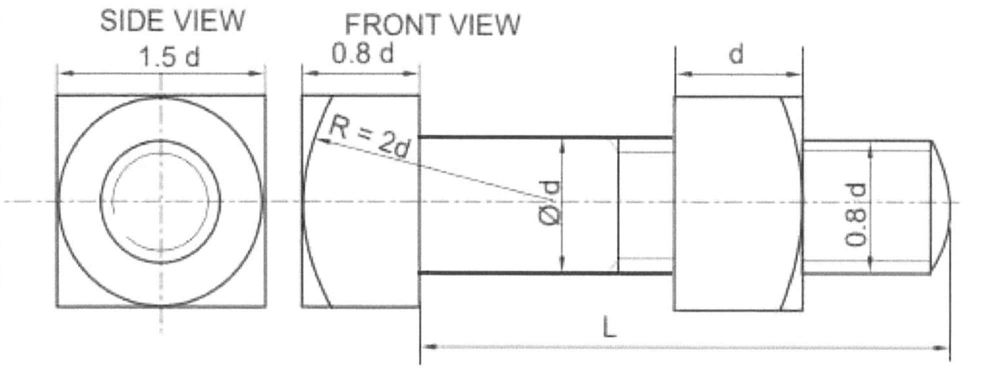

d	0.8d	1.5d	2d	2d+6	L
25	20	37.5	50	56	90

SQUARE BOLT AND SQUARE NUT IN POSITION

Figure 4.65

Exercises:

NOTE: Assume missing dimensions proportionately

1. Draw to scale 1:1, the front view, top view and side view of an assembly of hexagonal headed bolt of 30mm diameter with hexagonal nut and washer, keeping the axis parallel to V.P and H.P. Give standard dimensions.
2. Draw to scale 1:1, the front view and side view of an assembly of a hexagonal bolt of diameter 30mm and a hexagonal nut, keeping the axis parallel to V.P and H.P.
3. Draw to scale 1:1, the front view and side view of a square headed bolt of size M24, fitted with a square nut, keeping their common axis parallel to V.P and H.P.
4. Draw to scale 1:1, the front view and side view of the assembly of square headed bolt with a hexagonal nut and a washer, with the diameter of bolt as 30mm, keeping their axis parallel to V.P and H.P and two of the opposite sides of the square head of the bolt and of the hexagonal nut, parallel to V.P.

4.2 FREE HAND SKETCHES OF ORTHOGRAPHIC VIEW AND ISOMETRIC VIEW

Free Hand sketching of front view, top view and a suitable side view of simple components from their isometric views. Normal perspective of prism, pyramid, cylinder& cone in vertical position by visual ray method only.

PERSPECTIVE PROJECTION

The perspective projection, also sometimes called scenography projection or central projection, is the form of pictorial drawing which most nearly approaches the pictures as seen by the eyes.

Perspective projection is sometimes called scenography projection or central projection**,** since the lines of sight converge to a single point or centre. Perspective obtained will depend on the relative position of the object, picture plane and point of sight. In this projection, the eye is assumed to be situated at a definite position relative to the object. The picture plane (vertical plane) is placed between the object and the eye. Visual rays from the eye to the object pierce the picture plane and form an image on it. This image is known as **perspective** of the object.

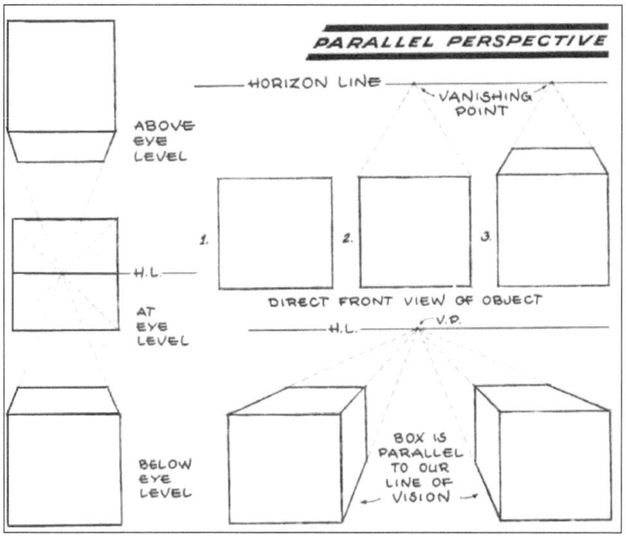

Figure 4.66

Figure 4.67

The points at which the visual rays joining the station point and the object pierces the picture plane in both the top and profile views, are projected to intersect each other to give points in the perspective. Since the perspective view is obtained by the intersection of the visual ray, this method is called Visual Ray Method.

IMPORTANCE POINT IN VISUAL RAY METHOD

1. Draw the PP line to represent the picture plane in the top view.
2. Draw the plan of the object based on its position with respect to the PP.
3. Draw the Ground Plane, the GP at any convenient distance from PP and project the front view based on the position of object with respect to gp.
4. Locate the position of the central plane with respect to the object and represent it as a line in both the views. On it mark the top view (s) and front view (s') of the station point based on this position with respect to PP and GP.
5. Join all plan points with s and note the intercepts of each line with PP line.
6. From each intercept, with PP, draw projector vertically till it meets the line joining the elevation of the corresponding point and s' to get the perspective.
7. Follow the above step to get the perspective of other points of the object.
8. Join all these points in proper sequence to get the perspective of the objective.

Perspective view of the point P

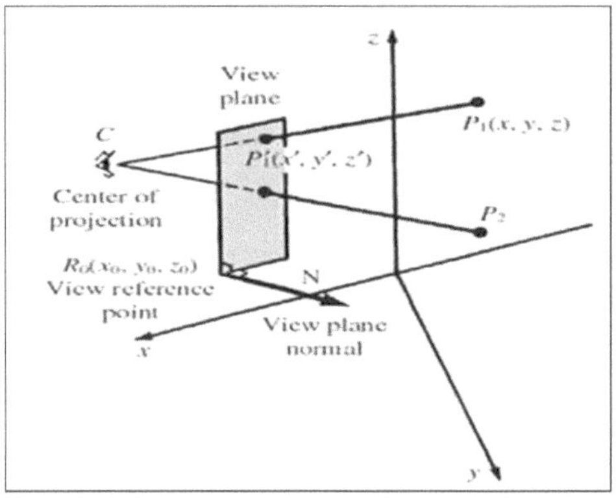

Figure 4.68

Perspective view of line

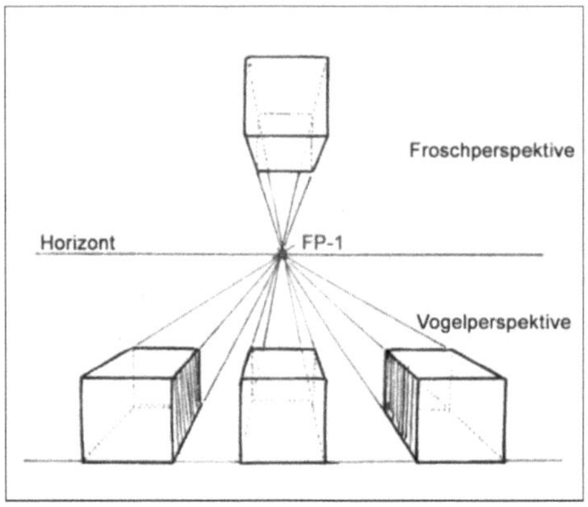

Figure 4.69

Assignment

1. The thread angle of B.S.F/B.S.W. screw threads are ___
a) 45°
b) 60°
c) 55°
d) 30°
2. B.S.F screw threads belong to coarse series.
a) True
b) False
3. Find the type of screw thread the following screw profile is indicating.

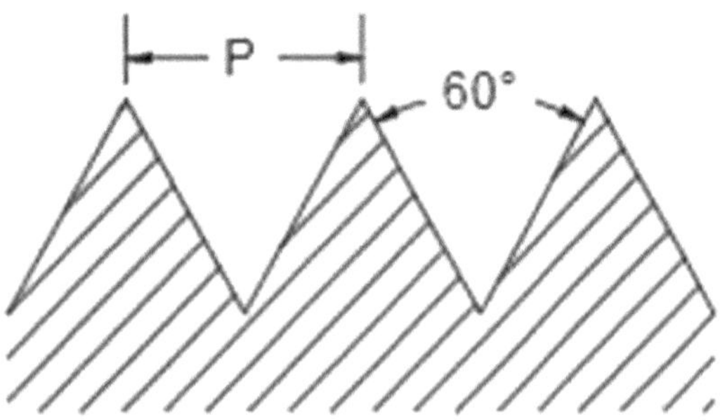

a) Worm thread
b) Buttress thread
c) V-thread

d) B.S.W thread

4. Which screw thread profile is the combination of square and V- threads?

a) B.S.F.

b) Buttress thread

c) ACME thread

d) B.S.W thread

5. The thread angle of V-threads ___________

a) 45°

b) 60°

c) 55° d) 30°

6. Find the type of screw thread the following profile is illustrating.

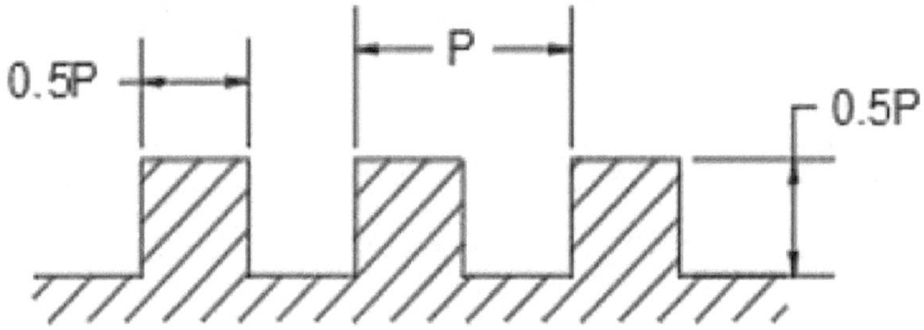

a) ACME thread

b) Buttress thread

c) Square thread

d) Worm thread

7. The thread angle of ACME thread __

a) 45°

b) 60°

c) 55°

d) 29°

8. Find the type of screw thread the following profile is illustrating.

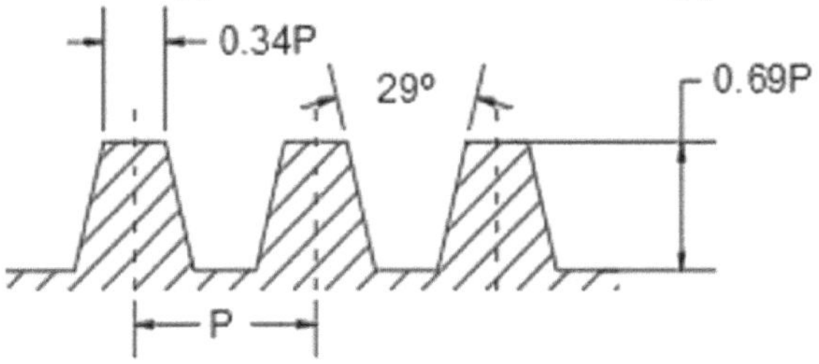

a) ACME thread

b) Buttress thread

c) Square thread

d) Worm thread

9. Worm threads are deeper than ACME threads.

a) True

b) False

10. Find the type of screw thread the following profile is illustrating.

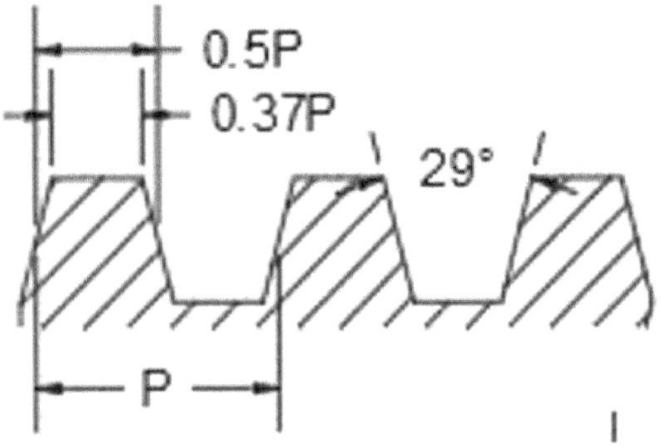

a) ACME thread
b) Buttress thread
c) Square thread
d) Worm thread

11. Find the type of screw thread the following profile is illustrating.

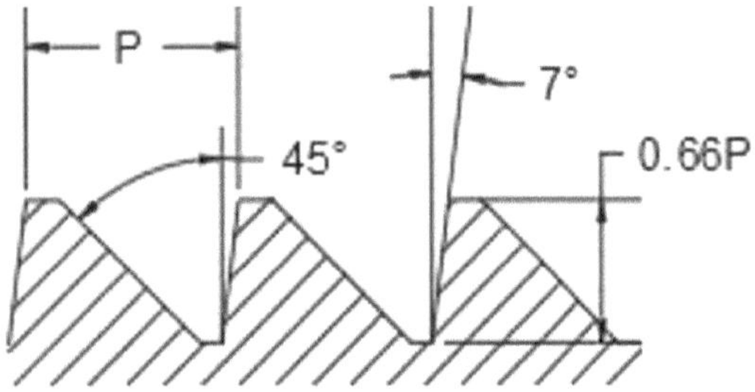

a) ACME thread
b) Buttress thread
c) Square thread
d) Worm thread

12. Find the type of screw thread the following profile is illustrating.

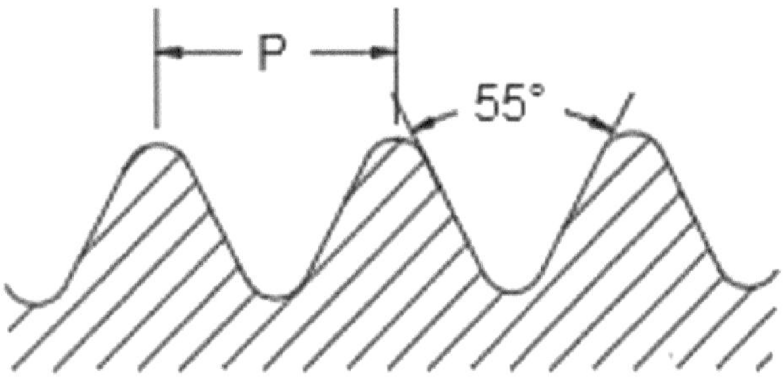

a) ACME thread
b) Buttress thread
c) Whitworth thread
d) Worm thread

13. Threads designed to handle high axial thrust in one direction is ___
a) V-thread
b) Buttress thread
c) Square thread
d) Worm thread

14. A thread designation is given as SQ 20 x 5, what does that indicate?
a) A square thread of pitch 5 mm and 20 mm is the nominal diameter
b) A square thread of pitch 20 mm and 100 mm is the nominal diameter
c) A thread of pitch 5 mm and 20 mm is the nominal diameter
d) A thread of pitch 20 mm and 5 mm is the nominal diameter
15. Screw thread profile adopted by B.I.S ____
a) Unified thread
b) Buttress thread
c) Square thread
d) Worm thread
16. Which of the following is not the method to prevent rotation of a bolt while screwing a nut on or off it?
a) By keeping a square neck

b) By providing a snug
c) By keeping a lock nut

d) By inserting a pin in shank
17. Which bolt is commonly used in bearing for shafts?
a) Hook bolt
b) Eye-bolt
c) T-headed bolt
d) Square-headed bolt
18. Which bolt is commonly used in big ends of connecting rods, eccentrics?
a) Hook bolt
b) Eye-bolt
c) T-headed bolt
d) Cheese-headed bolt
19. Which bolt is used in machine-tool tables in which T-slots are cut to accommodate the T-heads?
a) Hook bolt
b) Eye-bolt
c) T-headed bolt
d) Countersunk-headed bolt
20. The bolt which is used where the head of a bolt must not project above the surface of the connected piece is

____
a) Stud-bolt
b) Tap-bolt or cap screw
c) T-headed bolt
d) Countersunk-headed bolt
21. The bolt which is used when it is not possible to drill a hole in the piece adjoining the bolt-head is

____
a) Hook bolt
b) Tap-bolt or cap screw
c) T-headed bolt
d) Countersunk-headed bolt
22. The bolt which has a circular ring of rectangular cross-section as its head, which can be conveniently held to prevent its rotation is _______
a) Hook bolt
b) Eye-bolt
c) T-headed bolt
d) Stud bolt
23. Which bolt is used as an appliance for lifting heavy machines?
a) Hook bolt
b) Eye-bolt
c) Lifting eye-bolt
d) Cup-headed bolt

24. Which bolt is mainly used in marine shaft couplings?
a) Headless tapered bolt
b) Tap-bolt or cap screw
c) Lifting eye-bolt
d) Countersunk-headed bolt
25. Which bolt is used as a screw that is when it is not possible to accommodate the nut?
a) Headless tapered bolt
b) Tap-bolt or cap screw
c) Stud-bolt or Stud
d) Countersunk-headed bolt
26. Bolt which consists of only a cylindrical shank threaded at both ends is called _________
a) Headless tapered bolt
b) Tap-bolt or cap screw
c) Stud-bolt or Stud
d) Countersunk-headed bolt
27. The _____________ bolt is often made use of in gland and stuffing box arrangement in boiler mountings such as stop valve, feed-check valve etc.
a) Headless tapered bolt
b) Tap-bolt or cap screw
c) T-headed bolt
d) Countersunk-headed bolt
28. Screwed joints are ______________
a) Temporary joints
b) Permanent joints
c) Neither permanent nor temporary joint
d) Temporary and permanent joints
29. Angle turned by the spanner to get the alternative holds in a hexagonal nut is ______
a) 180°
b) 60°
c) 90°
d) 40°
30. Angle turned by the spanner to get the alternative holds in a square nut is ______
a) 180°
b) 60°
c) 90°
d) 40°
31. The angle of chamfering for the hexagonal nut with the base is ________
a) 30° to 45°
b) 50° to 65°
c) 10° to 25°
d) 60° to 75°
32. As per the approximate standard dimensions the relation between the thickness of the nut T and the nominal diameter of the bolt D for a hexagonal nut.
a) $D = 0.5T$
b) $T = 2D$
c) $T = D$
d) $D = 2T$
33. As per the approximate standard dimensions the relation between the width across flats W and the nominal diameter of the bolt D, in mm for a hexagonal nut.
a) $D = 0.5W + 3$
b) $W = 2D + 3$
c) $W = 1.4D$
d) $W = 1.5D + 3$
34. As per the approximate standard dimensions the relation between the radius of chamfer arc, R and the nominal diameter of the bolt D for a hexagonal nut.
a) $D = 0.5R$

b) R = 2D
c) R = 1.4D
d) D = 2R
35. As per the rough rule dimensions the relation between the thickness of the nut T and the nominal diameter of the bolt D for a hexagonal nut.
a) D = 0.5T
b) T = 2D
c) T = D
d) D = 2T
36. As per the rough rule dimensions the relation between the radius of chamfer arc, R and the nominal diameter of the bolt D for a hexagonal nut.
a) D = 0.5R
b) R = 2D
c) R = 1.4D
d) D = 2R
37. As per the rough rule dimensions what will be the distance across the opposite corners diagonally for a nut if the nominal diameter of the bolt D for a hexagonal nut.
a) D
b) 0.5D
c) 1.5D
d) 2D
38. As per the approximate standard dimensions the relation between the thickness of the nut T and the nominal diameter of the bolt D for a square nut.
a) D = 0.5T
b) T = 2D
c) T = D
d) D = 2T
39. A hexagonal nut with a flat circular disc attached to it is called ______
a) Dome nut
b) Cap nut
c) Flanged nut
d) Capstan nut
40. To prevent leakage through the thread and protect the ends of the bolt from corrosion ________ is used.
a) Ring nut
b) Cap nut
c) Flanged nut
d) Capstan nut
41. Nut in a cylindrical shape with holes drilled in the curved shape is called _________
a) Ring nut
b) Cap nut
c) Flanged nut
d) Capstan nut
42. Wing nuts are easily operated with the fingers.
a) True
b) False
43. A washer ensures that the nut screwed-on tightly.
a) True
b) False
44. When the nut diameter is D, what will be the approximate diameter of the washer?
a) D+0.5mm
b) D+1mm
c) 2D+3mm
d) 2D+0.5mm
45. When the nut diameter is D, the diameter of the hole in the washer is ______
a) D+0.5mm

b) D+1mm
c) 2D+3mm
d) 2D+0.5mm
46. When the diameter of the bolt is D, then the thickness of the washer is ______
a) 0.12D
b) 0.5D
c) 0.3D
d) D
47. Which of the following is true for the type-A washer?
a) It is a flat washer
b) It has a slot cut from its centre to its perimeter
c) It has a flat cut on its perimeter
d) They are also called as finishing washers
48. Which of the following washers are considered as SAE washers?
a) Type-A plain washers
b) Type-C
c) Type-D
d) Type-B

Answers

1. Answer: b
2. Answer: a
3. Answer: c
4. Answer: b
5. Answer: b
6. Answer: c
7. Answer: d
8. Answer: d
9. Answer: a
10. Answer: a
11. Answer: b
12. Answer: b
13. Answer: b
14. Answer: a
15. Answer: a
16. Answer: c
17. Answer: c
18. Answer: c
19. Answer: c
20. Answer: d
21. Answer: b
22. Answer: b
23. Answer: c
24. Answer: a
25. Answer: c
26. Answer: c
27. Answer: a
28. Answer: a
29. Answer: a
30. Answer: b
31. Answer: a
32. Answer: c
33. Answer: d
34. Answer: b
35. Answer: b
36. Answer: b
37. Answer: c

38. Answer: a
39. Answer: c
40. Answer: b
41. Answer: a
42. Answer: a
43. Answer: a
44. Answer: a
45. Answer: a
46. Answer: c
47. Answer: b
48. Answer: a

CHAPTER 5: COMPUTER AIDED DRAFTING INTERFACE

5.1 INTRODUCTION

Computer-aided design and drafting (CADD) is the process of using a computer with CADD software for design and drafting applications. Software is the program or instructions that enable a computer to perform specific functions to accomplish a task. CADD refers to the entire range of design and drafting with the aid of a computer, from drawing basic 2-D objects to preparing complex 3-D models and animations. CAD is the acronym for computer-aided design and a common reference to computer-aided drafting. Computer-aided design and computer-aided drafting refer to specific aspects of the CADD process.

CADD offers solutions to most engineering drawing and design problems, and it allows for increasingly complex projects. Several industries and most disciplines related to engineering and architecture use CADD. Most engineering firms and educational institutions that previously used manual drafting practices have evolved to CADD. Professionals have come to rely on the power and convenience of CADD in all aspects of design and drafting. CADD systems include tools to accomplish any drawing and design requirement, such as preparing the 3-D model of a home shown in Figure. 5.1

Figure 5.1

THE CADD WORKSTATION

The CADD workstation composed of a variety of computing devices. Hardware includes the physical elements of a device system, as an example the computer, monitor, keyboard, mouse, and printer. The figure illustrates a contemp orary CADD workstation. A CADD workstation relies on a device for information processing, calculations, and inte rcommunication with peripheral accessories. A peripheral is an external computing devices contrivance that utilizes the device to conduct functioning that the device cannot deal with.

Peripherals furnish input, output, and storage functioning and treatments. Input intends to put facts into the device th at the device behaves on in several way. Input happens from gadgets as an example the keyboard, a mouse or lookali ke input device, or a digitizer. Outcome consults to facts that the device forwards to a getting intention as an exampl e a monitor, a plotter, or a printer. Storage consults to disks and drives that allow the operator to store programs, files , symbols, and information.

5.2 CADD SOFTWARE AND HARDWARE PRODUCTS

The contemporary CADD workstation is powerful, inexpensive, and supports refined CADD tool. A few of CADD application manufacturers exist, andabundant cosmetics are accessible to encounter industry stands in need ofseveral CADD application is general objective and can implement to anypenalize. For example, Autodesk, Inc. creates Aut oCAD for 2-D and 3D construct and drafting. Other products focus on a specific CADD technology, industry, or discipline, such as drawings or models of mechanical parts and assemblies or those for architectural, civil, or structural engineering projects.

CADD SOFTWARE

A CADD program contains hundreds of functions that enable you to accomplish specific drawing tasks. A task may involve drawing an object, editing an existing drawing, displaying a view of the drawing, printing or saving it, or controlling any other operation of the computer. The functions contain a number of commands that enable you to specify exactly what you want to do and how you want to do it.

The functions are organized into modules that provide easy access to all the commands. The CADD program is divided into modules such as draw, edit, data output, function control, data storage and management. A program may also have a number of specialized functions such as layers, database and 3D. Let's have a look at the CADD modules:
1. Draw
2. Edit
3. Data output
4. System control
5. Data storage and management

SPECIAL FEATURES

DRAW

The draw module provides access to all the drawing functions of CADD. Whenever you need to draw something this group of functions is used. The draw module enables you to draw lines, arcs, circles, ellipses, text, dimensions, symbols, borders and many other drawing components.
Draw is CADD's most frequently used module because all drawing work is accomplished using it.

EDIT

The edit module lets you change existing drawing elements and manipulate them in a number of ways. You can move, copy or erase drawing components. You can enlarge or reduce the sizes of diagrams or change the colour and line type of drawing components. You can also change the size and style of text and dimensions, as well as edit a dimension to show different units of measurement. A good CADD program is designed to change the appearance of all drawing elements created with CADD.
The edit functions also act as convenient drawing-aid tools. They enable you to join missing corners of lines, trim drawing components along a line, stretch them to fit a new shape, etc. The list of editing capabilities goes on and on. The edit functions make CADD a dynamic drawing tool.

DATA OUTPUT

The data output module enables you to display drawings on the screen and then print them on paper. There are two separate sets of functions that help accomplish this:
* View-display functions
* Print/plot functions

The view-display functions allow you to display different views of a drawing on the screen. These functions are used quite often, because every time you need to draw something or edit something, you need to focus on that portion of the drawing. With the help of view-display functions, you can zoom in on a specific portion of the drawing.
The print and plot functions allow you to print drawings using a printer or a plotter. You can control many aspects of printing and plotting. You can print the same drawing in different sizes by applying the appropriate scale factor. You can plot the drawings with specific colours, pen thickness, and line types.

DATA STORAGE AND MANAGEMENT

The data storage and management module allow you to store and manage drawing data. Through the use of the functions in this module, you can store drawings as files on the hard disk. You can manage the files in directories and sub-directories, and move, copy or delete them as needed.
CADD data management functions also let you translate drawings created by other CADD programs. These functions convert drawing data to a generic format that can be read by any CADD program. Data exchange format (DXF) is one of the common data translation formats used by CADD programs. There are a number of data exchange formats available.

SYSTEM CONTROL

The system control module (also known as system defaults) allows you to control how CADD works. CADD programs are designed for a broad range of professionals, including architects, designers, engineers and surveyors. With the help of system control functions, you can set the working environment of CADD to suit your needs.
Example: You can set the type of units that you will be using, the accuracy of the units, a style for dimensions and text, colours, layers, line type in a drawing, etc. Additionally, you can customize screen menus, the display of colours on the screen, resolution of the screen, size, the speed of the cursor, etc.
You can also specify whether the selected defaults should apply to a single drawing, to a specific project, or to all the projects in a specific category. The defaults can be set on a temporary or permanent basis.

SPECIAL FEATURES

CADD programs usually offer a number of special features that make working with CADD easier and allow you to automate many drawing tasks. For example, you can create layers in a drawing that allow you to segregate drawing components. You can develop spreadsheets and databases that can be used to create many types of project reports. You can create three-dimensional (3D) drawings, such as isometrics and perspectives, with the help of 3D functions. You can also accomplish many other automated tasks with the help of macros.

The number of special features a CADD program has or how elaborate they varies from one program to another. Some vendors sell specialized features as separate packages, while others include them in a single package. It all depends how a program is written, how big or small it is, and how it is sold.

IMPORTANT TIP:

The functions of CADD are like the tools of a handyman. All of the tools are essential for working. Some are used more frequently than others are, but each has its own importance.

COMPUTER AIDED DESIGN AND DRAFTING

The user interface describes the items that allow you to input information to and get outputs from a device method. The GUI offers

the onscreen characteristics that allow you to interact with anapplication program. New cosmetics repeatedly emerge to reply to innovative tech and plan provisionbigger tool manufacturers, as an example Autodesk Inc., Dassault Systèms, Parametric tech Corporation, and Siemens PLM Solutions grasp the max-imal number of CADD users, and they traditionally have the aptitude to grow their cosmetics and capture smaller ap plication industries or available tool.

Some software manufacturers offer products intended to support various aspects of product development. For example, some software companies combine CADD and CAM tools for design, drafting, and manufacturing.

A few software companies offer specific applications or software packages to help manage all aspects of a project, known as product life cycle management (PLM). PLM systems include tools for CADD, product data management (PDM) to organize and monitor project data, computer-aided engineering (CAE) for simulation and analysis, CAM, and presentation.

ALIBRE INC.

Alibre offers application publicly for CADD in the fabricating industry. Alibre construct is a 3D opaque modeling a nd 2D drafting prog-ram. The expetmodel of Alibre construct includes tools for sheet metal construct and rendering. The specialist model of Alibre construct offers extra functioning as an example replication and FEA, PDM, CAM, a nd expanded translation tools.

Translation occurs when converting information from the fi le scheme of one CADD scheme to another, and it's freq uently appropriate when share with CADD information with others, as an example consultants, manufacturers, and vendors.

Most CADD software includes tools for several level of file translation. Separate translation software is available when necessary. Alibre also offers Alibre Personal Edition, which is a 3-D modelling and 2-D drawing software marketed to hobbyists.

ASHLAR-VELLUM

Ashlar-Vellum offers basic 2-D and 3-D CADD software. Graphite provides 2-D and 3-D wireframe drawing and modelling capabilities. Argon is a basic 3-D modelling software for conceptual design, visualization, and translation. Xenon and Cobalt, which include additional functions, are 3-D modelling programs with 2-D drafting capabilities.

AUTODESK, INC.

Autodesk provides a wide variety of software. AutoCAD is generalpurpose 2D and 3D CADD application and is the basic Autodesk creation.

Autodesk offers variations of AutoCAD for original markets, as an exampleAutoCAD LT for 2D drafting, AutoCA D Electrical for electrical comparison scheme design, and AutoCAD noncriminal 3D for noncriminal engineering pl an construct extra Autodesk cosmetics concentrate on specific CADD

tech and industries, as an example manufacturing, architecture, construction, infrastructure, media, and entertainmen t. Autodesk® Inventor® is a 3-D solid modelling and 2-D drafting program generally for CADD in the manufacturing industry.

Autodesk Inventor provides a comprehensive and flexible set of software for 3-D mechanical design, simulation, design visualization and communication, tooling creation, and 2-D documentation. Autodesk offers Autodesk Inventor Suites that combine Autodesk Inventor, AutoCAD Mechanical, and tools for specific applications, such as mould, tube and pipe, and cable and harness design.

Autodesk Revit is a 3D constructing construct program with 2D drafting and papers abilities. Models of Autodesk R evit concentrate on construct for architecture, mechanical, electrical, and plumbing(MEP), or constructing acts modelling (BIM) for structural engineering.

Autodesk fabricates abundant other application products, such as Autodesk Algor replication for opaque machine re plication and FEA, Autodesk

Vault for PDM, 3ds Max for 3D modeling, animation, and rendering, and application to confirm the ecological and e nvironmentally amicable const-ruct and growth.

BENTLEY SYSTEMS, INC.

Bentley Systems focuses on tool for engineering(lit) framework design, documentation, and technique. Framework i s the structures, facilities and

treatments demanded for an economy to function, as an example buildings, roads and bridges, water provide and se wer systems, and power-supply and telecommunication systems. Micro Station is a general

purpose 2D and 3D CADD software and is the most important Bentley Systems

creation.

Micro-Station additionally behaves as a program for other Bentley Systems application. For example, GEOPACK Civil Engineering Suite includes tools for civil engineering and transportation project design. Micro Station Power Draft is a version of MicroStation mainly for 2-D drafting. Bentley Systems manufactures other software, including Project Wise for PDM, and Pro Concrete for 3-D modelling, detailing, and scheduling of reinforced concrete structures.

DASSAULT SYSTEMS

Dassault Systèmes brands provide various software cosmetics commonly geared towards CAD and narrated tech for the fabricating

industry. CATIA is a plan advancement treat and is the essential Dassault Systèms sign device. CATIA offers tools f or 3D opaque modeling and 2D drafting and tools for specifi c applications, as an example mold, tube and pipe, wire and harness, and digital construct.

CATIA additionally provides replication and analysis, CAM, and PDM functioning. The extra Dassault Systèms sig n tool focuses on specifi c facets of PLM. SolidWorks is a 3D opaque modeling and 2D drafting program and is the basic Dassault Systems SolidWorks(www.Solidworks. com) sign item.

Dassault Systèms SolidWorks provides a classic model of SolidWorks and suites that incorporate SolidWorks with s imulation, analysisand PDM tools. SolidWorks replication includes tools for opaque system replication and FEA. So lidWorks movement offers fluid-

flow replication and thermal examination. Dassault Systèmes SolidWorks additionally fabricates application to confi rm the ecological and environmentally amicable construct and fabricating.

GOOGLE INC.

Google SketchUp is a tool purposed to have a simple to employ user interface for creating, sharing, and presenting 3 D versions. familiar applications for Google SketchUp include sketching and modeling for visualization in the cours e of the conceptual construct phase of a plan and building presentation drawings that look hand

sketched or photorealistic.

Google SketchUp additionally can lead to Google Earth for sketching relative to a physical location, as an example modelling a constructW5n on a factual lot.

GRAPHISOFT

GRAPHISOFT focuses on application for the architecture, engineering, and interpretation(AEC) industry. ArchiCA D is a 3D constructing construct program with 2D drafting and papers capacities it's the essential GRAPHISOFT device.

MEP Modeler includes 3D and 2D MEP functioning to ArchiCD. Virtual constructing is a 3D digital database that tr acks all components that compose a building, letting the designer to employ items as an example area place and volu me, thermal properties, area descriptions, costs, device information, and window, door, and finish schedules.

Virtual consults to something that seems to have the attributes of a genuine or factual thing or encounter. GRAPHISOFT also manufactures photo-realistic rendering software and software to support sustainable and environmentally friendly architectural design and construction.

IMSI/DESIGN, LLC

IMSI/Design offers basic CADD software for general-purpose and project-specific applications. Turbo CAD is the core IMSI/Design product. IMSI/Design provides variations of Turbo CAD for unique markets, such as Turbo CAD Designer for 2-D drafting, Turbo CAD Deluxe for 2-D drafting and 3-D modelling, and Turbo CAD Pro with additional 2-D and 3-D CADD functions. Additional IMSI/Design products focus on specific CADD technology and industries, such as Home & Landscape and Instant Architect for basic 2-D and 3-D home design.

INTERGRAPH

Intergraph fabricates application for specific industries and projects, containing the design, construction, and practic e of plants, shipsoffshore andtransportation and usefulness systems. For example, SmartMarine 3D is specific tool fo r 3D modeling, design, and paperof marine structures, as an example commercial ships. Intergraph additionally prov ides Smart Sketch for 2-D drafting.

IRON CAD

Iron CAD provides software generally for CADD in the manufacturing industry. IRONCAD is a 3-D solid modelling and 2-D drafting program with PDM functions. A third-party application offers simulation and FEA tools compatible with IRONCAD. INOVATE is a version of IRONCAD with fewer 3-D modelling functions and no 2-D drafting capabilities.

KUBOTEK CORPORATION

Kubotek fabricates CADD and CAM tool. KeyCreator is a 3D opaque modeling and 2D drafting program publicly f or CADD in the fabricating industry it's the chief Kubotek creation. Kubotek Validation software confirms construct precision wishlist or afterwards a specific activity, as an example a construct revision or information translation. Kubotek additionally fabricates KeyMachinest for CAM and preserves CADKEY for 3D wireframe modelling.

PARAMETRIC TECHNOLOGY CORPORATION

Parametric Technology Corporation, or PTC, offers several software products generally focused on CADD in the manufacturing industry. Pro/ENGINEER is a 3-D solid modelling and 2-D drafting program and is the core PTC software. PTC provides various additions to the Pro/ENGINEER platform, including tools for CAE, CAM, and PDM: for example, Pro/ENGINEER Mechanical for simulation and FEA, Pro/ENGINEER Piping and Cabling Extension for pipe and cable design, and Pro/ENGINEER Reverse Engineering for automating reverse engineering. Reverse engineering is the process of converting an existing physical product into drawings or digital models, and it involves discovering the technological principles of a device, object, or system by analysing its structure, function, and operation.

PTC manufactures other software, including Windchill for PDM, Co Create for CAD, CAE, and PDM, and Math CAD for engineering calculations.

SIEMENS CORPORATION

Siemens Corporation provides a wide variety of cosmetics and treatments. The Siemens PLM Solutions(www. plm. automation. siemens. com) sign fabricates PLM application. NX addresses each place of device development, and it' s the most important Siemens PLM Solutions application.

NX offers tools for 3D opaque modeling, 2D drafting, and specific applications as an example generator and fixture, routed system, and sheet metalcreation construct.

NX additionally provides simulation, FEA, CAM, and PDM functioningfurthermore NX, Siemens PLM Solutions cr eates SolidEdge for 3D opaque modeling and 2D drafting, publicly for CADD in the fabricating industry. The extra Siemens PLM Solutions sign tool focuses on specific aspect of PLM.

CAD SYSTEM HARDWARE

The hardware for a typical CAD system consists of the following components: (1) one or more design workstations, (2) digital computer, (3) plotters, printers. and other output devices. and (4) storage devices. The relationship among the components illustrated in Figure 5.2. In addition, the CAD system would have a communication interface to permit transmission of data to and from other computer systems. Thus enabling some of the benefits of computer integration

Design Workstations. The workstation is the interface between computer and user in the CAD system. Its functions are the following:
(1) communicate with the CPU.
(2) continuously generate a graphic image.
(3) provide digital descriptions of the image.
(4) translate user commands into operating functions, and
(5) facilitate interaction between the user and the system.
The design of the CAD workstation and its available features have an important influence on the convenience, productivity. and quality of the user's output. The workstation must include a graphics display terminal and a set of user input devices. The display terminal must be capable of showing both graphic and alphanumeric text. It is the principal means by which the system communicates with the user. For optimum graphics display, the monitor should have a large colour screen with high resolution.

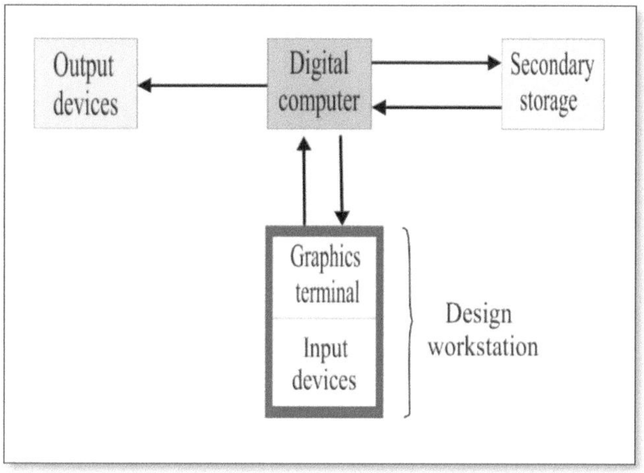

Figure 5.2

The user input devices permit the operator to communicate with the system. To operate the CAD system, the user must be able to accomplish the following:
(1) enter alphanumeric data,
(2) enter commends to the system (0 perform various graphics operations, and
(3) control the cursor position on the display screen.
To enter alphanumeric data, an alphanumeric keyboard is provided. A conventional type writer like keyboard allows the designer to input numerical and alphabetic characters into the system. The alphanumeric keyboard can also be used to enter commands and instructions to the system. However. other input devices accomplish this function more conveniently.
Special function keypads have been developed to allow entry of a command in only one or two keystrokes. These special keypads have from 10 to 50 function keys, depending on the system.
However, each key provides more than one function, depending on the combination of keys pressed or which software is being used. Another input device for entering commands to a CAD system is the electronic tablet, an electronically sensitive board on which an instruction set is displayed, arid commands arc entered using a puck or electronic pen.
Cursor control permits the operator to position the cursor on the screen to identify a location where some function is to be executed. For example, to draw a straight line on the screen, the endpoints of the line can be identified by locating the cursor in sequence at the two points and giving the command to construct the line.

There are various cursor control devices used in CAD, inducting pucks, mouse's, joysticks, trackballs, thumbwheels, light pens, and electronic tablets. An input device for entering coordinates from an existing drawing into the CAD system is a digit/t.er, which consists of a large flat board and an electronic tracking element such as a puck that can be moved across the surface of the board to record x and y-coordinate positions.
Digital Computer CAD applications require a digital computer with a high-speed central processing unit (CPU), math coprocessor to perform computation intensive operations, and large internal memory. Today's commercial systems have 32bit processors, which permit high-speed execution of CAD graphics and engineering analysis applications.

Several CAD system configurations are available within the general arrangement shown in Figure 5.3 Let us identify three principal configurations, illustrated in Figure 5.3 (a) host and terminal, (b) engineering workstation, and (c) CAD system based on a personal computer (PC).

The host and terminal - The original CAD configuration in the 1970s and early 1980s when the technology was first developing. For many years, it was the only configuration available. In this arrangement, a large mainframe computer or a minicomputer serves as the host for one or more graphics terminals.
These systems were expensive, each installation typically representing an investment of a million dollars or more. The powerful microprocessors and high-density memory devices that are so common today were not available at that time. The only way to meet the computational requirements for graphics processing and related CAD applications was to use a mainframe connected 10 multiple terminals operating on a timesharing basis. Host and terminal CAD systems are still used today in the automotive industry and other industries in which it is deemed necessary to operate a large central database.

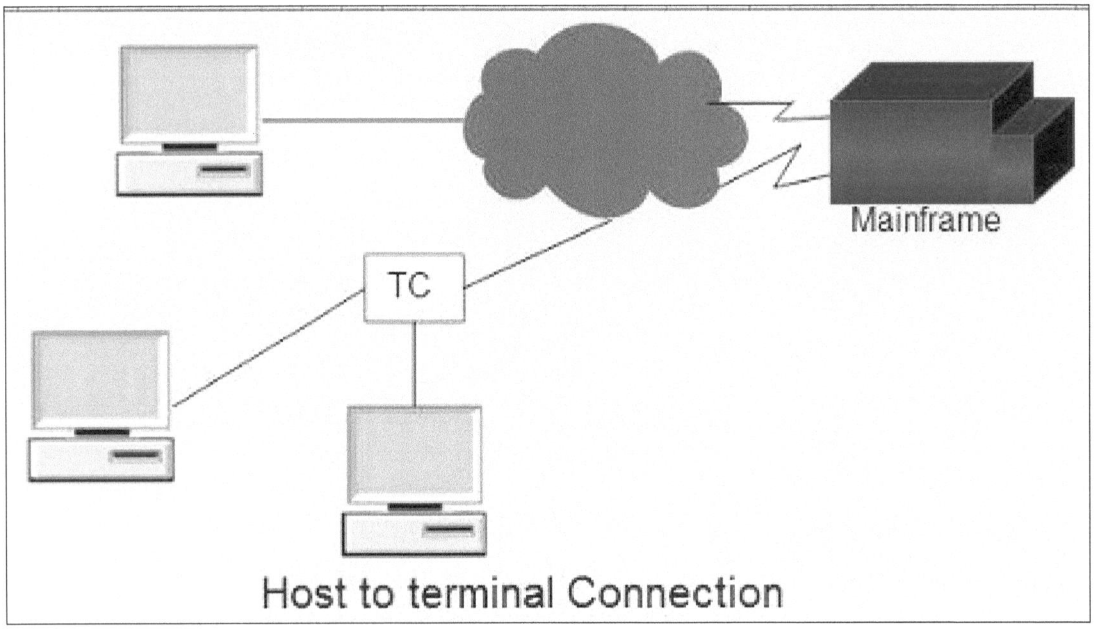

Figure 5.3

An engineering workstation is a standalone computer system that is dedicated to one user and capable of executing graphics software and other programs requiring high-speed computational power.
The graphics display is a high-resolution monitor with a large screen. As shown in our figure, engineering workstations are often networked to permit exchange of data files and programs between users and to share plotters and data storage devices.

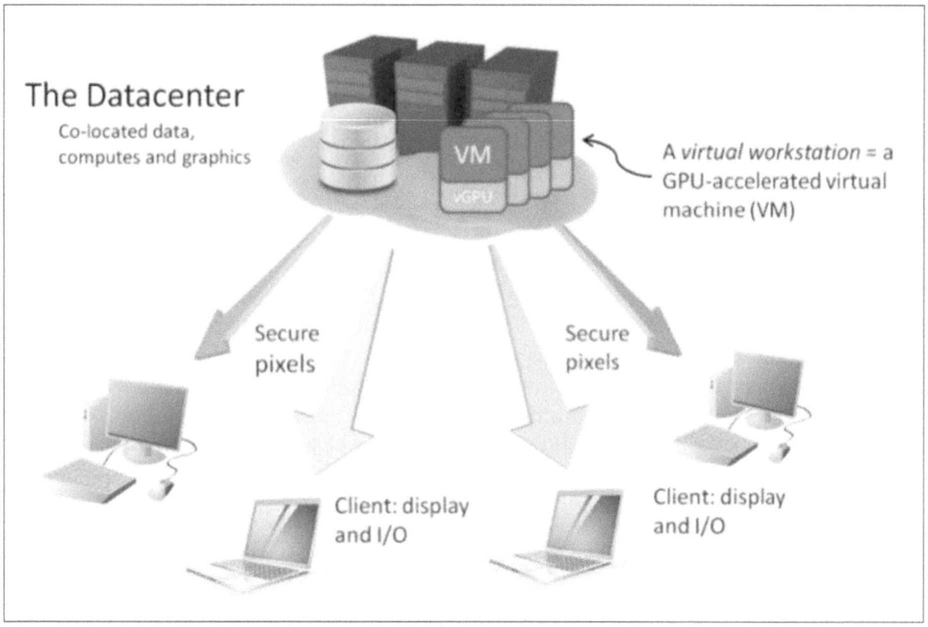

Figure 5.4

A PC-based CAD system is a PC with a high-performance CPU and medium to high resolution graphics display screen. The computer is equipped with a large random access memory (RAM), math coprocessor, and large capacity hard disk for storage of the large applications software packages used for CAD. PC-based CAD systems can be networked 10 share file output devices, and for other purposes. Starting around 1996, CAD software developers began offering products that utilize the excellent graphics environment of Microsoft Windows NT, thus enhancing the popularity and familiarity of PC-based CAD.

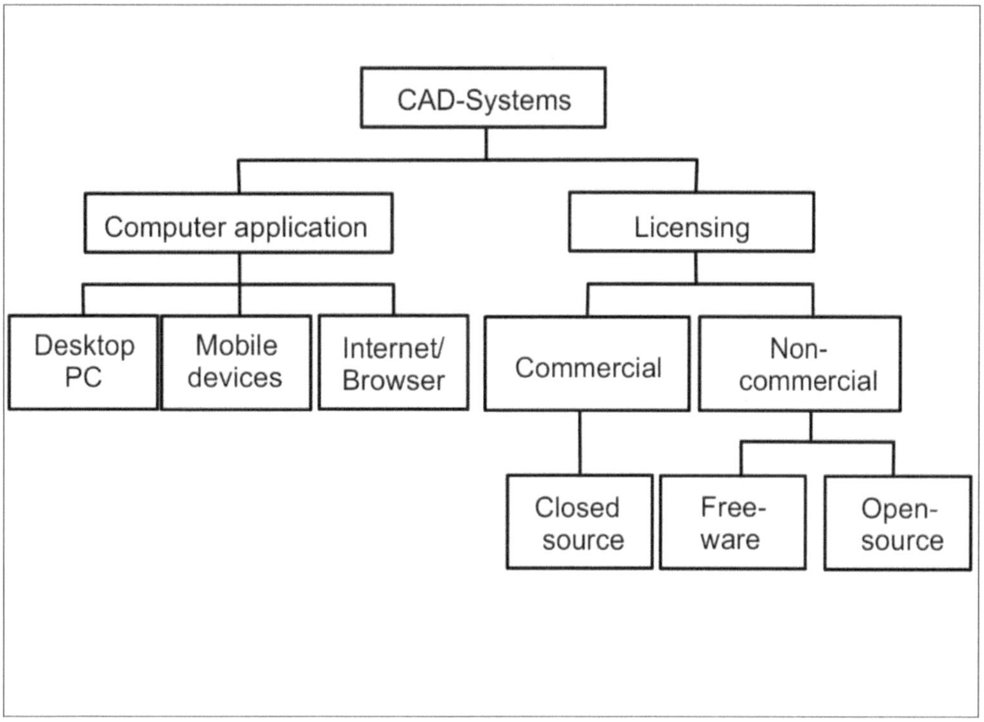

Figure 5.5

When the engineering workstation is compared with the PC-based system, the former is superior in terms of most performance criteria. Its capacity to efficiently accomplish 3D geometric modelling and execute other advanced software exceeds that of a PC, and this makes the workstation more responsive and interactive than a PC-based CAD System. However, the performance characteristics of PCS are improving each year, and the prices of engineering workstations are dropping each year, so that the distinction between the two types is becoming blurred.

CAD COMPUTER PARTS

DEFINE HARDWARE IN CAD SYSTEM

It is used to describe the physical equipment (electronic circuitry) and its peripherals (mouse, printer). The term "hardware" is used to cover all those pieces of equipment which are used to implement computer programs. These includes the computer with its memory and CPU, means of storing programs on disk or magnetic tape, the display screens, various devices for inputting instructions to the computer i.e., keyboards, etc. and output devices which can give "hard copy", i.e., printers and plotters. Hardware consists of the graphics workstations, graphics input and output devices. Computer software include: The operating system which controls the basic house-keeping operations in the computer, software packages used for geometric modelling, and application software for design analysis and synthesis.

INPUT/OUTPUT DEVICES

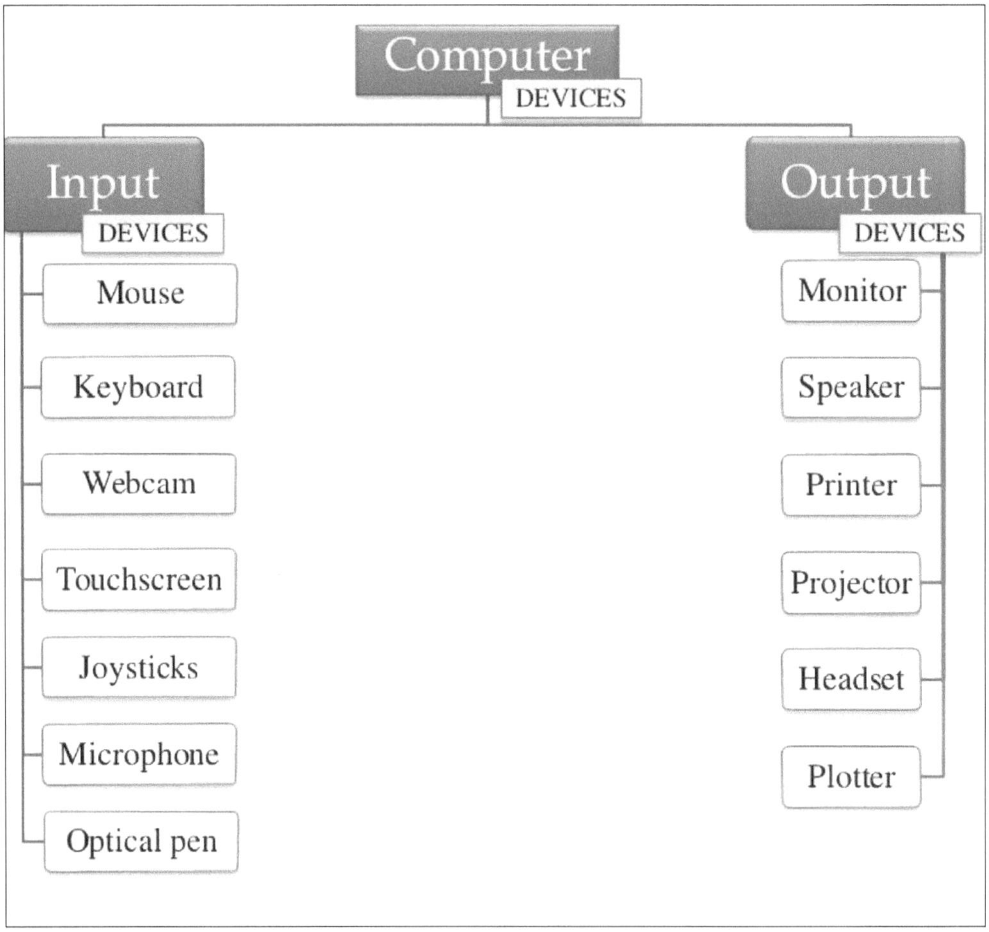

Figure 5.6

INPUT DEVICES:

KEYBOARD:

The computer keyboard is used to enter text information into the computer, as when you type the contents of a report. The keyboard can also be used to type commands directing the computer to perform certain actions.

KEYPAD:

Most laptop computers today have a touch pad pointing device. You move the onscreen cursor by sliding your finger along the surface of the key pad.

DIGITIZER:

Digitizer is defined as device for converting analogue signals into digital signals or to convert data to digital form for use in a computer. It is a term used by the American speaking countries and mostly used by physicists.

TRACKBALL:

The trackball is sort of like an upside-down mouse, with the ball located on top. You use your fingers to roll the trackball and internal rollers sense the motion which is transmitted to the computer.

JOYSTICK:

Joysticks and other game controllers can also be connected to a computer as pointing devices. They are generally used for playing games and not for controlling the onscreen cursor in productivity software.

TABLET:

A tablet consists of an electronic writing area and a special "pen" that works with it. Graphics tablets allows artists to create graphical images with motions and actions similar to using more traditional drawing tools.

MOUSE:

The Mouse pointing device sits on your work surface and is moved with your hand. In order mice, a ball in the bottom of the mouse rolls on the surface as you move the mouse, and internal rollers sense the ball movement and transmit the information to the computer via the cord of the mouse.

OPTICAL MOUSE:

A Cordless or wireless mouse communicates with the computer via radio waves.

OUTPUT DEVICES:

MONITOR:

Monitor commonly called as Visual Display Unit (VDU) is the main output device of a computer. It forms images from tiny dots, called pixels, that are arranged in a rectangular form. The sharpness of the image depends upon the number of the pixels.

PRINTERS:

Printer is the most important output device, which is used to print information on paper.

SPEAKERS:

Computer speakers, or multimedia speakers, are speakers external to a computer, that disable the lower fidelity built-in speaker.

PROJECTOR:

An output device that can take the display of a computer screen and project a large version of it onto a flat surface. Projectors are often used in meetings and presentations to help make sure everyone in the room can view the presentation.

AUTOCAD SYSTEM REQUIREMENTS

AUTOCAD 2024

Operating System: 64-bit Windows 11 or 10 version 1809 or above

Processor: 2.5 GHz (3+ GHz recommended), ARM processors not supported

Memory: 8 GB (32GB recommended)

Disk space: 10.0 GB

Display: 1920 x 1080 resolution with True Colour

Display Card: 2 GB GPU (8 GB recommended) with 29 GB/s bandwidth (106 GB/s recommended) and DirectX 11 compliant (DirectX 12 recommended)

AUTOCAD 2023

Operating System: 64-bit Windows 11 or 10 version 1809 or above

Processor: 2.5 GHz (3+ GHz recommended), ARM processors not supported

Memory: 8 GB (16GB recommended)

Disk space: 10.0 GB

Display: 1920 x 1080 resolution with True Colour

AUTOCAD 2022
Operating System: 64-bit Windows 10
Processor: 2.5 GHz (3+ GHz recommended)
Memory: 8 GB (16GB recommended)
Disk space: 10.0 GB
Display: 1920 x 1080 resolution with True Colour

AUTOCAD 2021
Operating System: 64-bit Microsoft Windows 10 or 8.1
Processor: 2.5 GHz (3+ GHz recommended)
Memory: 8 GB (16GB recommended)
Disk space: 7.0 GB
Display: 1920 x 1080 resolution with True Colour

AUTOCAD 2020
Operating System: Microsoft Windows 10 (64-bit only), 8.1 (64-bit only), or 7 SP1 (64-bit only)
Processor: 2.5 GHz (3+ GHz recommended)
Memory: 8 GB (16GB recommended)
Disk space: 6.0 GB
Display: 1920 x 1080 resolution with True Colour

AUTOCAD 2019
Operating System: Microsoft Windows 10 (64-bit only), 8.1 (32-bit & 64-bit), or 7 SP1 (32-bit & 64-bit)
Processor: 2.5 GHz (3+ GHz recommended)
Memory: 8 GB (16GB recommended)
Disk space: 6.0 GB
Display: 1920 x 1080 resolution with True Colour

AUTOCAD 2018
Microsoft Windows 10 (64-bit only), 8.1 (32-bit & 64-bit), or 7 SP1 (32-bit & 64-bit)
1 GHz or faster processor
RAM
- 32-bit: 2 GB
- 64-bit: 4 GB

Disk space: 4.0 GB
1360 x 768 display resolution with True Colour

AUTOCAD 2017
Microsoft Windows 10, 8.1 or 7 SP1
Intel® Pentium® 4 or AMD Athlon™ 64 processor
RAM
- 32-bit: 2 GB
- 64-bit: 4 GB

Disk space: 6.0 GB
1360 x 768 display resolution with True Colour

AUTOCAD 2016
Microsoft Windows 8 or 7
Intel® Pentium® 4 or AMD Athlon™ 64 processor
RAM
- 32-bit: 2 GB
- 64-bit: 4 GB

Disk space: 6.0 GB
1024 x 768 display resolution with True Colour

AUTOCAD 2015
Microsoft Windows 8 or 7
Processor
- 32-bit: Pentium 4 or AMD Athlon Dual Core, 3.0 GHz or greater with SSE2 technology
- 64-bit: Athlon 64 or Opteron with SSE2 technology, or Intel Xeon or Pentium 4 with EM64T support and SSE2 technology

2 GB RAM
Disk space: 6.0 GB
1024 x 768 display resolution with True Colour

AUTOCAD 2014
Microsoft Windows 8, 7, or XP SP3
Processor
- 32-bit XP: Pentium 4 or AMD Athlon Dual Core, 1.6 GHz or greater with SSE2 technology
- 32-bit Windows 7 or 8: Pentium 4 or AMD Athlon Dual Core, 3.0 GHz or greater with SSE2 technology
- 64-bit: Athlon 64 or Opteron with SSE2 technology, or Intel Xeon or Pentium 4 with EM64T support and SSE2 technology

2 GB RAM
Disk space: 6.0 GB
1024 x 768 display resolution with True Colour
AUTOCAD 2013
Microsoft Windows 7 or XP SP3
Processor
- 32-bit XP: Pentium 4 or AMD Athlon Dual Core, 1.6 GHz or greater with SSE2 technology
- 32-bit Windows 7: Pentium 4 or AMD Athlon Dual Core, 3.0 GHz or greater with SSE2 technology
- 64-bit: Athlon 64 or Opteron with SSE2 technology, or Intel Xeon or Pentium 4 with EM64T support and SSE2 technology

2 GB RAM
Disk space: 6.0 GB
1024 x 768 display resolution with True Colour
Microsoft Internet Explorer 7.0 or later

AUTOCAD 2012
Microsoft Windows 7, Vista SP2 or XP SP3
Processor
- 32-bit XP: Pentium 4 or AMD Athlon Dual Core, 1.6 GHz or greater with SSE2 technology
- 32-bit Vista or 7: Pentium 4 or AMD Athlon Dual Core, 3.0 GHz or greater with SSE2 technology
- 64-bit: Athlon 64 or Opteron with SSE2 technology or Xeon or Pentium 4 with EM64T support and SSE2 technology

2 GB RAM
Disk space
- 32-bit: 2.0 GB
- 64-bit: 2.0 GB

1024 x 768 display resolution with True Colour
Microsoft Internet Explorer 7.0 or later

AUTOCAD 2011
Microsoft Windows 7, Vista SP1 or XP SP2
Processor
- 32-bit XP: Pentium 4 or AMD Athlon Dual Core, 1.6 GHz or greater with SSE2 technology
- 32-bit Vista or 7: Pentium 4 or AMD Athlon Dual Core, 3.0 GHz or greater with SSE2 technology
- 64-bit: Athlon 64 or Opteron with SSE2 technology or Xeon or Pentium 4 with EM64T support and SSE2 technology

2 GB RAM
Disk space

32-bit: 1.8 GB
64-bit: 2.0 GB
1024 x 768 display resolution with True Colour
Microsoft Internet Explorer 7.0 or later

AUTOCAD 2010
Microsoft Windows Vista SP1 or Windows XP SP2
Processor
32-bit XP: Pentium 4 or AMD Athlon Dual Core, 1.6 GHz or greater with SSE2 technology
32-bit Vista: Pentium 4 or AMD Athlon Dual Core, 3.0 GHz or greater with SSE2 technology
64-bit: Athlon 64 or Opteron with SSE2 technology or Xeon or Pentium 4 with EM64T support and SSE2 technology
2 GB RAM
Disk space
32-bit: 1 GB
64-bit: 1.5 GB
1024 x 768 VGA with True Colour
Microsoft Internet Explorer 7.0 or later

AUTOCAD 2009
Microsoft Windows Vista or Windows XP SP2
Pentium 4 processor or AMD Athlon, 2.2 GHz or greater or Intel or AMD Dual Core processor, 1.6 GHz or greater
1 GB RAM (Windows XP), 2 GB RAM (Windows Vista)
750 MB free disk space
1024 x 768 VGA with True Colour
Microsoft Internet Explorer 6.0 (SP1 or later)

AUTOCAD 2008
Microsoft Windows Vista, Windows XP Home and Professional (SP2), or Windows 2000 Professional (SP3)
Microsoft Internet Explorer 6.0 (SP1 or later)
Pentium 4 processor
512 MB RAM
750 MB free disk space
1024 x 768 VGA with True Colour

AUTOCAD 2007
Microsoft Windows XP (Professional, Home Edition, or Tablet PC Edition) or Windows 2000 Professional (SP3 or SP4)
Microsoft Internet Explorer 6.0 (SP1 or later)
Pentium IV processor recommended
512 MB RAM
750 MB free disk space
1024 x 768 VGA with True Colour
Mouse, trackball, or compatible pointing device

AUTOCAD 2006
Microsoft Windows XP (Professional, Home Edition, or Tablet PC Edition) or Windows 2000 Professional (SP3)
Microsoft Internet Explorer 6.0 (SP1 or later)
Pentium III or later or compatible, 800 mhz or higher
256 MB RAM
300 MB free disk space
1024 x 768 VGA with True Colour
Mouse, trackball, or compatible pointing device

AUTOCAD 2005
Microsoft Windows XP (Professional, Home Edition, or Tablet PC Edition) or Windows 2000 Professional

Microsoft Internet Explorer 6.0 (SP1 or later)
Pentium III or later or compatible, 800 mhz or higher
256 MB RAM
300 MB free disk space
1024 x 768 VGA with True Colour
Mouse, trackball, or compatible pointing device

AUTOCAD 2004
Microsoft Windows XP (Professional, Home Edition, or Tablet PC Edition), Windows 2000 Professional, or Windows NT 4.0 (SP6a or later)
Microsoft Internet Explorer 6.0
Pentium III or later or compatible, 500 mhz or higher
128 MB RAM
300 MB free disk space
1024 x 768 VGA with True Colour
Mouse, trackball, or compatible pointing device

AUTOCAD 2002
Microsoft Windows XP, Windows 2000 Professional, Windows NT 4.0, or Windows 98
Microsoft Internet Explorer 6.0
Pentium III or later or compatible
64 MB RAM
300 MB free disk space
1024 x 768 VGA with True Colour
Mouse, trackball, or compatible pointing device

AUTOCAD 2001
Microsoft Windows XP, Windows 2000 Professional, Windows NT 4.0, or Windows 98
Microsoft Internet Explorer 6.0
Pentium III or later or compatible
64 MB RAM
300 MB free disk space
1024 x 768 VGA with True Colour
Mouse, trackball, or compatible pointing device

AUTOCAD 2000
Microsoft Windows XP, Windows 2000 Professional, Windows NT 4.0, or Windows 98
Microsoft Internet Explorer 6.0
Pentium III or later or compatible
64 MB RAM
300 MB free disk space
1024 x 768 VGA with True Colour
Mouse, trackball, or compatible pointing device

UNDERSTANDING AUTOCAD USER INTERFACE

Whether you just started using AutoCAD, or you are already an experienced user, you are likely not fully aware of all user interface elements. The same applies to AutoCAD commands. Let us review the user interface elements in AutoCAD 2018 as well as some AutoCAD commands that can give you a shortcut that will improve your workflow.

AUTOCAD INTERFACE

APPLICATION MENU

This is the first thing you see on launching AutoCAD, located in the left corner at the top of the screen. It is represented by a big red "A" button with a downward facing black arrow. Single-clicking this button opens a drop-down menu which contains some number of the same options in the quick access toolbar, and more. More importantly, it contains links to most recently opened drawings and sheet sets. With the click of button you can change this to a list of currently open drawings. Double-clicking the application menu is not advised unless you mean to close AutoCAD application.

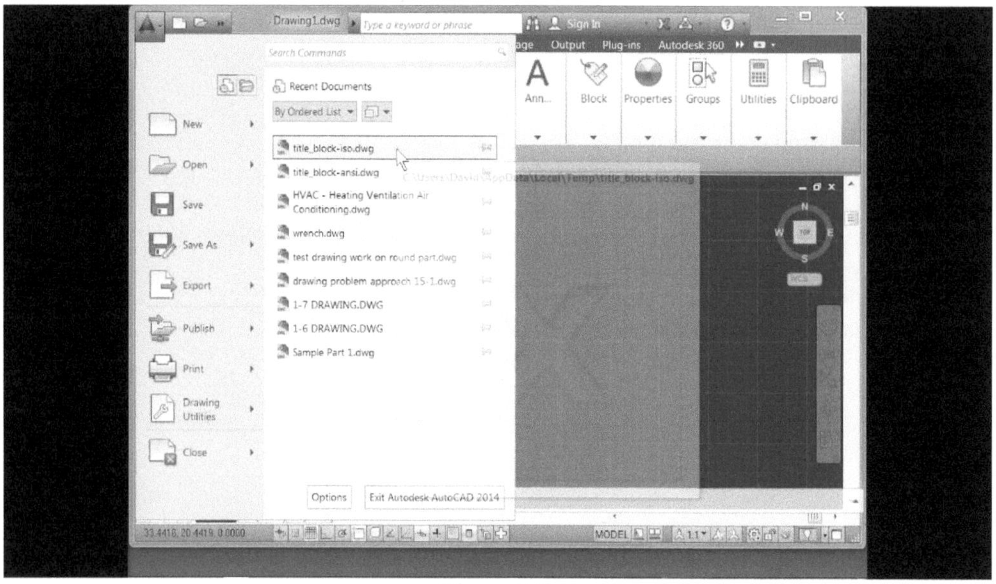

Figure 5.7

Another feature of the Application Menu is the Search Commands part. Here you may type the full name or partial name of AutoCAD commands you are not sure of, and AutoCAD will carry out a search for related commands in the menu. The figure below shows the results you get when you type 'poly' in the Search Commands part.

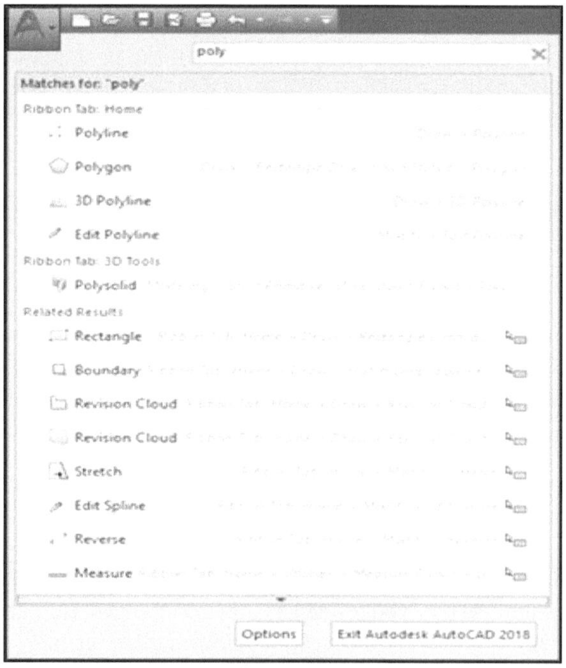

Figure 5.8

RIBBON

The ribbon is a user interface element that contains various AutoCAD commands arranged in panels and tabs. The tabs are quite powerful. They have panels of commands as well as options that relate to any task at hand. For instance, if you select a HATCH object, a contextual tab appears. The tab contains a Hatch editing command that appears automatically.

Figure 5.9

IN-EDITOR ELEMENTS

These are elements found inside the drawing panel. The user can control their visibility. In the upper left corner are the view port controls that allow the user to alter the view and visual styles. The View Cube, Navigation bar and the UCS icon are positioned as shown in the figure below 5.10

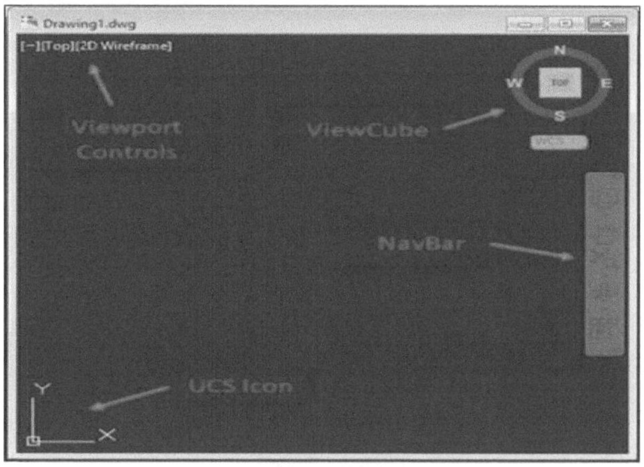

Figure 5.10

STATUS BAR

At the bottom of the screen, is the Status Bar. It contains only icons. Several icons, by default, are not enabled, but the user can decide to turn on some of them.

The figure below shows a status bar, with an expanded control icon on the far right, to show you the other available icons.

Figure 5.11

FILE TABS

File tabs give access to all open documents. Each document that is open has a tab. Left-clicking on a tab makes it the current open document. Right-clicking on a tab gives access to more options, including Save As and Close. The tab that has a plus sign allows the user to open a new document. File Tabs are located just below the Ribbon. The figure below shows three File Tabs.

Figure 5.12

LAYOUT TABS

Layout tabs give access to different Model Spaces of each Layout in any given document. Left-clicking on a tab to makes it the current open layout. Right-clicking on a tab gives you access to more options that include Print, Move, or Rename. The tab that has a plus sign allows the user to create a new layout. Layout Tabs, by default, are located at the bottom of the drawing panel below the command line, in the area around status bar.

Figure 5.13

DROP-DOWN MENUS

A drop-down menu is a top level name that is clickable. On left-clicking the menu reveals a list of commands or sub menus below it. When a command is selected, the menu vanishes. An example of a drop-down menu is shown in the figure below.

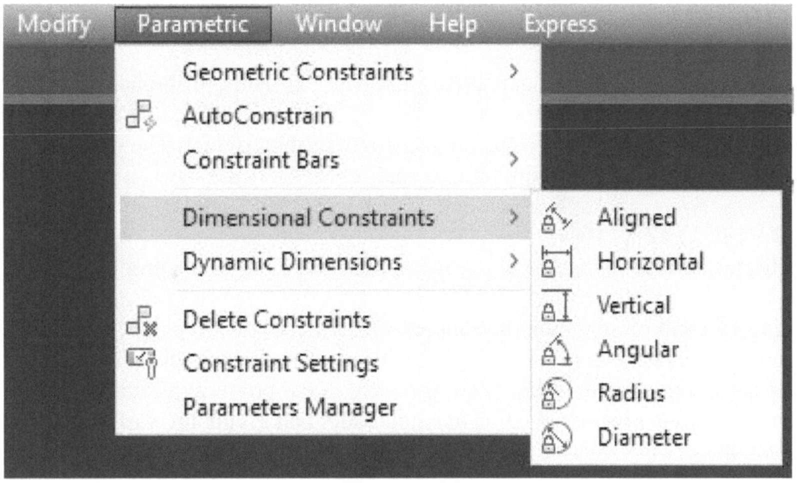

Figure 5.14

TOOLBARS

Toolbars are flat panels containing buttons or drop-down lists. Each button has an icon and macro that performs a either a command or a macro. The figure below shows two toolbars.

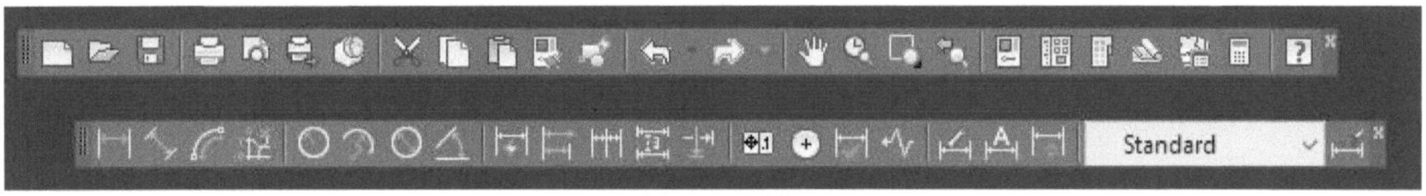

Figure 5.15

COMMAND LINE

This is a palette in which you can input command names. When docked, it has some fixed number of visible lines. When floating, the command line can expand to show more data. The figure below shows an example of a floating command line:

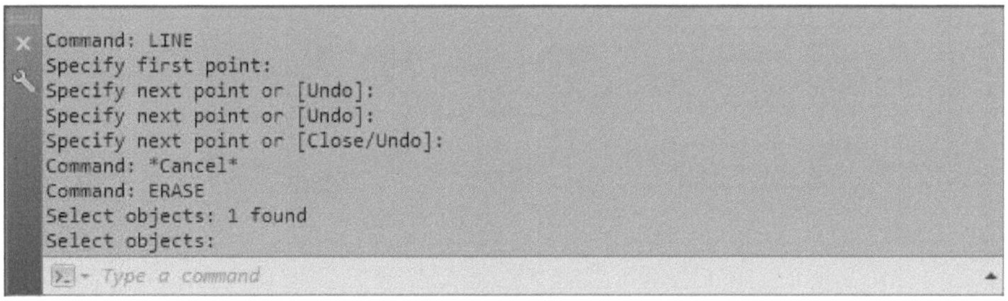

Figure 5.16

ELEMENTS OF CAD

CAD consist of 3 levels namely drafting modelling analysis and blended of two main factors i.e., machine and human factor, working together to achieve the optimum design. The computer graphic capability and computer power allows the designer to test their ideas interactively in real time without creating a real prototype as in traditional method of design.

A). **Human Factor: -** The human factor is the key factor in their success of CAD system. This concern the training of the designer or the engineer, without whom the CAD system becomes redundant. The trained designer may be actively involved in CAD process due to the following factors.

§ **Computer Aided Analysis**

§ **Computer Aided Visualization**

§ **Computer Aided Synthesis**

B). **System Factor: -** The system of CAD consists of two components namely hardware and the software.

COMPUTER AIDED ANALYSIS:

The computer is highly useful in solving of analytical problems consisting of higher order of variables. This is also useful to solve numerous equations in a short span of time. This not only helps design engineer to work on the analytical part but also saves time to work on the other aspects of the product. The result interpretation will also get improved if the designer is left with some more time to analyse it.

COMPUTE AIDED VISUALIZATION:

A designer need a better visualization to analyse the overall development of the product. A designer can visualize the product by:

An automated drafting tool and presentation of graphics

A means for presenting numerical data in form of graphical sheets such as graphs and pie charts

Animating a product in virtual space to visualize the working of the product in actual cycle

Visualizing the aesthetic look by creating a 3Dimensional model and giving the various material properties.

The effecting visualization not only saves time money effort but also improves the quality of work. The designer not only looks at product from one point but is free to look for all the possible directions.

COMPUTER AIDED SYNTHESIS:

The synthesis process is carried out with the help of the computer. The preliminary design ideas for similar products are created in a common database. This common database can be used by other elements of CAE to perform the computer aided synthesis. The evaluation of the various preliminary design needed to develop the various model for the problem and thus computer aided synthesis helps to develop the model in real time and evaluate the performance of the design.

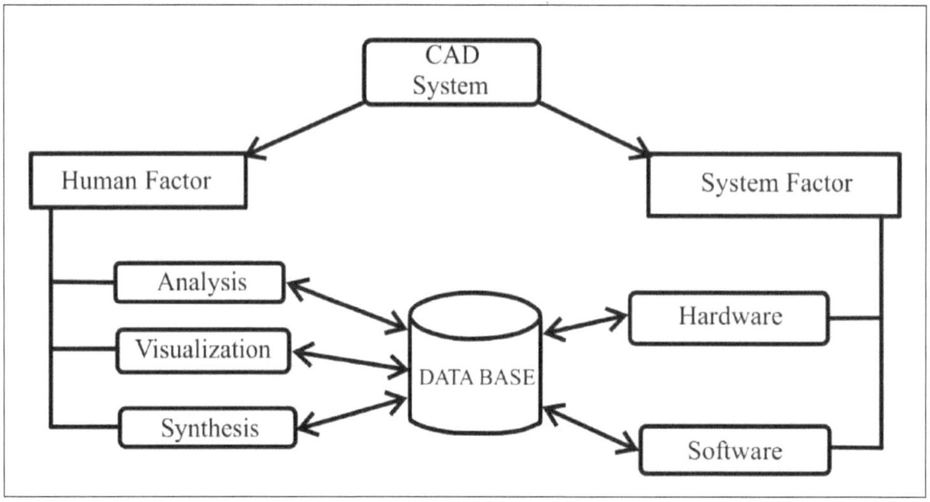

Figure 5.17

5.3 COMPONENTS OF AUTO CAD SOFTWARE

AutoCAD is a software application for 2D and 3D computer-aided design (CAD) and drafting available since 1982 as a desktop application and since 2010 as a mobile web- and cloud-based app, currently marketed as AutoCAD 360. Developed and marketed by Autodesk, Inc. AutoCAD was first released in December 1982 having been purchased a year prior in its original form by Autodesk founder John Walker. The software is currently marketed in its eighteenth generation.

The AutoCAD 3D has two 3D displays.

3D Modelling

3D Basics

The **3D Modelling** includes some advance icons, while **3D Basics** includes basic icons.

The display of AutoCAD 3D Modelling is shown in the image below:

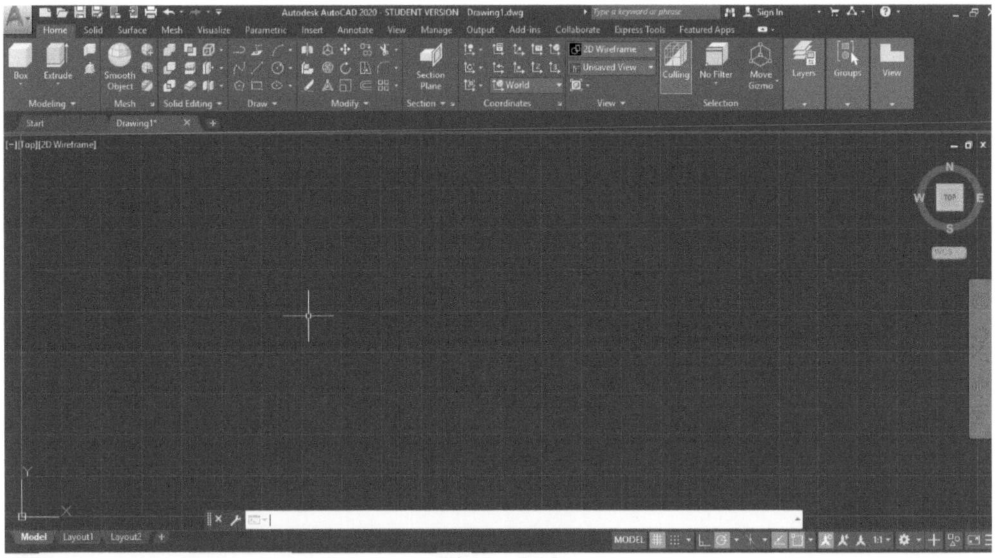

Figure 5.18

The display of AutoCAD **3D Basic** is shown in the image below:

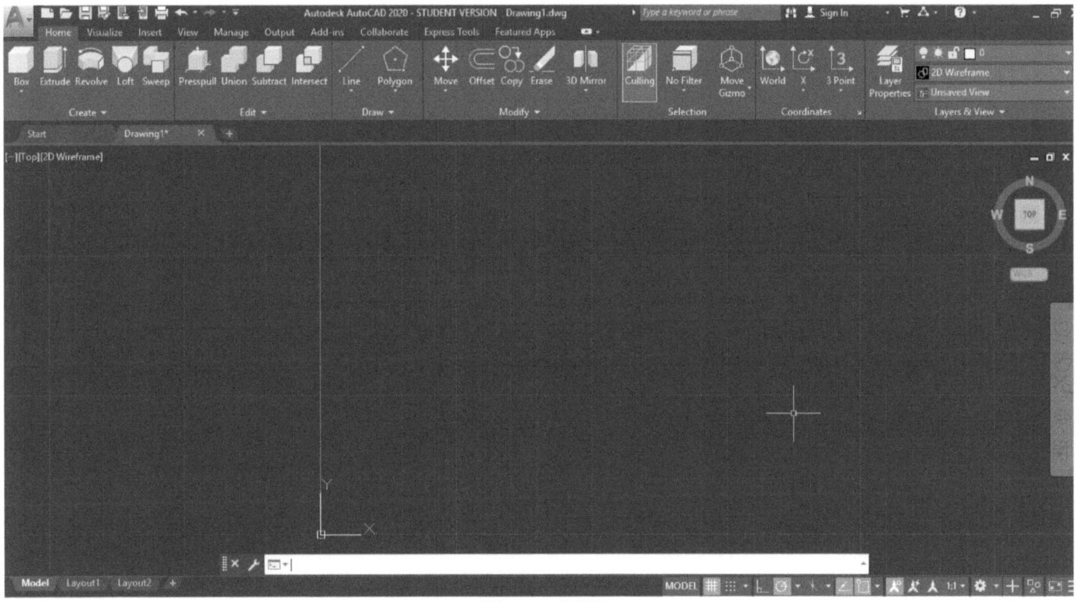

Figure 5.19

The parts of the screen are numbered for better understanding.
The numbered screen is shown in the below image:

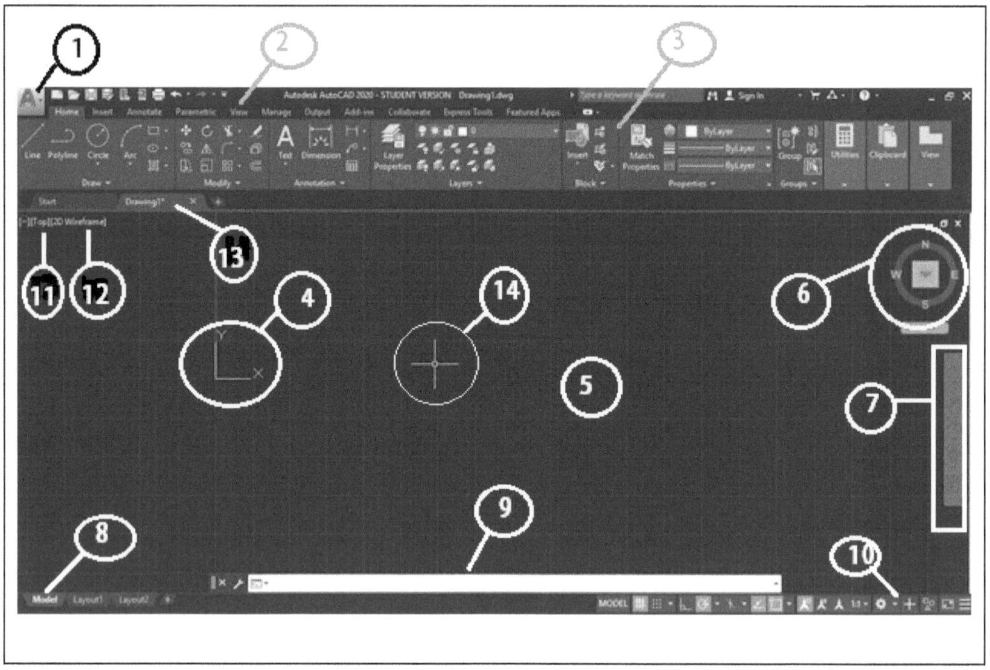

Figure 5.20

The parts of the AutoCAD 2020 display are displayed in the numbers.
The names of the numbered parts are listed below:

1. APPLICATION MENU/BUTTON.

It is present at the upper-left corner of the workspace. To close the application menu, we can click anywhere outside the application button or window. The button is shown in the image marked as number 1.

2. QUICK ACCESS TOOLBAR

The Quick Access Toolbar is located at the top of the application window and right of the application menu. It consists of the set of frequently used commands. We can add and remove the commands according to the requirements.

3. RIBBON PANEL

It provides access to the dialog box related to that panel. If the ribbon panel disappears, then we need to enter the 'ribbon' on the command line to display the ribbon panel on the workspace.

4. USER COORDINATE SYSTEM (UCS)

The UCS is the active coordinate system that represents the XY plane in 2D and XYZ planes in 3D. It acts as the direction in the X, Y, and Z-axis for drawing and modelling. We can control the origin and orientation of the UCS to make drawings according to the specific points and coordinates. We can also work with drawing aids such as Grid and the ortho mode.
The 2D UCS is shown in the given image:

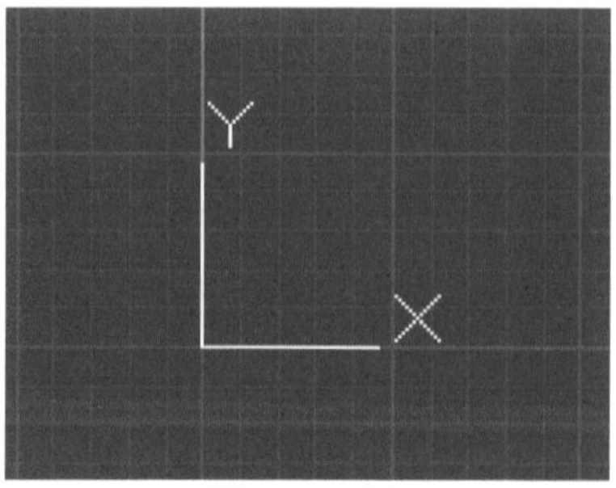

Figure 5.21

The 3D UCS is shown in the given image:

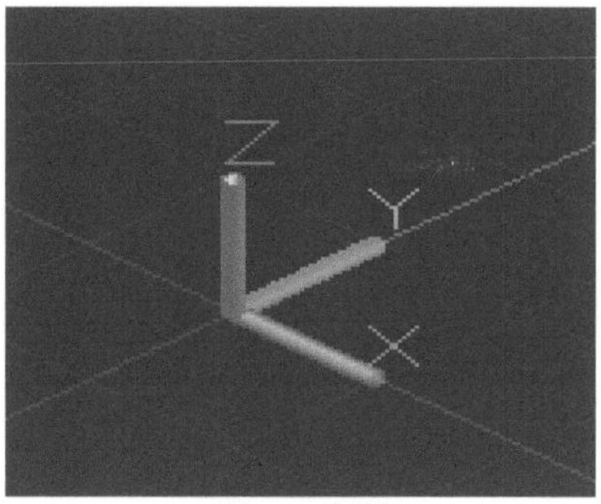

Figure 5.22

5. MODEL SPACE / WORK SPACE / DRAWING WINDOW

It is defined as the area to create 2D and 3D drawings, models, and objects. We can create using different commands according to the requirements.

6. VIEW CUBE

The View Cube is termed as the navigation tool that is displayed when we are working on a 2D or 3D model space. We can switch between the isometric view and the standard view of our drawings or model.

7. NAVIGATION BAR

The Navigation Bar is used to access the navigation tools. It is a user interface element, where we can access both unified (common tools) and product-specific tools (unique product tools).
The navigation bar is shown in the below image:

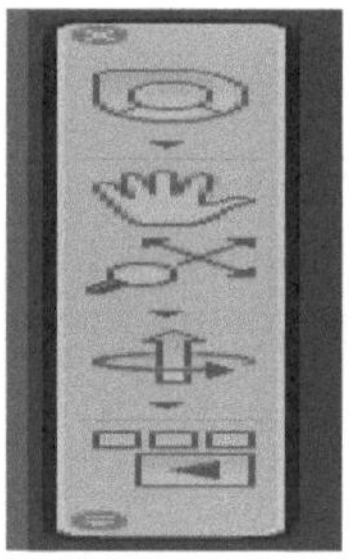

Figure 5.23

8. MODEL LAYOUT TAB

A layout is defined as a 2D working environment for creating the drawing sheets.
The model tab is the screen where we create the 2D and 3D drawings.
If the model and layout tab is not visible, then follow the steps given below:

a. We need to click on the AutoCAD option displayed on the top left corner of the screen, shown as 
b. Click on the 'Options' button at the bottom.
c. Then on the 'display' option on the top, select the option 'Display Layout and Model tabs' and then click 'Ok' as shown below:

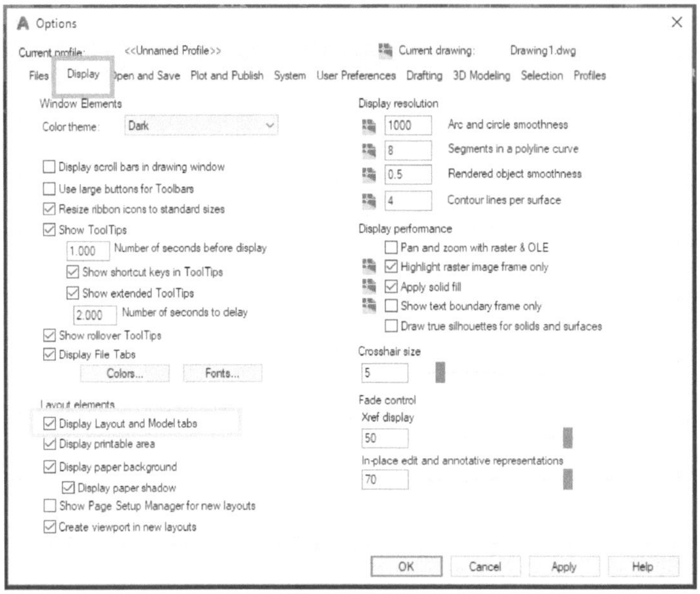

Figure 5.24

9. COMMAND LINE WINDOW (SHORTCUT KEY- CTRL+9)

The command line window is used to write the commands. We need to press 'Enter' after typing any particular command. The AutoCAD also displays the steps after each command on the Command line. We need to select the option and press 'Enter' after each step.

10. STATUS BAR

The status bar displays the drawing tools that affect the drawing environment. It provides quick access to most of the commonly used drawing tools. It includes options such as ISODRAFT, ORTHOMODE, AUTOSNAP, OSNAP, etc.

11. VIEW CONTROL

The View Control is displayed on the left corner of the viewport. It provides a suitable way to change views, visual settings, and styles. It includes options such as top, bottom, left, right, etc.

12. VISUAL STYLE CONTROL

The Visual Style Control is the customized setting that controls the structure view of the models or 3D drawings created on the Viewport. The options include 2D Wireframe, Realistic, Shaded, Conceptual, Hidden, etc.

We can also enter the visual styles on the command line to display the options of the visual style control.

After entering on the command line, the box will appear displaying clear information with figures for each option. It is shown in the below image 5.25

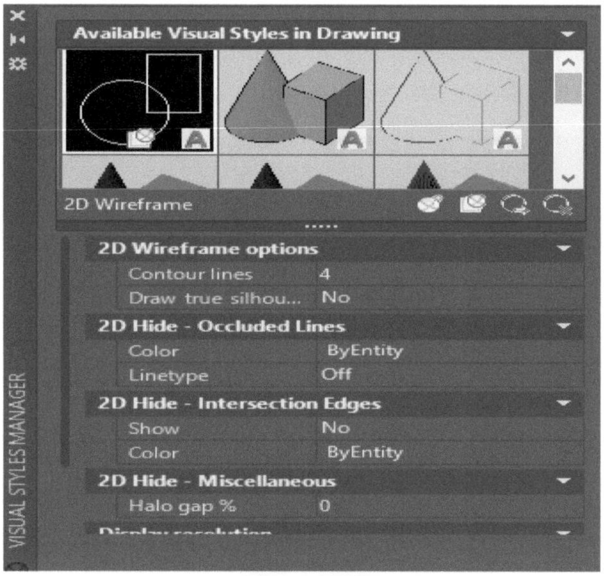

Figure 5.25

13. FILE TAB

The file tab consists of the current drawing files opened on the screen.

We can also type FILETAB command in the command line to turn on turn on the file tabs. To close the file tab, we can type FILETABCLOSE command in the command line window.

We can click on the '+' sign to add a new tab on the File Tab.

14. MOUSE CURSOR

The cursor used to draw figures, etc. is known as the mouse cursor.

The cursor in 2D is shown in the given image:

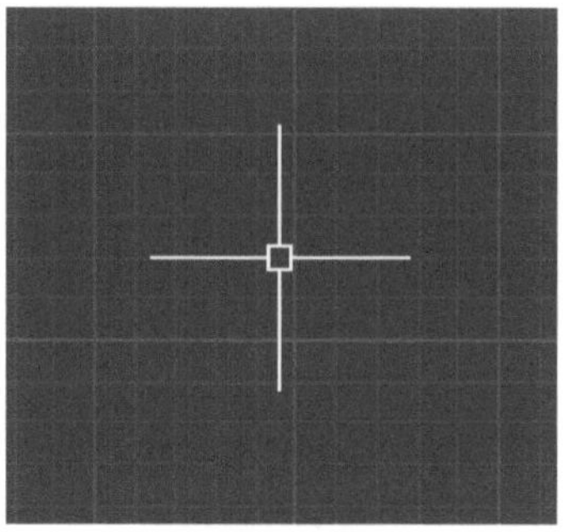

Figure 5.26

The cursor in 3D is shown in the given image:

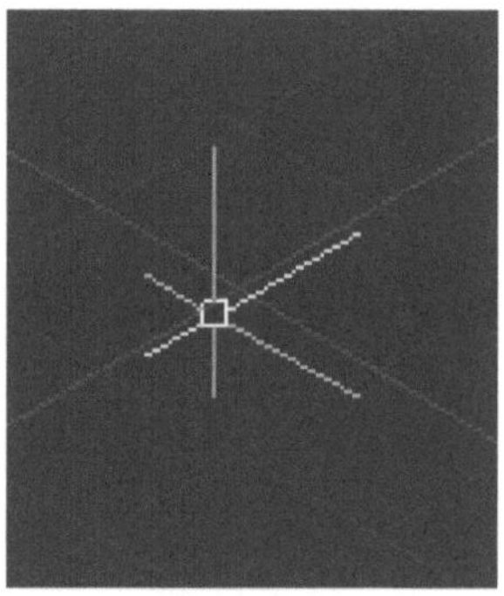

Figure 5.27

2D TO 3D

The steps to convert the 2D display of AutoCAD into 3D are listed below:

- Click on the button on the status bar as shown in the below image.

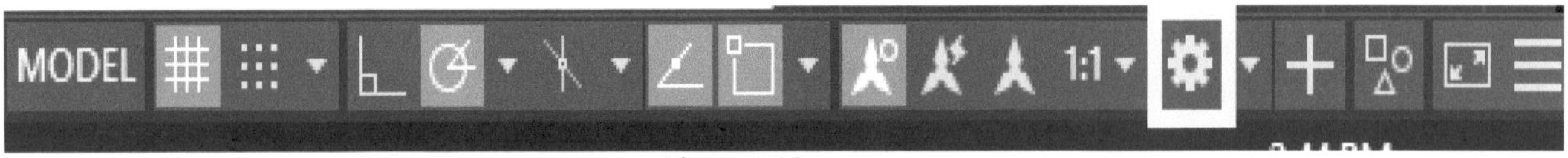

Figure 5.28

The drop-down list will appear. Select the 3D Modelling or 3D Basics option from the drop-down menu as shown in the below image:

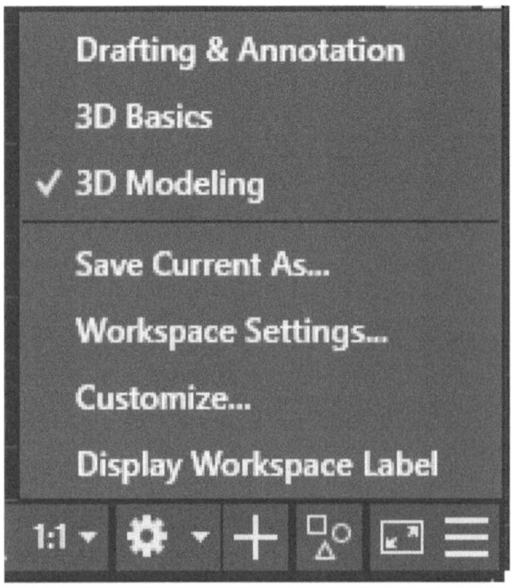

Figure 5.29

We can select 3D Modelling or 3D basics according to the requirements.
The 3D modelling will include advance commands to create 3D models.
The 3D Basics will include basic icons.
Introducing the AutoCAD Graphics Window At the top of the graphics window sit the Ribbon, the Quick Access toolbar to the left, and the Info centre and related tools on the right.

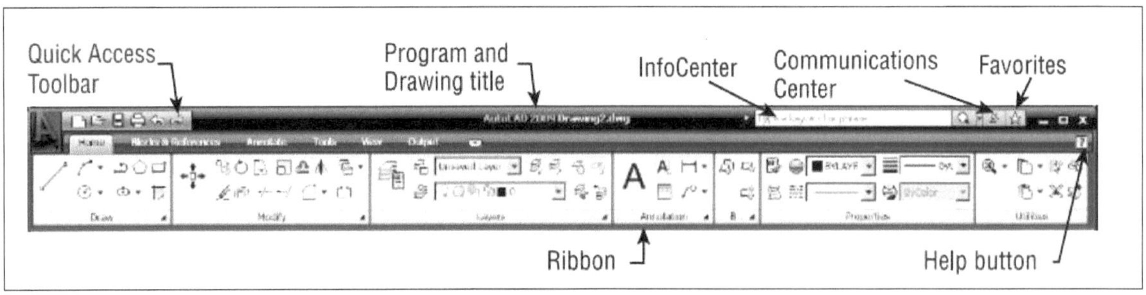

Figure 5.30

15. THE TITLE BAR

The title bar is analogous to the title bar in any Windows program. It contains the program name (AutoCAD or AutoCAD LT) and the title of the current drawing with its path, as long as any drawing other than the default Drawing is open. Below the title bar is the Ribbon, where you'll find most of the AutoCAD commands and tools needed to complete any drawing task. Related tasks are found under the different tabs, which are further segmented into panels containing similar tools.
To the far right of the title bar are the Info Centre, Communications Centre, Favourites, and Help buttons. You can enter a question in the field to the left of the Info Centre button to quickly access information from the Help system

through the Info Centre drop-down panel. With the Communications Centre you can determine what type of information, such as software updates, product support, or RSS feeds, Autodesk sends directly to your system. With the Favourites tool, you can define a list of help or informational topics that can be quickly accessed whenever you need them. The Help button is a direct link to the AutoCAD help system.

The blank middle section of the screen is called the drawing area. Notice the movable crosshair cursor. The crosshairs on your cursor might extend completely across the screen. Later in this chapter, I will show you how to modify the length of the crosshairs as well as make a few other changes. Notice the little box at the intersection of the two crosshair lines. This is one of several forms of the AutoCAD cursor. When you move the cursor off the drawing area, it changes to the standard Windows pointing arrow.

5.4 FEATURES OF AUTO-CAD:

AUTO-CAD is used to refer CAD which is any type of computer aided design/drafting. Basically AUTO-CAD is the name of the software which is develop by the company named **Autodesk**. The auto in Auto-CAD doesn't mean anything but is just a reference to company's name. As, it is ca software for computer sided design so it is used in construction or designing of cars/ boats/ airplanes/ small-buildings and in the design of tools and machinery.

All CAD systems uses basic geometric entities to construct design. The construction used in technical design are based upon 2-dimensional planar geometry. The method and number of operations are required to accomplish the instructions are different from one system to another. In order to become an expert in Auto-CAD, we must learn to create geometric entities quickly and accurately.

Auto-CAD provides various several different mechanisms or a tool with the help of which different drawings can be made easily and can be viewed properly. That is why, designers or architect engineers mostly used it. It gives the facility with which we can enter the commands through the command-prompt or selecting them from pull-down menu or from the tool-bar. Most commands are either execute immediately after it has been involved or prompt from further information.

Multiple window environment – Auto-CAD provides the facility of drawing and having each window in different drawing pad. The user can create top-view, idea-view and can save it in different windows. This option is available in layout menu.

Three dimensional view – Auto-CAD has an option for viewing any object in its three dimensional form. It has several 3D tools to perform this function. Two of these are as follows:

3D surface: This option can be used to create 3D cones, cylinders, polygons etc on a normal view.

3D orbit: By using this option, we can change and relate the object in any view.

GUI features – Auto-CAD supports graphical user interface features as it is loaded under windows operating system. Some of its features are:

Interaction with peripheral devices: Several peripheral devices like printer, plotter, mouse etc can be used. There is no provision for page setting during plotting line, also several commands like isometric view, different structures designing can be performed using mouse.

Auto-CAD 2000 supports 24 bit true colours.

Intelligent mouse functions.

DB connect manager – This is an advance feature of Auto-CAD 2000 where we can connect drawings with database. Auto-CAD can connect with SQL or with any access database during the designing of any type of structure. During DB connect manager; a new window is opened for connecting the database.

AUTOSLIP editor – AUTOSLIP editor is basically a programming language embedded in AUTO-CAD. It lets the user automate tasks and add custom commands to AUTO-CAD. This is useful to connect a C program to AUTO-CAD.

Object properties manager – Like usual basic AUTO-CAD 2000 is also designed with the object-oriented features. Every drawing element whether it is circle, ray, line. rectangle or any type of shape has its own properties. The properties box shows the colour, x and y co-ordinates, width, height, length and several other properties.

Command-specific shortcut menus – AUTO-CAD provides the facility of drawing objects by entering different commands through the command line.

Colours – We can set different background colours for different windows. We can make objects of different colours. We can change/slash the different settings of whatever is viewed on the screen e.g., screen menu, scroll bars, layout-tabs, size of fonts etc. It supports 24-bit true colours.

Zoom – We can use zoom command to any part of any structure to have a close look of particular part of interest.

Layers – To make the complimented drawings in an easier manner, we can divide the structure of drawing in different layers. In layers, the changes are made only on current layers. To make changes on others we have to select the layer first.

5.5 SETTING UP NEW DRAWINGS

AutoCAD is an industry-leading commercial CAD software. AutoCAD is used by AEC(Architecture, Engineer, and Construction) to generate and optimize 2D and 3D designs. AutoCAD is a widely used software program that can help you draft construction documentation, explore design ideas, visualize concepts through photorealistic renderings, and simulate how a design performs in the real world(Autodesk).

AutoCAD was first released in December 1982 as a desktop app. In 2010, AutoCAD was released as a mobile- and web app, marketed as AutoCAD 360.

5.5.1 UNITS

While you should consider carefully which unit convention to choose — cm, mm, or inch — the unit of a drawing has no effect on the measured numerical values in that drawing. For example, if you use a dimension to measure a segment of 1000, it means that even if your CAD drawing is set to cm or mm, the measurement value of Dim is still unchanged at 1000.

UNIT OPTIONS

The following units can be used when drawing in AutoCAD:

Inches
US Survey Feet
Feet
Miles
Nanometres
Centimetres
Millimetres
Meters
Kilometres
Yards
Microinches
Mils
Angstroms
Microns
Decimetres
Dekametres
Hectometres
Light years
Gigametres
Astronomical
Parsecs
Unitless

LENGTH UNITS OVERVIEW

DECIMAL UNITS

Any unit, such a meter, millimetre, inch, or other, can be used to represent a decimal unit.

Decimal units in AutoCAD don't refer to any specific unit by name, and the software permits dimensions in the form of units, such as 1.85 or 3.365. A maximum of eight decimal places can be entered in AutoCAD.

Architectural Units: Feet and inches serve as the foundation for architectural components. The fractional inches are expressed using fractions. It presents feet and inches in decimal and fractional form. For instance, 4′-5 1/2 ". Here, feet are represented by a single apostrophe (') and inches are represented by a double apostrophe (").

Scientific Units: A measurement yields 1.5500E+01 when the length of 15.5 units is included. Dimensions may be entered in the format of a real number divided by 10, for example, I25E+02. Scientific notation can be used to depict the measurement system. Using scientific notation is mainly beneficial when dealing with extremely large numbers.

Engineering Units: 1′-3.5 is produced from a measurement that has a length of 15.5 units. Dimensions may be represented in feet, inches, and inch-decimal places. Important to bear in mind is that 12 inches is automatically

converted to 1 foot and is displayed as such. To enter the feet, use a single apostrophe ('). Each engineering unit is equivalent to one inch in AutoCAD's units. Building plans and topographic maps used in the design and construction of highways and harbours are a few examples of common civil drafting tasks.
Fractional Units: Fractional units are unitless and display values as fractions rather than decimal numbers, similar to decimal units. A measurement yields 15 ½ units of length, or 15.5 units total. Dimensions may be expressed as a collection of units or as fractional segments of units. The types of fractional units include feet, miles, and inches. The unit of measurement can be represented using fractions.

ANGLE UNITS OVERVIEW
Various angle measurements can also be specified in a similar way.
DECIMAL DEGREES
Decimal degrees measure angles and display them as Real numbers with up to eight decimal places. This form of angle measurement is the default in AutoCAD.
Radians: The common unit of angular measurement used in different branches of mathematics is the radian, abbreviated as rad. The angle subtended from a circle's centre that intercepts an arc with a length equal to the circle's radius is defined as a radian.
Surveyor's Units: Angle measurement is shown along with quadrant bearings in the form of degrees, minutes, and seconds. In essence, the circle is divided into four quadrants, with the result that no angle can be higher than 90 degrees.
Grads: The gradian, also known as the grade, or grad, is a unit of measurement for angles used in trigonometry. It is equivalent to one hundredth of a straight angle, or 100 gradians, or 90 degrees.
Deg/Min/Sec: The format of degrees, minutes, and seconds is used to display angular measurement.

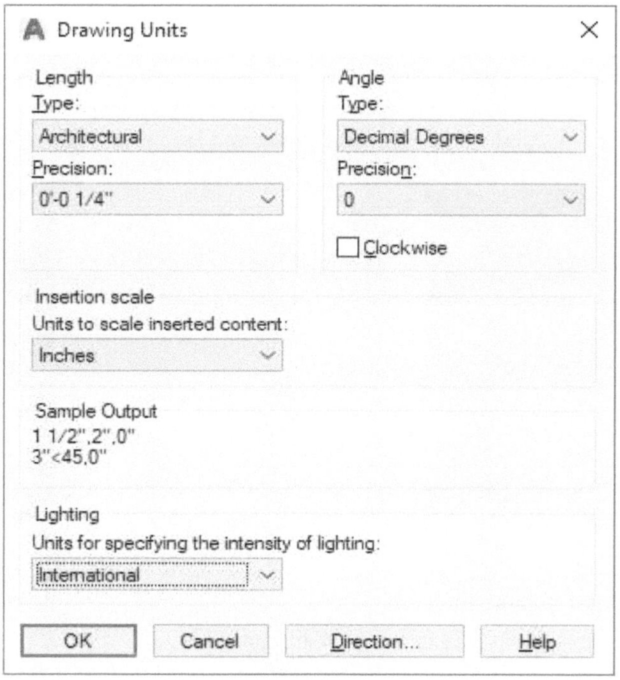

Figure 5.31

HOW TO CHANGE UNITS IN AUTOCAD
Hereafter, this article will show you 3 ways to do it in CAD that only takes a few minutes.
USING THE UN COMMAND

QUICK GUIDE
Select any point.
Enter the UN command.
Press Enter.
After the Drawing Units tab appears In the Length section, select the Precision section to adjust the unit.

In the “Units to scale inserted content,” select the appropriate unit for the drawing.
Select OK to finish.

DETAILED INSTRUCTIONS

After entering the CAD drawing, select any point and enter the UN command, then press Enter.

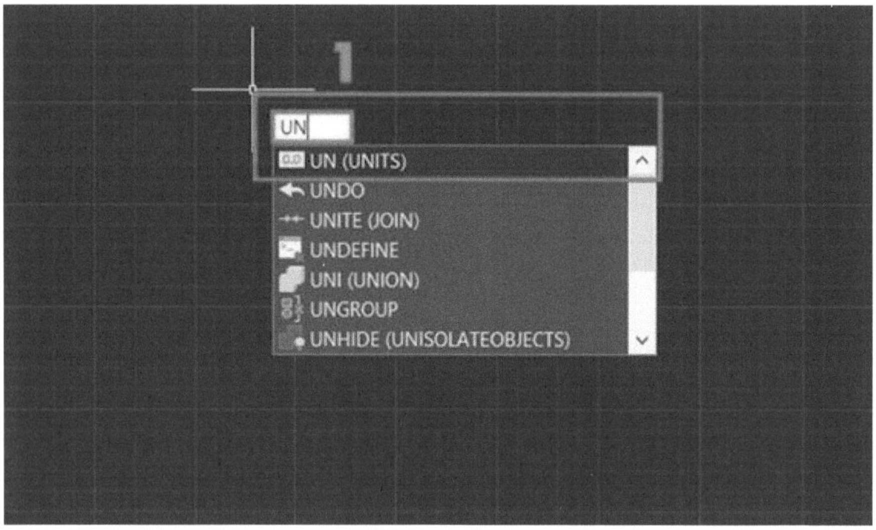

Figure 5.32

2. After the Drawing Units tab appears, you need to care about items such as Length (Type and Precision), Angle (Type and Precision), and Insertion Scale (Units to scale inserted content).
Length (Type and Precision), Angle (Type and Precision), and Insertion Scale (Units to scale inserted content).

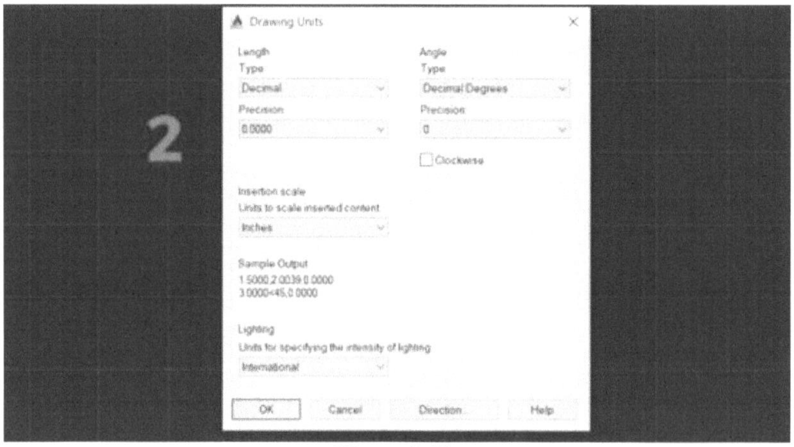

Figure 5.33

3. In the Precision options of the Length section, you can adjust the unit size, including scientific, decimal (according to TCVN standards, you should choose this option), engineering (for scientists or engineers), architectural, and fractional.

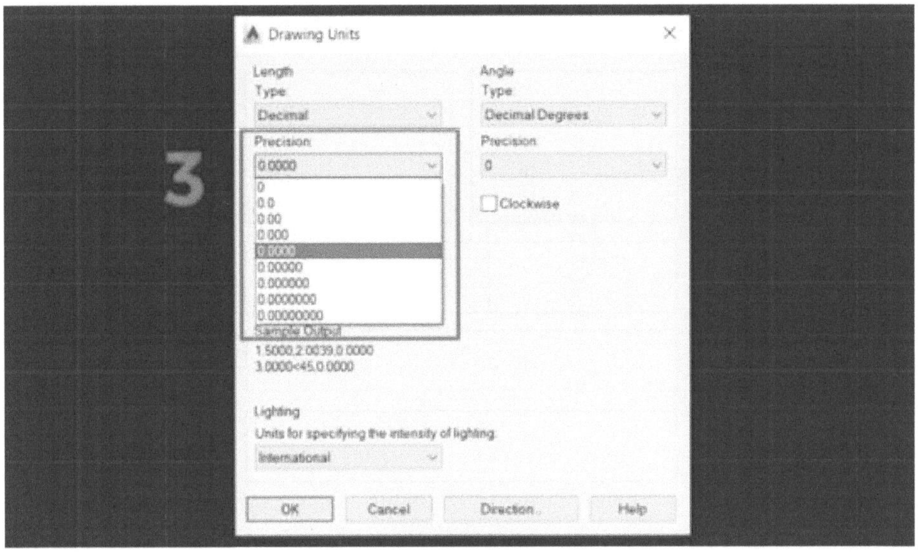

Figure 5.34

Precision options of the Length section
4. In the Units to scale inserted content section, select the unit that matches the drawing.

Insertion scale
5. After selecting the unit, you need to select OK to complete.

ADJUSTING UNITS IN AUTOCAD BY TURNING ON THE ORIGINAL TEMPLATE FILE
QUICK GUIDE
Select the file you want to adjust the unit of.
Select New item or press Ctrl + N.
The Selection tab will appear with 4 sample files.
Select file.
Click Open.

5.5.2 LIMITS
LIMITS COMMAND
The Limits command in AutoCAD is used to set an invisible rectangular boundary in the drawing area or viewport. It limits the grid display and the point locations.
We are required to specify the coordinates of the opposite corners of the rectangular window. The opposite corners are named as Upper right and Lower left corners.
The opposite corners are shown in the below image.

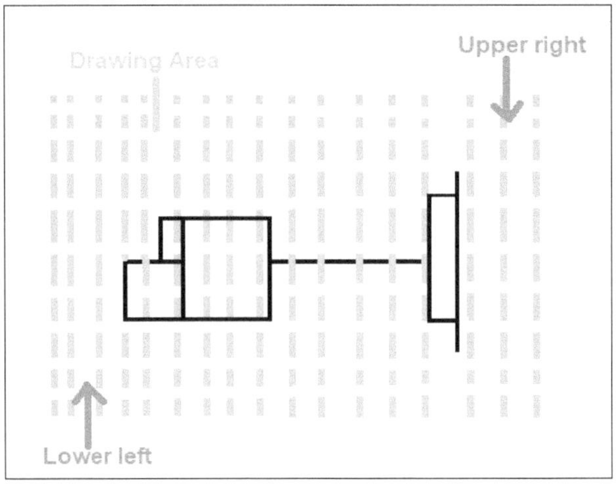

Figure 5.35

WHY WE NEED LIMITS?
Some dimensions in a drawing are represented in hundreds and thousands of meters. For example, bridge designing in AutoCAD, Architecture designing, etc.
The representation of such large dimensions is difficult in a regular drawing area. For such purposes, limits play a vital role in setting the boundary of the drawing area. We can specify the limits according to our requirements.

LIMITS CHECKING
The two options under limits checking are:
ON
It turns **ON** the limit checking. It tests only the points we enter, so we cannot specify the points outside the grid limits. But portions of circles and other objects can extend outside the grid limits.
OFF
It turns **OFF** the limit checking. The current value of limits is maintained for the next time when we turn ON the limits.

STEPS TO SET THE LIMITS
The steps to set the limits are listed below:
Open the AutoCAD software.
Type LIMITS on the command line or command prompt.
Press Enter or spacebar.
Write the coordinates of the lower-left corner. For example, (0,0). The coordinates for the lower-left are usually the coordinates of origin for better understanding. We can also modify the coordinates according to the requirements.
Press Enter.
Write the coordinates of the upper-right corner. For example, (200, 200).
Press Enter.
Write Z. Here Z signifies Zoom.
Press Enter.
Write E. Here, E signifies Extend.
Press Enter.
The **Z** and **E** are mandatory to activate the limits. It is also called a regenerating mode of limits.
Let's understand **limits** with few examples.
EXAMPLE 1: DRAW A RECTANGLE OF DIMENSION 100 X 200.
Here, the length of the rectangle is 100, and the width is 200.
The steps are listed below:
Type LIMITS on the command line or command prompt.
Press Enter.
Write the coordinates of the lower-left corner (0, 0).
Press Enter.
Write the coordinates of the upper-right corner (300, 400). You can also increase the limits according to your choice.

Press Enter.
Write **Z**. Here Z signifies Zoom.
Press Enter.
Write **E**. Here, E signifies Extend.
Press Enter.
Type Rec or Rectangle on the command line.
Press Enter.
Specify the first corner point on the viewport.
Specify the length and breadth of the rectangle in the form of @length, width. For example, @100,200
Press Enter.
When you start a new drawing, its template has settings for units and limits that are used by default. You can use the settings used in the template or make your own settings for units and limits. In this example, we will open a blank drawing then set the units and limits in it and then we will save the drawing as a template file. To start a new drawing in AutoCAD, you can click on the start panel of the Start tab or select one of the readymade templates under the start panel.
The list of ready made templates will show up underneath the Start Drawing panel when you click the arrow pointing downward in the panel:

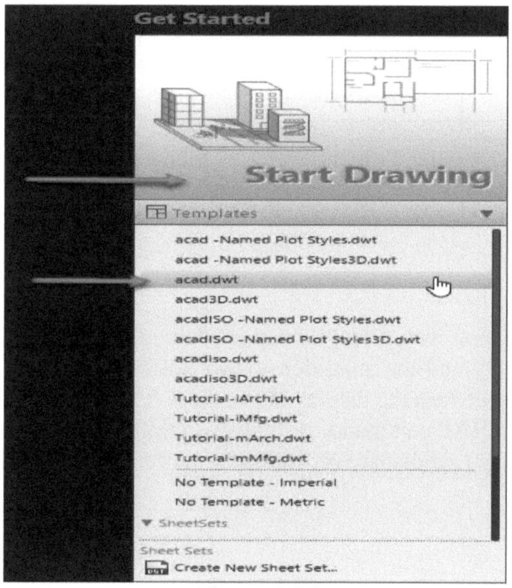

Figure 5.36

5.5.3 GRIDS
Grids are a series of dots or lines that display and overlay the Graphic window in a rectangular or square shape similar to overlaying a paper drawing with grid paper. Grids do not plot and are not part of the drawing's database, they are simply a drawing aid.
The grid can be set to display using dots or lines. See Figure 5.27. The grid's density can be adjusted and their display can be enabled or disabled, when required. The horizontal and vertical spacing between the grid dots or lines can be different, but by default are equal. The grid display can be toggled with the F7 key, the icon on the Status bar, with the GRID command, or by using the Drafting Settings dialogue box. The size of the rectangle that the grid fills can be set to the size of the drawing limits or it can be set to fill Model space. Grids can also be displayed in a layout in Paper space.

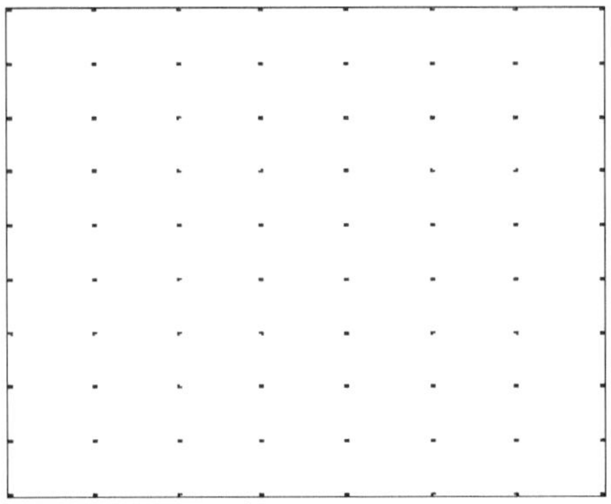

Figure 5.37

AUTOCAD COMMAND: GRID

The GRID command is used to display a series of dots or lines in either an equal or unequal horizontal and vertical spacing. The function key F7 toggles grid mode.

Shortcut: **F7**

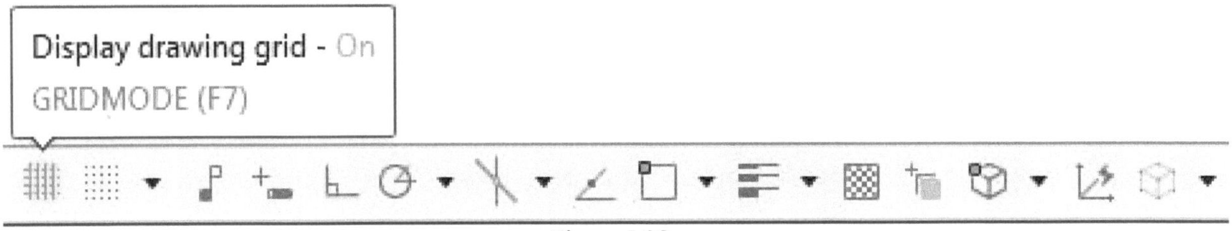

Figure 5.38

5.5.4 SNAP

The snap feature works in conjunction with grids. If snap is enabled, the crosshairs snap only to grid dots, similar to object snap mode. The grid density and the snap density do not have to be set at the same value. Therefore, the crosshairs could snap to locations between the displayed grid dots. The snap can be toggled using the F9 key, with the icon on the Status bar, using the SNAP command, or with the Drafting Settings dialogue box. If snap is enabled, it snaps to the snap density even if the grid display is disabled.

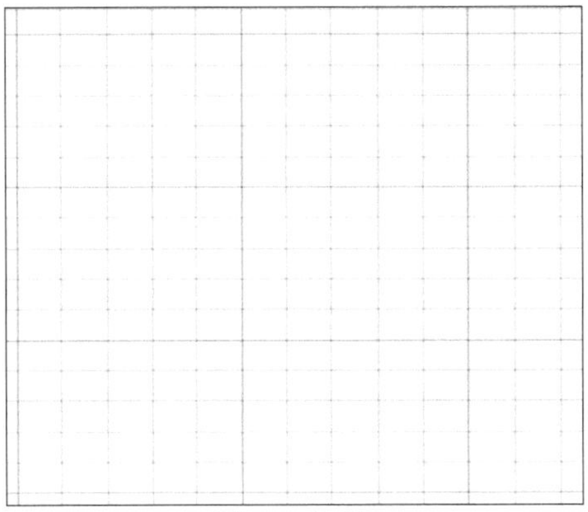

Figure 5.39

AutoCAD Command: SNAP

The SNAP command is used to control and enable the snap feature. The function key F9 toggles snap mode.

Shortcut: **F9**

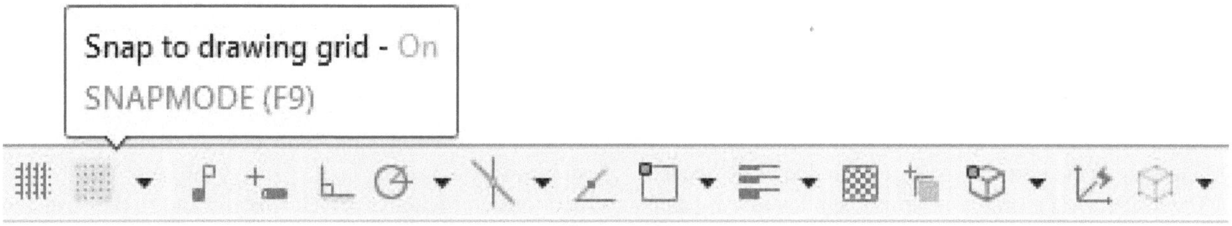

Figure 5.40

CHARACTERISTICS OF AUTOCAD

In the Above Section, we have learned about Introduction to AutoCAD; there are some main characteristics that one needs:

AutoCAD is a global application. It is being used globally by product development teams, manufacturing facilities, medical professionals, educational institutions, professionals, and engineers.

3D modelling and visualization are two main important features of the program. AutoCAD allows the modelers to create powerful 3d models, wireframes, meshes & surfaces using various 3D tools & commands.

AutoCAD is a professional application with flexibility in design changes and an auto-specification check feature. The mistakes in the designs or the product can be avoided and re-edited as and when required.

AutoCAD can create section planes. These section planes help in achieving cross-sectional views of 3D objects. The users can change, select, or move the section planes to inspect the inner details of 3D objects.

AutoCAD helps the user to develop, modify, and design better infrastructure, deliver scalable and feasible building assignments, supervise production finances, and foresee project results.

Recreating and editing 2D images with their text properties can also be achieved with AutoCAD.

AutoCAD drawings can also be linked directly to Microsoft Excel Spreadsheets. With this technique, formulas, and data from Excel can be imported directly to AutoCAD. This feature is an Auto-Update command.

AutoCAD also has an SHF text recognition tool that helps users convert text with SHF fonts into text objects.

With AutoCAD, the students can even import PDF files for any project reference.

3D models from other CAD applications can also be imported to AutoCAD. The designers can also work on imported online maps or locations.

APPLICATIONS OF AUTOCAD

After learning the Introduction to AutoCAD, its components, and the Characteristics of AutoCAD, we will now learn the application of AutoCAD. Digital literacy is a boon to students and a skill essential for career growth. AutoCAD is a gigantic program that delivers desired outputs and is used in many ways.

The students can use AutoCAD for the following:

AutoCAD is compatible with other applications like 3D Max. The designers can import or export DWG & DXG Files to the application. Also, these files can even be exported from AutoCAD to other applications for creating more improved projects & presentations.

3D printers use various file formats not supported by all the CAD software. On the other hand, AutoCAD supports these file formats preferred by most 3D printers.

AutoCAD is adaptive to the CAD interface and its usage. AutoCAD allows professionals to check the framework's workings within the design stage. Designers can use this framework for projects & presentations.

AutoCAD comprises integrated design layouts of various templates specially designed for architectural planning and infrastructure constructions; the users can work on plans that include creating architectural arrangements for construction purposes without having to master the software.

AutoCAD is not only a powerful application but also has a simpler user interface. It comes with many functions and commands that are to the point and without any confusion. Designers also use AutoCAD in the fashion industry as they can design difficult shapes and designs without any problems.

Engineers also use AutoCAD, which incorporates easier drafting tools and helps model engineering designs, blueprints, and other components. With AutoCAD, engineers can design realistic models with minimum errors. AutoCAD is a medium for users to design mechanical parts and components for their projects.

ADVANTAGES AND DISADVANTAGES OF AUTOCAD

After going through all the components, characteristics, and the wide Introduction to AutoCAD, we are going to explore the advantages and disadvantages of AutoCAD:

With AutoCAD, users can link their files to multiple platforms; they benefit from importing and exporting all kinds of files.
The users can expand the base program through programming; thus, AutoCAD is a versatile application standardized in computer designs.
AutoCAD also has project managing tools that allow users to share information simultaneously & effectively.
The cost of the program is quite high due to its license.
The software is a time-consuming application. It has many functions and features which specific and to the point.
A powerful computer with good RAM and Hard disk is required for the application to work with processing speed.

WHAT IS THE PURPOSE OF AUTOCAD SOFTWARE?

The most basic purpose of AutoCAD is drafting programs. Several industries that develop or design products use different computer-aided designing software and the most commonly used computer-aided designing software is AutoCAD.
AutoCAD creates realistic 2D and 3D sketches of numerous things like buildings, automobiles, products, bridges, clothes, landscapes, and furniture. It assists engineers and designers by:
Saving time taken to create things manually.
Reducing the cost of material in trial phases.
Creating precise and accurate yet detailed sketches.
Helping in detecting errors in designs and drafts.
Allowing for testing the durability of products through life-like simulations rather than theoretical calculation.
Providing several analysis tools that help optimize efficiency.

WHAT ARE SOME REASONS TO CHOOSE AUTOCAD OVER OTHER CAD SOFTWARE?

Some of the primary reasons to pick AutoCAD over any other CAD software are:
Friendly Interface: AutoCAD has a very beginner-friendly user interface, that is, people who are using AutoCAD for the very first time will not have a problem finding their way around the tools.
Multipurpose: AutoCAD is one of the first CAD software and has gained massive popularity. It is the most commonly used CAD software and the uses of AutoCAD are infinite. Therefore, every industry requires people who know how to operate AutoCAD. Certain industries practically rely on AutoCAD to fulfill their component design requirements.
Saves Time: AutoCAD saves you a lot of design time because it has built-in workflows, several pre-designed parts and features. AutoCAD's robust library of APIs, LISP functionality, and specialized toolsets also allow users to automate tasks and save time.
Easy collaboration: You can share your designs and collaborate with other designers/engineers/architects using the share feature in AutoCAD. This feature enables many people to work on the same project at a time.
Secure & safe: AutoCAD Web App uses cloud storage to store your designs and therefore you do not run the risk of losing your files or filling up your PC's memory.
Functionality: AutoCAD has an array of features and tools and can almost do everything other software can. If you learn the features and uses of AutoCAD first, you will be able to make a shift to any domain-specific software later.

WHERE IS AUTOCAD USED?

Well, let us learn about the different fields of studies where AutoCAD is used. Uses of AutoCAD in civil engineering AutoCAD know-how is a must-have skill for civil engineers. Almost all the work of civil engineers like designing, drafting, and editing the blueprints of a bridge or building is done with the help of AutoCAD.
Other significant applications of AutoCAD in civil engineering are:
A graphic design tool to plan and map extra space in designs
3D Max for representing available space in floor plans
3D printing feature to have proffered file compatibility with the diversified format in civil engineering study.
Easy editing features like copy, stretching, scale, rotate, mirror, etc.

USES OF AUTOCAD IN MECHANICAL ENGINEERING

Mechanical engineers use AutoCAD in several aspects of machinery design. They primarily use it for drawing, drafting, editing ideas, and then analysing the drawings to make changes or optimize the drawing at the preliminary stages of a project.

This is done to remove any inadequacies or flaws in the design before it goes into production.
Following are the applications of AutoCAD in the field of mechanical engineering:
Helping reduce the cost of physical materials by creating life-like digital simulations.

Providing the digital prototyping feature that enables engineers to make changes as and when required without having to redo the entire design.

Providing high speed and flexibility in delivering outputs saves time and provides corporations with a competitive advantage.

A vital area where AutoCAD software plays a key role in mechanical engineering is by allowing users to simulate a variety of conditions (e.g.: stress, temperature, repetitiveness) to examine the durability of a product, building or system over time, thus aiding in defining structural and functional specifications.

USES OF AUTOCAD IN ELECTRICAL ENGINEERING

Electrical engineers, too, use AutoCAD to design flawless circuits and simulate such circuits before they go into production.

Drawing single power line diagrams and motor-controlled circuit schematics

Figuring out general switchboard arrangements

Communicating with mechanical engineers regarding electrical design plans and where to embed them.

Detecting and eradicating errors from circuits

Some of the more specific applications of AutoCAD by electrical engineers are optimizing and analysing the circuits without having to change the entire configuration each time.

Significantly reduces overhead costs by removing the need to purchase primary testing components.

USING AUTOCAD FOR LANDSCAPE DESIGN AND INTERIOR DESIGN

Professional landscaping artists or interior decorators widely use AutoCAD these days because of the several features of AutoCAD which helps them visualize and pitch their designs to their clients much better than on conventional pen and paper.

Specific applications of AutoCAD in the landscaping and interior designing industries are:

It helps in streamlining workflow

Makes designs and landscapes look realistic with gradient features and gives the customers as well as the designers a clear idea of what to expect

Saves a lot of time for designers as basic components like doors and windows are already in-built in the software and need not be created from scratch.

USES OF AUTOCAD IN THE AUTOMOBILE AND AEROSPACE ENGINEERING INDUSTRY

Automobile and aerospace engineers over the globe widely use AutoCAD to create designs for cars and aircraft and test their efficiencies via simulation.

Some of the specific applications of AutoCAD in automobile and aerospace engineering are:

AutoCAD designs are self-explanatory to production engineers.

Saves a lot of time and money by reducing the cost of materials.

High precision upholstery, engine and wheels are designed using AutoCAD

The 3D design of models made using AutoCAD also helps in detecting and correcting errors easily.

Assignment

1. The computer-aided design (CAD) hardware doesn't include.

a) Graphic display terminals
b) Computer
c) Computer programmes
d) Keyboard

2. How many types of CAD are there?

a) 6
b) 4
c) 2
d) 5

3. Modem CAD systems are based on:

a) ICG
b) GCI
c) GIF
d) IFG

4. The computer communicates with the user via:
a) CPU
b) CRT
c) Graphics
d) Display button
5. The process of designing consists of __________ identifiable steps.
a) 8
b) 5
c) 4
d) 6
6. Implementing CAD improves communications.
a) True
b) False
7. The functionality areas of CAD application can be grouped into __________ categories.
a) 2
b) 3
c) 4
d) 5
8. The colour on CRT screen is obtained by the combination of:
a) Red, yellow, blue
b) Red, green, blue
c) Green, black, yellow
d) Red, black, yellow
9. Who formulated the six ground rules to be considered in designing graphics software?
a) Newman and Sproull
b) Schaffer G
c) Fitzgerald W
d) Krouse and Lerro
10. The input devices in CAD can be divided into:
a) 2
b) 5
c) 3
d) 4
11. The typical ICG system consists of software component.
a) True
b) False
12. Which is the latest version of AutoCAD software?
a) 2016
b) 2017
c) 2018
d) 2019
13. Which key is used to obtain properties palette in AutoCAD?
a) Control+1
b) Control+2
c) Control+3
d) Control+4
14. AutoCAD was first released in the year:
a) 1858
b) 1966
c) 1898
d) 1982
15. How many units are available in AutoCAD?
a) 4
b) 5
c) 7
d) 6

16. Which mode allows the user to draw 90° straight lines:
a) Osnap
b) Ortho
c) Linear
d) Polar tracking
17. To obtain parallel lines, concentric circles and parallel curves; _________ is used.
a) Array
b) Fillet
c) Copy
d) Offset
18. The default grid spacing in both X and Y directions is:
a) 10
b) 20
c) 5
d) 15
19. How many workspaces are available in AutoCAD?
a) 2
b) 4
c) 3
d) 5
20. Scale command can be accessed easily by typing:
a) SL
b) S
c) SC
d) C
21. Which command is used to divide the object into segments having predefined length?
a) Divide
b) Chamfer
c) Trim
d) Measure
22. How many grip points does a circle have?
a) 5
b) 4
c) 3
d) 2
23. In AutoCAD 2D Modelling, which axis is not accessible for drafting?
a) X
b) Y
c) Z
d) WCS
24. A Polyline can be broken into individual lines and arcs using which of the following command?
a) BREAK
b) TRIM
c) EXPLODE
d) OVERKILL
25. Origin of the AutoCAD drawing space is,
a) 0,0
b) 1,0
c) 0,1
d) 1,1
26. Scrolling of mouse can perform which following action?
a) Zoom in / zoom out
b) pan & scan
c) extents / all
d) scale
27. In AutoCAD, the number of workspaces is,

a) 1
b) 2
c) 3
d) 4
28. Is there any difference between Command Plot and Print?
a) plot command prints only big plans
b) The plot command for CNC (CAM)
c) No difference
d) print command can print up to A3 size paper
29. Find the false statement in regards with "If an array is made associative".
a) You can't use Boolean operations on the array elements
b) To make the array non associative you need to use explode command on it
c) The array can be edited again by selecting it
d) The array can't be modified in any way once it is made associative
30. An open area can be filled with which of the following gradient:
a) Gap
b) Tolerance
c) Transparency
d) Open
31. The shortcut to start a table is,
a) T
b) TA
c) TBL
d) TABLE
32. What is the shortcut for zoom command?
a) Z Command
b) Rotate mouse wheel
c) ZOOM Command
d) All of the above
33. To select a set of objects in the workspace, what should be done?
a) By a crossing window drawn from right to left
b) By a crossing window drawn left to right
c) Shift + clicking on the objects
d) None of the above
34. A line shows its dimension as 14.52. What is the precision of this line?
a) [0.0]
b) [0.00]
c) [0.000]
d) Display of length has nothing to do with precision
35. Model Tab and Layout Tab can be majorly distinguished as,
a) The Model tab is used for drawing in 3D and a Layout is used for drawing in 2D
b) The Model tab is where you create the drawing, and a layout tab represents the sheet that you will plot or print on
c) The colour of the background
d) The Model tab displays the drawing you are copying from and the Layout tab is where you lay out the new drawing
36. Shortcut key for Multileader in AutoCAD 2022,
a) ML
b) MLN
c) MLEADER
d) MULTI
37. To bring a particular area of the workspace in focus, what command should be used?
a) Zoom Extents
b) Zoom Window
c) Zoom Dynamic
d) Zoom All
38. What are the various options from left to right and the opposite direction?

a) Choose a different category of objects
b) Select objects according to their colour
c) Select objects according to their position
d) No difference
39. Shortcut for Trim command is,
a) T
b) TR
c) TI
d) X
40. A Boolean operation that is used to select interfering region between solids,
a) Union
b) Intersect
c) Subtract
d) None of the above
41. What command allows us to select objects based on some status?
a) Properties
b) Qselect
c) Pselect
d) Attributes
42. If you change the scale list a project that I have started from 1:50 1:10 then,
a) You will have to start over
b) You should not raise the objects already exist (scale) by 5
c) You will not need to change anything in hitherto methodology
d) Should be converted into new items that will add based on the new scale.
43. Print scale 1:100 means that
a). The draft is 100 times less expensive than the original
b). A 3 cm corresponds to a meter
c). A measure corresponds to 100 cm
d). None of the above
44. AutoCAD file is saved in which format?
a). DWG
b). DWT
c). DWF
d). DXF
45. The shortcut command for a rectangle
a). R
b). RE
c). REC
d). RET

Answers

1. Answer: c
2. Answer: b
3. Answer: a
4. Answer: b
5. Answer: d
6. Answer: a
7. Answer: b
8. Answer: b
9. Answer: a
10. Answer: d
11. Answer: a
12. Answer: d
13. Answer: b
14. Answer: d
15. Answer: d

16. Answer: b
17. Answer: d
18. Answer: a
19. Answer: c
20. Answer: a
21. Answer: a
22. Answer: c
23. Answer: c
24. Answer: c
25. Answer: a
26. Answer: b
27. Answer: b
28. Answer: b
29. Answer: d
30. Answer: a
31. Answer: b
32. Answer: b
33. Answer: a
34. Answer: c
35. Answer: b
36. Answer: c
37. Answer: b
38. Answer: c
39. Answer: b
40. Answer: b
41. Answer: b
42. Answer: d
43. Answer: b
44. Answer: a
45. Answer: a

CHAPTER 6: COMPUTER AIDED DRAFTING

6.1 INTRODUCTION TO BASIC ENTITIES

The story of CAD/CAM was accelerated in early 1950s. Up to year 2012ithas become one of the supreme technology available on Planet earth. CAD is used to create precise drawings and models of objects and structures, either in 2D or 3D. CAD is used to create precise drawings and models of objects and structures, either in 2D or 3D. CAD software takes into account how the various materials involved in a project interact, allowing designers to consider every element in a project, from plumbing to electricity. This results in fewer revisions and a more efficient workflow.

Today, CAD software incorporates cloud technology, providing entire teams with instant, remote access to projects. These advantages mean CAD has had a huge impact on architecture, engineering, and construction, among other industries.

There are many types of CAD software out there, and while they all operate differently, all are based on geometry. Every CAD program has X (horizontal), Y (vertical), and Z (depth) coordinates which allow users to create 2D or 3D models. CAD programs typically use either vector-based graphics or raster graphics to represent drawings and models. CAD software takes into account how the various materials involved in a project interact, allowing designers to consider every element in a project, from plumbing to electricity. This results in fewer revisions and a more efficient workflow.

Today, CAD software incorporates cloud technology, providing entire teams with instant, remote access to projects. These advantages mean CAD has had a huge impact on architecture, engineering, and construction, among other industries.

There are many types of CAD software out there, and while they all operate differently, all are based on geometry. Every CAD program has X (horizontal), Y (vertical), and Z (depth) coordinates which allow users to create 2D or 3D models. CAD programs typically use either vector-based graphics or raster graphics to represent drawings and models. CAD software takes into account how the various materials involved in a project interact, allowing designers to consider every element in a project, from plumbing to electricity. This results in fewer revisions and a more efficient workflow.

It is being used in almost all the fields of engineering but primarily in mechanical engineering branches. The development in the field is still gaining speed.

- CAD Technology = Design Techniques + Computers.
- The CAD Process is the subset of the Design process.
- The CAM Process is a subset of Manufacturing Process.
- Integration of CAD and CAM leads to automation.

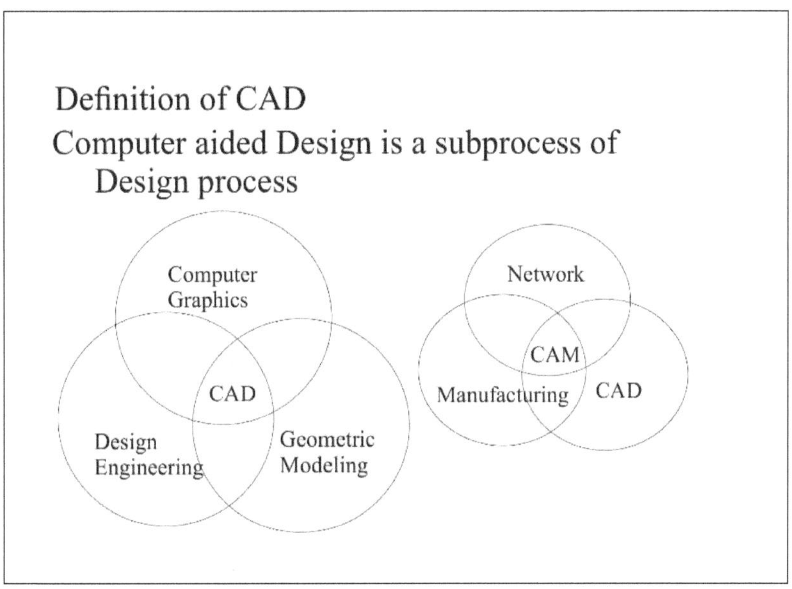

Figure 6.1

In general, a Computer Aided Design (CAD) package has three components:
a) Design
b) Analysis and
c) Visualization
A brief description of these components follows:

a) Design: Design refers to geometric modelling, i.e., 2-D and 3-D modelling, including, drafting, part creation, creation of drawings with various views of the part, assemblies of the parts, etc.

b) Analysis: Analysis refers to finite element analysis, optimization, and other number crunching engineering analyses. In general, a geometric model is first created and then the model is analysed for loads, stresses, moment of inertia, and volume, etc.

c) Visualization: Visualization refers to computer graphics, which includes: rendering a model, creation of pie charts, contour plots, shading a model, sizing, animation, etc.

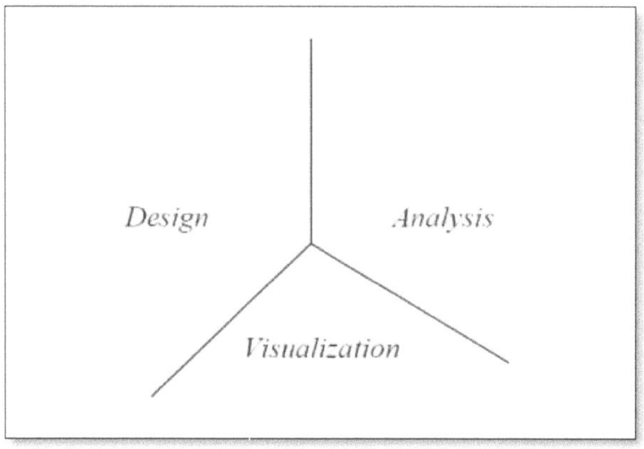

Figure 6.2

Today, CAD software incorporates cloud technology, providing entire teams with instant, remote access to projects. These advantages mean CAD has had a huge impact on architecture, engineering, and construction, among other industries.

There are many types of CAD software out there, and while they all operate differently, all are based on geometry. Every CAD program has X (horizontal), Y (vertical), and Z (depth) coordinates which allow users to create 2D or 3D models. CAD programs typically use either vector-based graphics or raster graphics to represent drawings and models.

2D & 3D CAD

CAD can be used to create both 2D drawings and 3D models. Some CAD software specialize in one or the other, while other programs are capable of producing high quality 2D and 3D designs.

2D CAD models are flat, two-dimensional drawings and provide the information, such as dimensions and layouts, needed to reproduce or build the subject. These 2D models are commonly used in industries such as architecture, civil engineering, automotive, interior design, landscaping, cartography and fashion.

For example, CAD software is used by architects to create floor plans of buildings and houses, and these floor plans contain most of the information that would be needed to construct the inside of the building. As well as floor plans, CAD can be used to create technical drawings and blueprints, piping and instrumentation diagrams, HVAC diagrams, site and plot plans, electrical schematics, and wiring diagrams.

2D & 3D CAD.

CAD can be used to create both 2D drawings and 3D models. Some CAD software specialize in one or the other, while other programs are capable of producing high quality 2D and 3D designs.

2D CAD models are flat, two-dimensional drawings and provide the information, such as dimensions and layouts, needed to reproduce or build the subject. These 2D models are commonly used in industries such as architecture, civil engineering, automotive, interior design, landscaping, cartography and fashion.

For example, CAD software is used by architects to create floor plans of buildings and houses, and these floor plans contain most of the information that would be needed to construct the inside of the building. As well as floor plans,

CAD can be used to create technical drawings and blueprints, piping and instrumentation diagrams, HVAC diagrams, site and plot plans, electrical schematics, and wiring diagrams.

CHARACTERISTICS OF CAD3

Efficiency: An Efficient software is that which can use less resources to give a better output.
Simplicity: A software must be easy to understand, simple to use and must be user friendly.
Flexibility: The software must be able to integrate the design modification without much of difficulty.
Readability: This provides the capability within the software to help the user as and when required.
Portability: The software must have the capacity to get transferred from one system to other.
Recover ability: A Good software must be able to give warnings before getting crashed and must be able to recover.

BASIC CONCEPT OF CAD

ATTRIBUTES

Every entity has certain attributes such as its color, line type and line width.

LAYERS

A basic concept of computer aided drafting is the use of layers to organize a drawing. Every entity of a drawing is on exactly one layer and a layer can contain lots of entities. Typically, entities with a common 'function' or common attributes are put on the same layer. E.g., you might want to put all axes in a drawing on a layer named 'axes' (see Figure 1). Layers can have attributes (color, line width, line style). Each entity can have its own attributes or have its attributes defined by the layer it is placed on. In the latter case you can change for example the color of all 'axes' entities by changing the color of the layer 'axes'. In manual drafting, a similar approach was used. Different building systems, such as wiring and air conditioning were often drawn on separate transparent sheet of paper. These sheets were then overlaid on one another to produce final drawings.

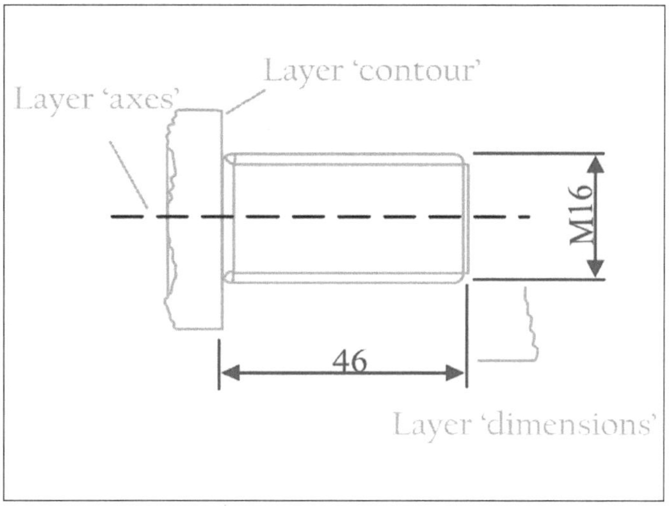

Figure 6.3

BLOCKS

A block is a named group of entities. Blocks can be inserted into the same drawing more than once at different locations, with different attributes, scaling factors and rotation angles (see Figure 6.4). Such instances of a block are usually called 'inserts'. Inserts have attributes just like other entities. An entity that is part of an insert can have its own attributes, inherit its attributes from the layer it is placed on, or inherit them from the insert it is part of. Once created, inserts are still dependent on the block they instantiate. The power of inserts is, that you can modify a block and all inserts that were previously made from that block will be updated to reflect the changes.

Figure 6.4

DRAFTING IN CAD

In many ways, CAD is similar to traditional drafting. When drawing a plan or a view of an object on a paper, you would use tools such as a ruler to draw lines. CAD systems offer many tools to achieve the same goal. The big advantage of a CAD system is the fact that you can change every entity of your drawing easily after you have created it. This might be one of the more difficult things to learn when moving from paper to CAD. When working with a CAD system you will very often create lines that will not be on the final printout or which don't have the correct length and need to be trimmed later (see Figure 6.5). A common mistake of CAD beginners is wanting to create the final drawing right away. Never hesitate to create an auxiliary construction if it helps you to define or verify a part of your drawing.

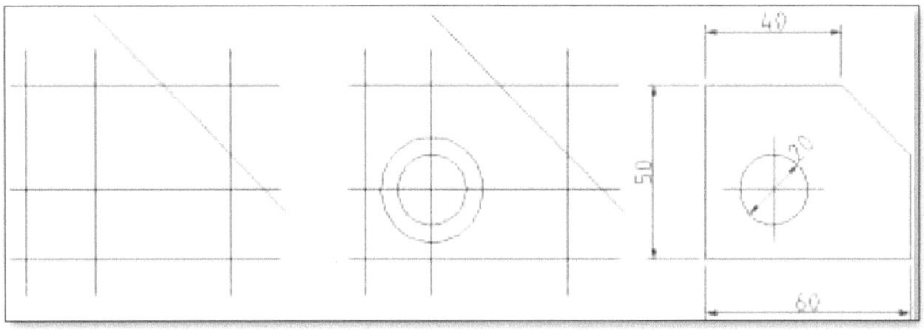

Figure 6.5

COORDINATE SYSTEMS

A good understanding of how the most common coordinate systems work is absolutely essential if you are to make the best use of any CAD program. If you are not familiar with coordinates it is highly recommended that you take some time to familiarize yourself with this fundamental concept.

ORIGIN

The origin (or origin point) of the drawing is the point where the X and Y axis cross each other. It is also the absolute zero point of the drawing.

RELATIVE ZERO POINT

In addition to the absolute zero point, there is also a relative zero point in QCad. The relative zero point can be moved to any location by the user as a temporary reference for a local construction.

CARTESIAN COORDINATES

The Cartesian coordinate system is the standard coordinate system that is usually used if no other system is specified. In the Cartesian coordinate system, the position of a point can be described by its distance from two axes, X and Y. Cartesian coordinates are usually written in the format:

x-ordinate, y-ordinate

For example, the coordinate 3.5,7 is located 3.5 units to the right and 7 unit to the top of the origin.

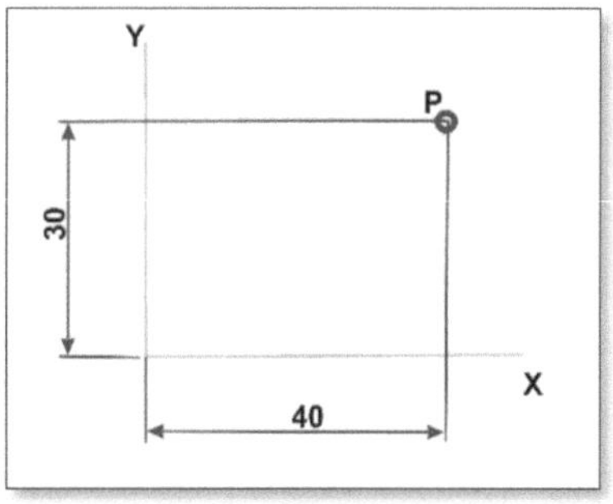

Figure 6.6

Cartesian coordinates can also originate in a position different from the origin. In that case we talk about 'relative coordinates'. There is no standard notation for relative Cartesian coordinates, but in QCad they are written in the format:
@x-ordinate,y-ordinate.

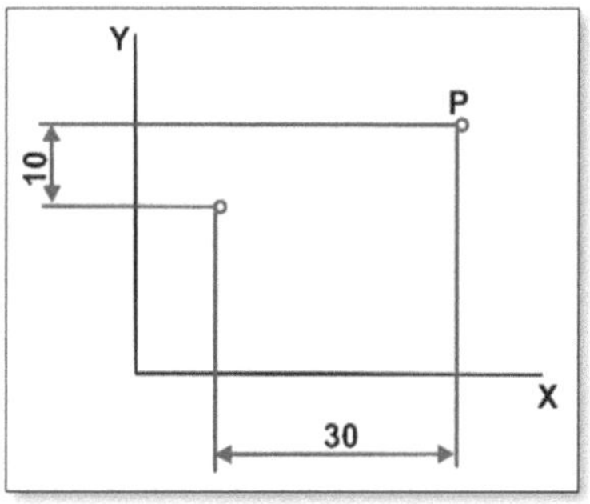

Figure 6.7

POLAR COORDINATES
Polar coordinates use a distance and an angle to describe the position of a point. The angle 0 is always pointing to the right on your screen (east). In QCad, polar coordinates are noted in the format:
distance<angle

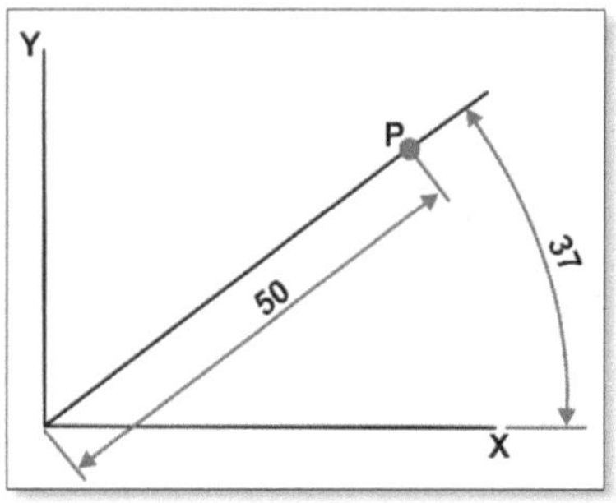

Figure 6.8

OBJECT AND GRID SNAPPING
Whenever you need to specify a coordinate in QCad, you can use the snap functions. They allow you to precisely select grid points or significant points on existing objects such as endpoints, midpoints, centre or intersections between entities.

BASIC EDITING
ENTITY CONSTRUCTION
Entities can be added to a drawing with various drawing tools or by duplicating existing entities. To draw an entity means to define all points and parameters that define the entity, such as the endpoints of a line.

ENTITY SELECTION
Before an entity can be deleted, duplicated, or transformed, it must be selected. Entity selection is one of the most basic of CAD operations. However, selecting the right entities for an operation is not always trivial. QCad offers a wide variety of selection tools to quickly select groups of entities, entities within a range, connected entities, etc.

DELETION
Deleting an entity means to remove it from the drawing. In QCad, all features that have to do with deleting entities are grouped with the modification features.

MODIFICATIONS
Existing entities can be modified in many ways. Basic modifications include translation, rotation, reflection, and scaling. These operations do not alter the characteristic geometry of the affected entities. Other modifications change an entity more fundamentally (for example trimming, extending or stretching).

ENTITIES
Entities are graphical objects in a CAD system. Typical entities which are supported by most CAD systems are: points, lines and circular and elliptical arcs. More complex and CAD specific entities include polylines, texts, dimensions, hatches and splines.
A circle is defined as a closed curve in which all points are the same distance from its centre point. The centre point is a single XY coordinate.
A circle is 360 degrees and can be divided into four quadrants. All points on a circle are at a given distance from is centre point. The distance between any of the points and the centre is called the radius.
Study the drawings in Figure 6.8, 6.9,6.10 for a description of the geometry of a circle.

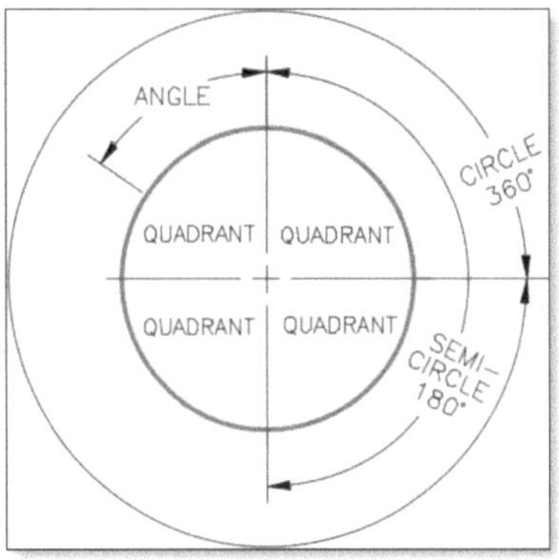

Figure 6.8

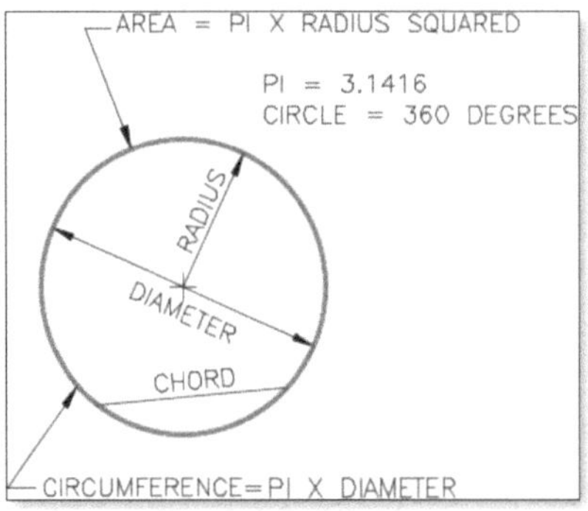

Figure 6.9

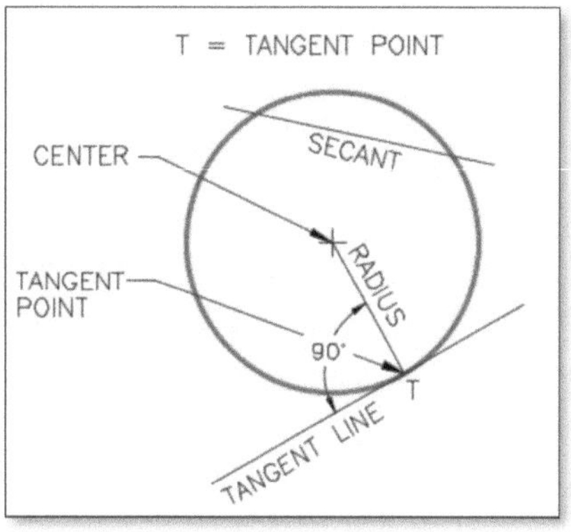

Figure 6.10

An arc is defined as an open curve in which all points are the same distance from its centre point. Study the drawings in Figure 6.11 and 6.12 for a description of the geometry of an arc.

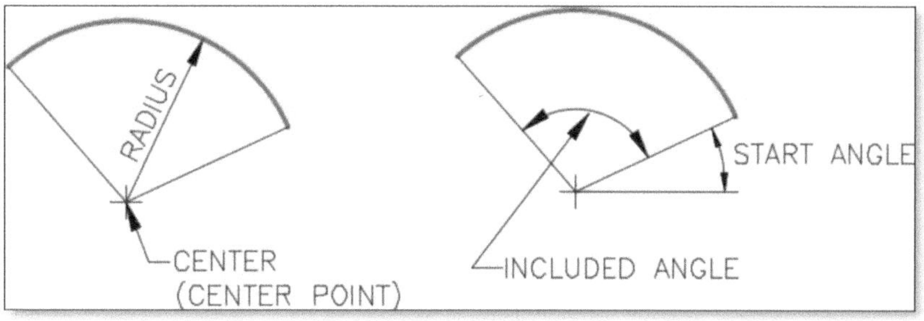

Figure 6.11

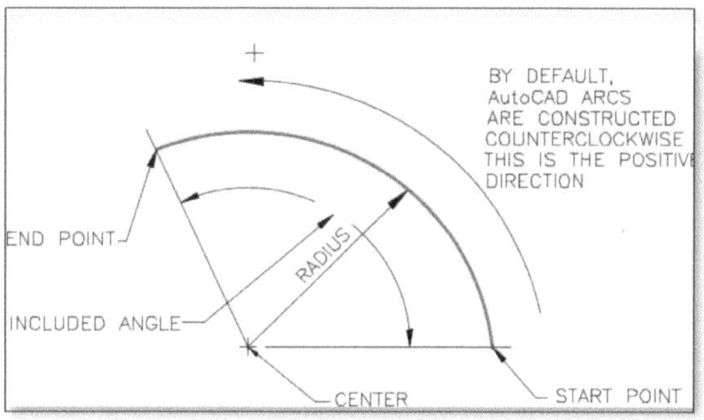

Figure 6.12

AutoCAD Command: CIRCLE
The CIRCLE command is used to draw circles.
Shortcut: **C**

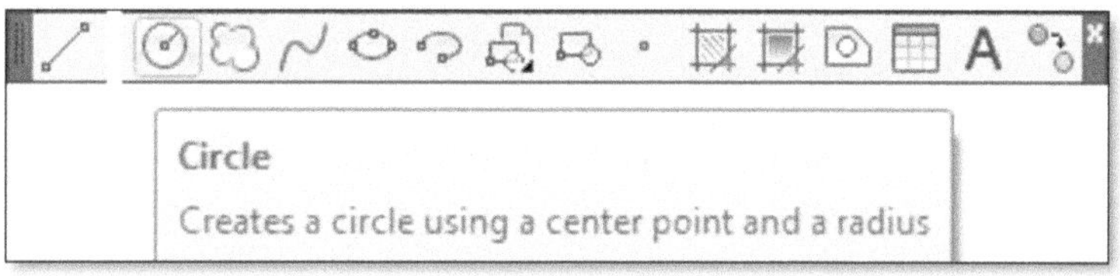

Figure 6.13

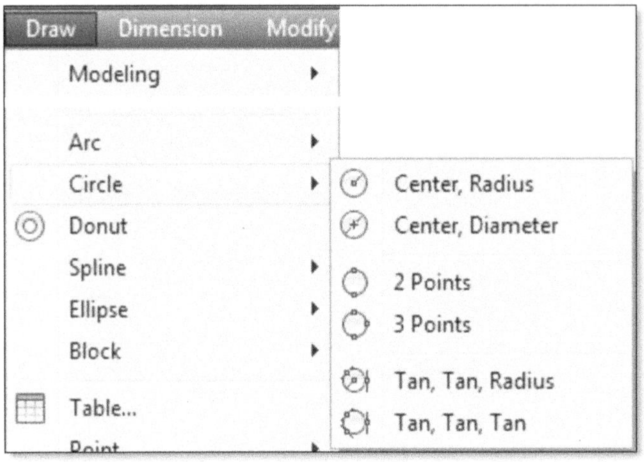

Figure 6.14

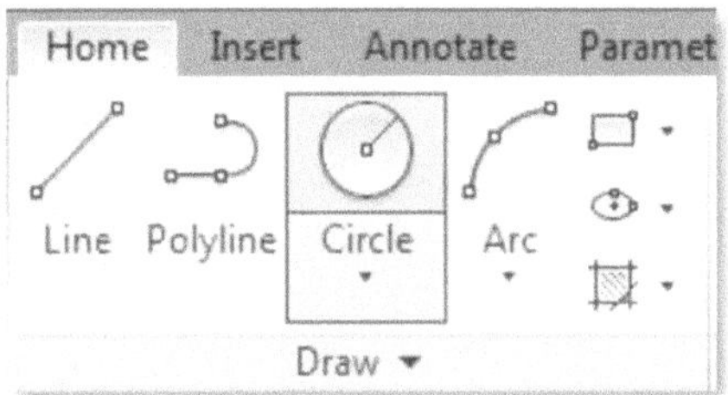

Figure 6.15

AutoCAD Command: ARC
The ARC command is used to draw arcs.
Shortcut: **A**

Figure 6.16

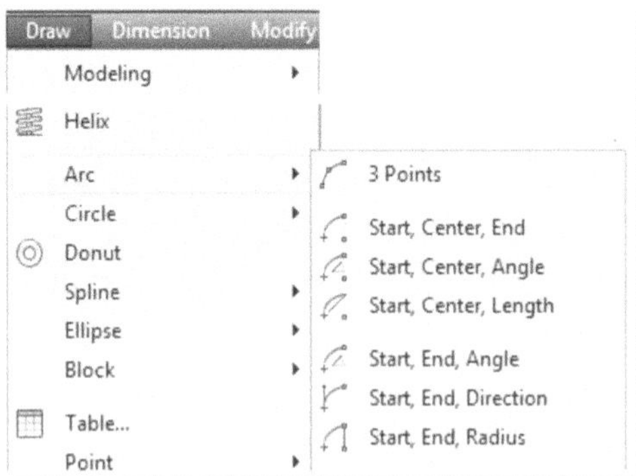

Figure 6.17

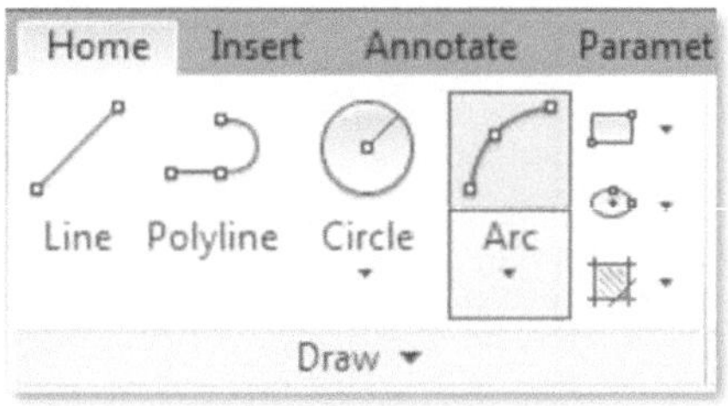

Figure 6.18

Draw the arc and circles using CAD.
STEP 1
Start a new drawing using the template
STEP 2
Save and name the drawing
STEP 3
Create layer: Object, colour red. Set it as the current layer.

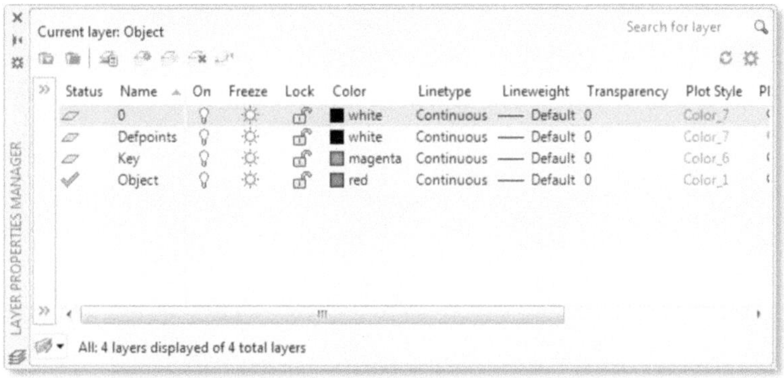

Figure 6.19

STEP 4
Using the figures and what you already leaned, draw the two lines shown in Figure 6.20. (Figure Step 6.20A and 6.20B)

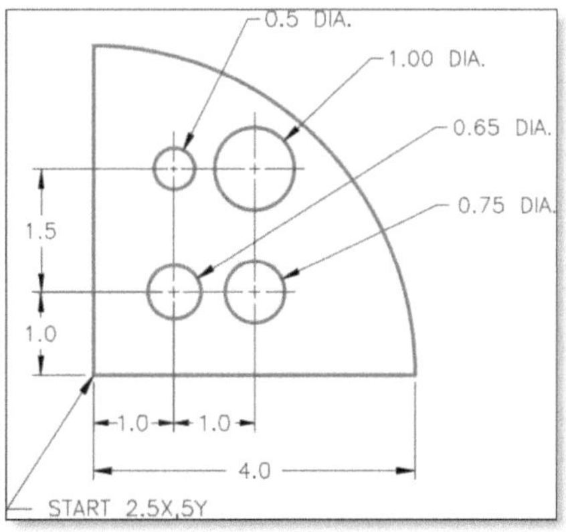

Figure 6.20A

Figure 6.20B

STEP 5

Enter the ARC command, as shown below, to draw the arc. (Figure Step 5)

Command: **ARC**

Specify start point of arc or [Centre]: **C**

(If possible, always draw arcs starting with the centre point first.)

Specify centre point of arc: **2.5, 5**

*(*Enter the centre point*.)*

Specify start point of arc: **@4,0**

*(*Enter the start point. Here the trick is to give AutoCad the start point and the radius of the arc at the same time)

Specify end point of arc or [Angle/chord Length]: **A**

Specify included angle: **90**

(Enter the angle of the arc. In this case, it is 90 degrees.)

Command:

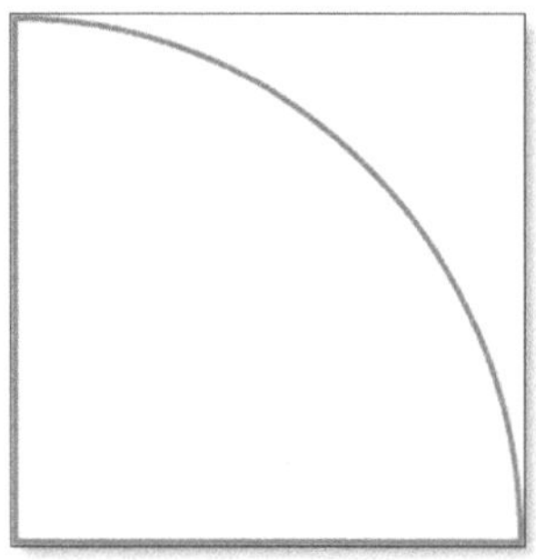

Figure 6.21

STEP 6

Enter the CIRCLE command, as shown below, to draw the four circles to complete the drawing. (Figure Step 6A and 6B)

Command: **CIRCLE**

Specify centre point for circle or [3P/2P/Ttr (tan tan radius)]: **3.5,6**

(Since this is the first circle, enter an absolute coordinate for the centre point.)

Specify radius of circle or [Diameter]: **D**

(Circles are usually dimensioned with diameters rather then a radius, enter D)

Specify diameter of circle: **0.65**

Command: **CIRCLE**

(Press Enter or Space bar as that will repeat the last command.)

Specify centre point for circle or [3P/2P/Ttr (tan tan radius)]: **@1,0**

(Notice here, since AutoCAD always remembers the last point, use an @. The last point was the centre of the first circle placed.)

Specify radius of circle or [Diameter] <0.3250>: **D**

Specify diameter of circle <0.6500>: **0.75**

Command: **CIRCLE** (Press Enter or Space bar as that will repeat the last command.

Specify centre point for circle or [3P/2P/Ttr (tan tan radius)]: **@0,1.5**

(Use an @ from the centre of the last circle drawn.)

Specify radius of circle or [Diameter] <0.3750>: **D**

Specify diameter of circle <0.7500>: **1**
Command: **CIRCLE**

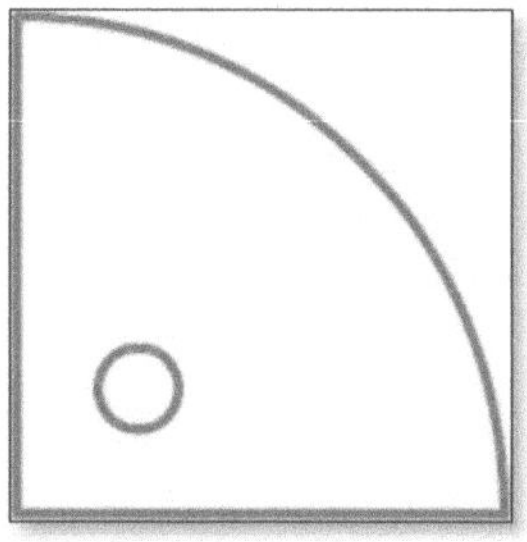

Figure 6.22

(Press Enter or Space bar. That will repeat the last command and save you entering it again.)
Specify centre point for circle or [3P/2P/Ttr (tan tan radius)]: **@-1,0**
Specify radius of circle or [Diameter] <0.5000>: **D**
Specify diameter of circle <1.0000>: **0.5**
Command:

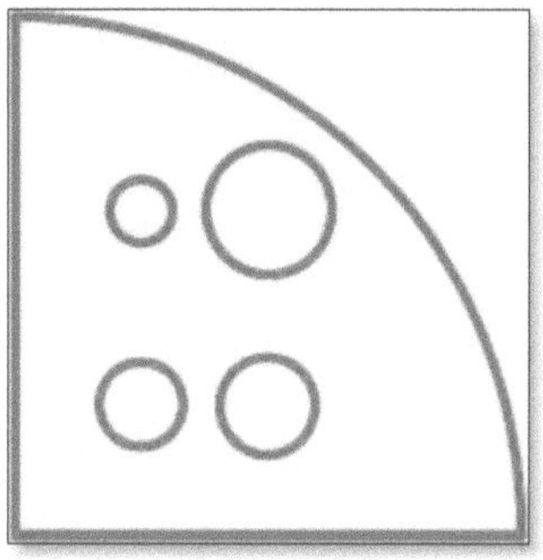

Figure 6.23

STEP 7
Save and close the drawing.

6.3 BASIC COMMANDS
MEASURING COMMANDS
GRID: Displays a grid of dots at a desired spacing on the screen.
Command: GRID (enter)
On/Off/Tick spacing(x)/Aspect: (enter value) (enter)

SNAP: Specifies a "round off" interval so that points entered with the mouse can be locked into alignment with the grid spacing.
Command: SNAP (enter)
On/Off/Value/Aspect/Rotate/Style: (enter value) (enter)

BASIC DRAW COMMANDS
CIRCLE: Draws circles of any size.
Command: Circle (enter)
3P/2P/TTR/<center point>: (pick a center point)
Diameter or: (Pick a point on the circle)

LINE: Draws straight lines between two points
Command: LINE (enter)
From Point: (pick a point using the mouse)
To Point: (Pick a point using the mouse)
To Point: (Press return to end the command)

ARC: Draws an arc (any part of a circle or curve) through three known points.
Command: ARC (enter)
Center/ < Start point >: (pick the first point on the arc)
Center/End/ < Second point >: C
Center: (pick the arc's center point)
Angle/Length of chord/: (pick the arc endpoint)

DISPLAY COMMANDS

LIMITS: Sets the size of the drawing paper. For size "A" drawing paper the limits should be set for 10.5 x 8.
Command: LIMITS (enter)
On/Off/Lower left corner (enter)
Upper right corner: 10.5,8 (enter)

ZOOM: Enlarges or reduces the display of a drawing.
Command: ZOOM (enter)
All/Center/Dynamic/Extents/Left/Previous/Vmax/Window/<Scale(x/XP)>: (pick a point to define one corner of a rectangular viewing window then pick a point to define the second point to define the opposite diagonal corner of the viewing window)

Note: To return the picture to its original viewing size enter ALL and press the enter key when prompted instead of defining a window.

PAN: Allows you to move your view point around the drawing without changing the magnification factor.
Command: PAN (enter)

EDITING COMMANDS

CHANGE: Alters properties of selected objects
Command: CHANGE (enter)
Select objects or window or Last (select objects to be changed)
Properties/<Change Point>: (type P) Change what property (Color/Elev/Layer/LType/Thickness)? (type Layer)
New Layer: (enter new layer name and press enter)

ERASE: Erases entities from the drawing.
Command: ERASE (enter)
Select objects or Window or Last: (Select objects to be erased and press enter when finished)

EXTEND: Lengthens a line to end precisely at a boundary edge.
Command: Extend (enter)
Select boundary edge(s)...
Select Objects (pick the line which represents the boundary edge which lines will be extended to)
(press enter when finished selecting cutting edges)
<Select object to extend>/Undo: (pick the line(s) that need to be extended)

TRIM: Trims a line to end precisely at a cutting edge.
Command: Trim (enter)
Select cutting edge(s)...
Select Objects (pick the line which represents the cutting edge of line in which objects will be trimmed to)
(press enter when finished selecting cutting edges)
<Select object to trim>/Undo: (pick the line(s) that need to be trimmed)

GRIPS You can edit selected objects by manipulating grips that appear at defining points on the object. Grips is not a command. To activate grips simply pick the object. Small squares will appear at various entity-specific positions. By selecting an end grip you can stretch the entity to change its size. By selecting the center grip you can move the entity to a new location. To remove grips press CTL-C twice. You can perform the following using grips: Copy, Multiple Copy, Stretch, Move, Rotate, Scale, and Mirror.

CREATING LAYERS

LAYER: Creates named drawing layers and assigns color and line type properties to those layers.

Command: LAYER (enter)

A Layer & Line type Properties dialog box will be displayed. To add a new layer, pick the New button. A new layer listing appears, using a default name of Layer1. the layer name can be changed by highlighting the layer name. Colors and Line type can be assigned to each new layer by picking the color box to assign a color and picking the line type box to assign a line type.

Standard AutoCAD Colours

1 = Red	2 = Yellow	3 = Green	4 = Cyan
5 = Blue	6 = Magenta	7 = White Standard	

Standard AutoCAD line types

Hidden2 = hidden lines
Center2 = center lines
Phantom2 = phantom or cutting-plane lines

CONSTRUCTION COMMANDS

ARRAY: Makes multiple copies of selected objects in a rectangular or circular pattern

Command: ARRAY (enter)
Select objects or Window or Last: (select object to array)
Rectangular or Polar array (R/P) <current>: (P)
Center point of array: (pick the point around which to form the array)
Angle to fill (+=CCW, -=Cw)<360>: (enter)

COPY: Draws a copy of selected objects.

Command: COPY (enter)
Select objects or Window or Last: (select objects to be copied)
Base point or displacement: (pick a point on the object to be use as a reference point)
Second point of displacement: (pick a point which represents the new location of the copied object)

MIRROR: Makes mirror images of existing objects.

Command: MIRROR (enter)
Select objects or Window or Last: (select objects to be mirrored)
First point of mirror line: (pick a point on top of the mirror line)
Second point: (pick a point on the bottom of the mirror line)
Delete old objects? <N> y or n (enter)

MOVE: Moves designated entities to another location.

Command: MOVE (enter)
Select objects or Window or Last: (select objects to move)
Base point or displacement: (pick a point on the object to be use as a reference point)
Second point of displacement: (pick a point which represents the new location of the object)

OFFSET: Constructs an entity parallel to another entity at a specified distance. Offset can be used with lines, circles, arcs, and polylines.

Command: OFFSET (enter)
Offset distance or Through: (enter a distance value)
Select object to offset: (select object to offset)
Side to offset: (Pick any point on the side of the object you wish to offset)

FILLET: Changes any corner to a rounded corner.

Command: FILLET
Polyline/Radius/Angle/Trim/Method/ <Select first line>: (pick the first line)
Select second line: (pick the second line)

CHAMFER: Changes any corner to an angled corner.

Command: CHAMFER

Polyline/Distance/Angle/Trim/Method/ < Select first line >: (pick the first line)
Select second line: (pick the second line)

OSNAP: Instantly locates exact points relative to existing objects (points). Object Snap Modes: Endpoint, Midpoint, Center, Quadrant, Intersection, Insertion, Perpendicular, Tangent, Nearest, Node, and None.

PLACING LETTERING ON A DRAWING

TEXT: Draws text characters of any size.

Command: TEXT (enter)
Justify/Style/: (pick a starting point or enter a justification letter)
Height (0) (enter the height of the lettering)
Rotation Angle (0) (enter)
Text: (enter the desired lettering) (enter)

CROSSHATCHING A DRAWING

BHATCH: Allows the user to crosshatch areas of a section view.

Command: BHATCH (enter)
The Boundary Hatch dialogue Box will be displayed. Select the Hatch Options box.
The Hatch Options box will be displayed. Select the Patterns box.
The Choose Hatch Pattern box will be displayed. Select the desired hatch pattern.
The Hatch Options box will be displayed again. You can select a scale and rotation angle for the crosshatch pattern.
Select the OK box when finished.

The Boundary Hatch dialogue box will be displayed again. Select the Pick Points box. When prompted select the internal point of the are to be crosshatched. Press the enter key when finished.

The Boundary Hatch dialog box will be displayed again. Select the Apply box to add the crosshatching to the drawing.

Summary of Options

<Start Point>	Left-Justifies text along its baseline
Justify	Justifies text according to the alignment options
Style	Enters a new text style
Null reply	Enters a new line of text below the previous text.

6.4 DIMENSIONING

A technical drawing is incomplete without annotation. Notes on a drawing are important to convey design information or clarify intent which can't be portrayed by linework alone. Measuring distances from a scaled, printed drawing is inaccurate. Dimensions allow exact values to be shown on a drawing which can then be used for construction.
You can access the dimensioning tools from the Ribbon **Annotate** tab > **Dimension** panel. The required dimension tool can be chosen from the Dimension drop-down.

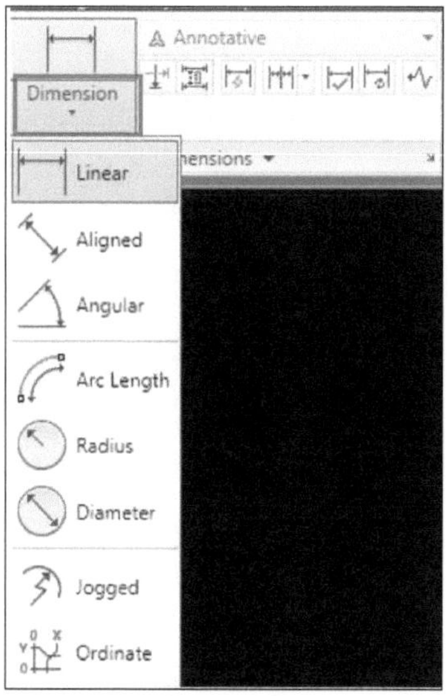

Figure 6.25

DIMENSION TYPES IN AUTOCAD

CONCEPT:

This tutorial will explain the different types of dimensions which are commonly used in AutoCAD and help you understand their differences. A good AutoCAD drawing will convey the needed information clearly and accurately. Dimensioning is used to let the drawing reader know what size each length or curve is, without having to open the drawing and measure it. If your dimensions are inaccurate, you will not be working in CAD for very long, no matter how accurate your drawing is. Learn when each dimension is needed and how to use it.

Dim Type	Keystroke	Icon	Description
Linear	DIMLINEAR / DIMLIN		Dimensions for straight lines either horizontal or vertical to the UCS.
Aligned	DIMALIGNED / DIMALI		Dimensions for straight lines that are angled to the UCS and provides the line's length.
Angular	DIMANGLUAR / DIMANG		Dimensions between 2 points and gives the resulting angle measurement.
Arc Length	DIMARC		Dimensions the length of an arc or arc segment of a polyline.
Radius	DIMRADIUS / DIMRAD		Dimensions the radius of an arc, circle or arc segment of a polyline.
Diameter	DIMDIAMETER / DIMDIA		Dimensions the diameter of an arc, circle or arc segment of a polyline.
Center Mark	DIMCENTER		Places a center mark at the center point of an arc or circle
Rotated	DIMROTATED	-	A modifier to a dimension to make it parallel to a selected line.
Spacing	DIMSPACE		Aligns the selected dimensions to an even spacing.
Continue	DIMCONTINUE / DIMCONT		Continues dimensioning from a base dimension
Baseline	DIMBASELINE / DIMBASE		Continues dimension from a base dimension to give an overall dim.

Figure 6.26

LINEAR

The Linear dimension tool measures horizontal or vertical distances. To create Linear dimensions click the points 1, then 2 to identify the element to be dimensioned. Click point 3 to locate the dimension line. The dimension distance will show the vertical or horizontal distance depending on whether you move your cursor in a horizontal or vertical direction when placing point 3.

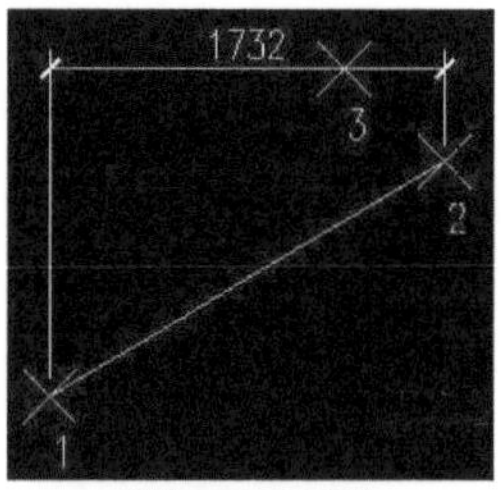

Figure 6.27

COMMAND LINE: To start the Linear Dimension tool from the command line, type "DIMLIN" and press [Enter]

Note that there are some options at the command line:

OPTION	DESCRIPTION
M-Text/Text	You don't need to use these, it doesn't really affect your dimension
Angle	This will place your text at the angle you specify
Horizontal	This will create a horizontal dim, but you select this when you place your dimension line. Rarely needed.
Vertical	This will create a vertical dim, but you select this when you place your dimension line. Rarely needed.
Rotated	This will create a 'rotated' dimension.

Figure 6.28

ANGULAR

The Angular dimension tool measures an angle between two objects.

To create an Angular dimension select object 1, and then object 2 to identify the elements creating the angle. Click point 3 to locate the dimension line.

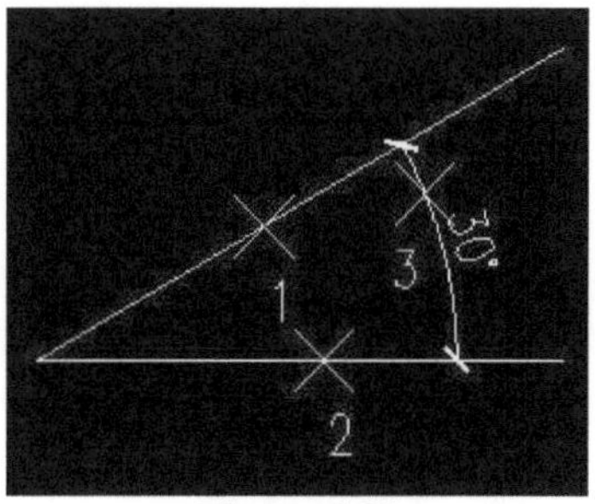

Figure 6.29

COMMAND LINE: To start the Angular Dimension tool from the command line, type "DIMANG" and press [Enter].

ALIGNED

The Aligned dimension tool measures the length along an object.

To create an Aligned dimension click the points 1, then 2 to identify the element to be dimensioned. Click point 3 to locate the dimension line.

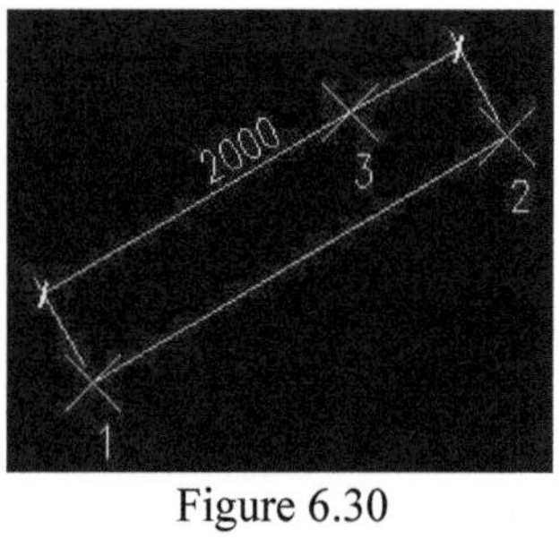

Figure 6.30

COMMAND LINE: To start the Aligned Dimension tool from the command line, type "DIMALI" and press [Enter].

ARC LENGTH
The Arc Length dimension tool measures the length of an arc.
To create an Arc Length dimension, select object 1 then click point 2 to place the dimension line.

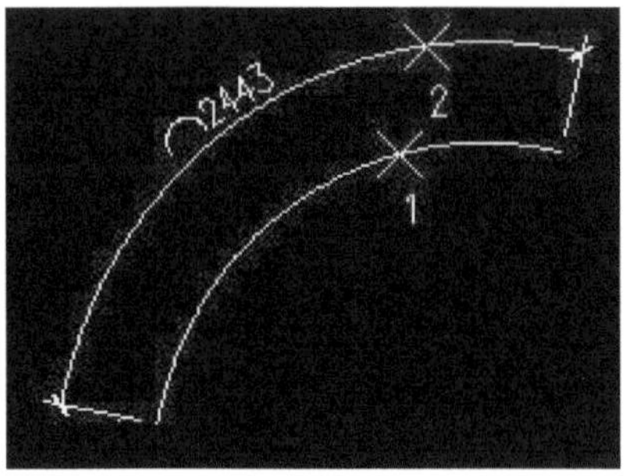

Figure 6.31

COMMAND LINE: To start the Arc Length Dimension tool from the command line, type "DIMARC" and press [Enter].

DIAMETER
The Diameter dimension tool measures the diameter of an arc or circle.
To create a Diameter dimension, click object 1 then, click point 2 to locate the dimension line.

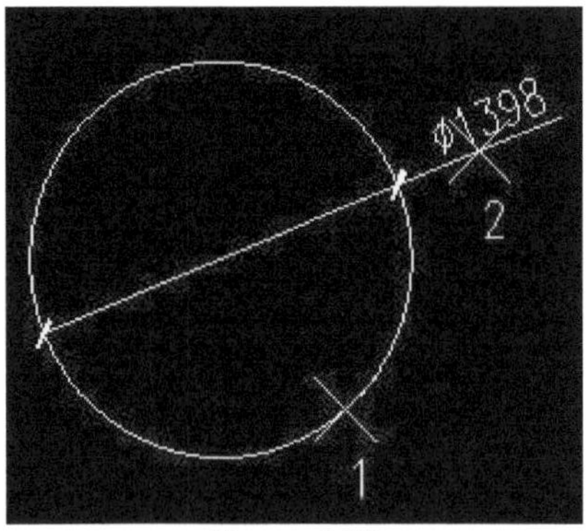

Figure 6.32

COMMAND LINE: To start the Diameter Dimension tool from the command line, type "DIMDIA" and press [Enter].

JOGGED
The Jogged dimension tool measures the radius of an arc or circle but places the leader line in an alternative place (inferred point) to the arc or circles centre point. The jog indicates the dimension line doesn't point to the centre of the circle. This tool can be used when space is limited or the arc radius is large.
To create a Jogged dimension, click on the arc you want to dimension (point 1). Click point 2 to identify the inferred centre point. Click point 3 to locate the dimension line and text. Click point 4 to locate the jog within the dimension line.

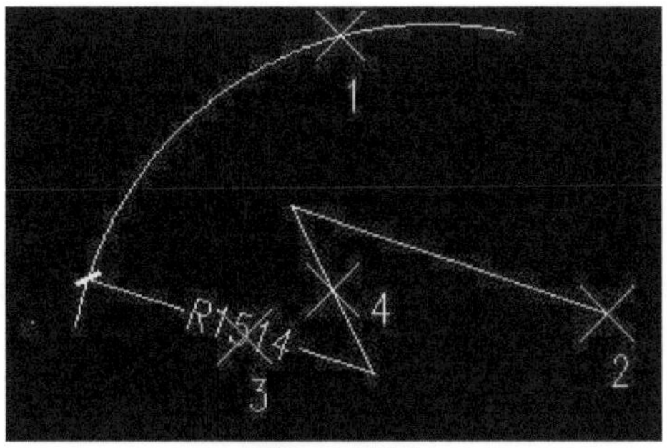

Figure 6.33

COMMAND LINE: To start the Jogged Dimension tool from the command line, type "JOG" and press [Enter].

RADIUS

The Radius dimension tool measures the radius of an arc or circle.
To create a Radius dimension, select object 1 then, click point 2 to locate the dimension line.

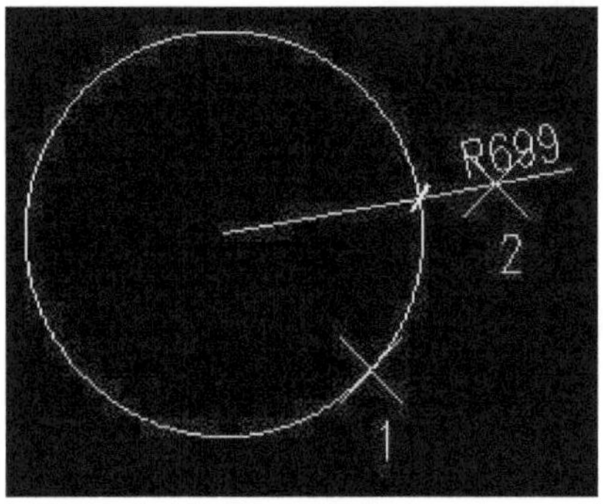

Figure 6.34

COMMAND LINE: To start the Radius Dimension tool from the command line, type "DIMRAD" and press [Enter].

ORDINATE

Ordinate dimensions are used to locate the X or Y coordinate value of a point. Click point 1 to identify the point location, click point 2 to locate the dimension line. Ordinate dimensions are especially useful for elevations or OS grids which use specific coordinates.
Note: The ordinate dimension will display the X value if your second point is above or below the first point. The Y value will be displayed if your second point is to the side of the first point.

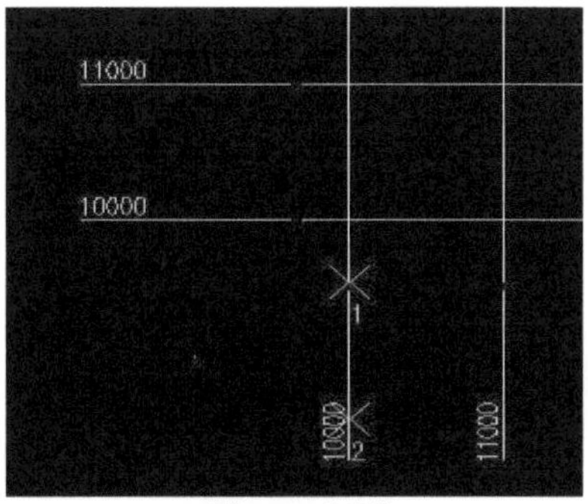

Figure 6.35

COMMAND LINE: To start the Ordinate Dimension tool from the command line, type "DIMORD" and press [Enter].

QUICK DIMENSION

The Quick Dimension tool is used for dimensioning multiple elements.

Quick Dimensions can be added by clicking the Quick Dimension tool from the Ribbon **Annotate** tab > **Dimensions** panel.

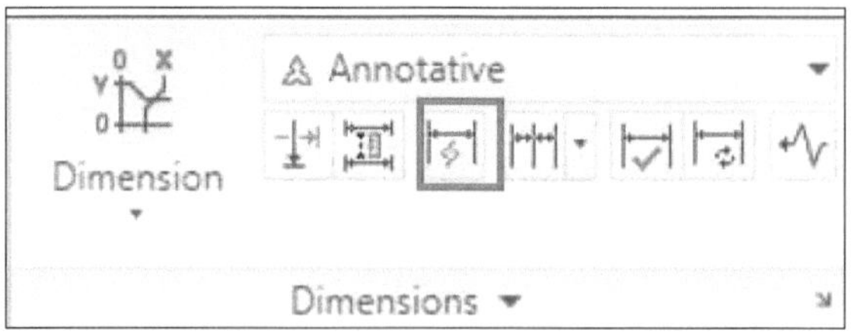

Figure 6.36

COMMAND LINE: To start the Quick Dimension tool from the command line, type "QDIM" and press [Enter].

Select all the objects you wish to dimension (objects 1). You can either select these 1 at a time or use a crossing window selection to select multiple objects.

Click point 2 to locate the dimension line.

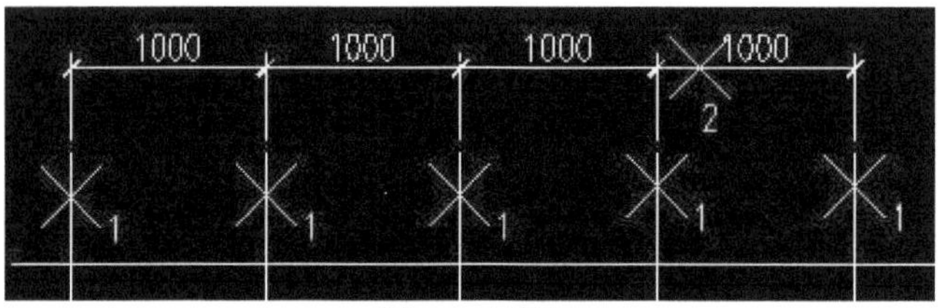

Figure 6.37

CONTINUE

The Continue dimension tool is used to add additional dimensions to a linear, aligned Angular or Arc length dimension.

Continue only works on an existing Linear or Aligned dimension. It takes the settings and height from the existing dimension chosen.

Note: The direction is based on the extension line clicked. If you select the right-side extension line it will start to measure from the right side, if you select the left side extension line, it will measure from the left side.

You can create a Continue dimension by clicking the Continue tool from the Ribbon **Annotate** tab > **Dimensions** panel.

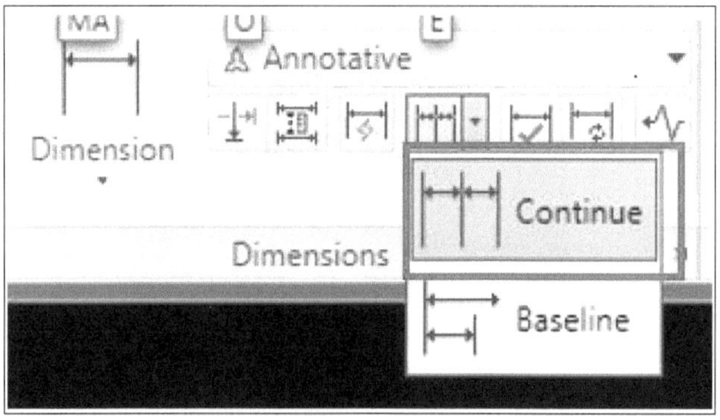

Figure 6.38

COMMAND LINE: To start the Continue Dimension tool from the command line, type "DIMCONT" and press [Enter].

A Continue dimension is created by selecting the already created Linear or Aligned dimension (1) then by clicking the subsequent points (2, 3 and 4).

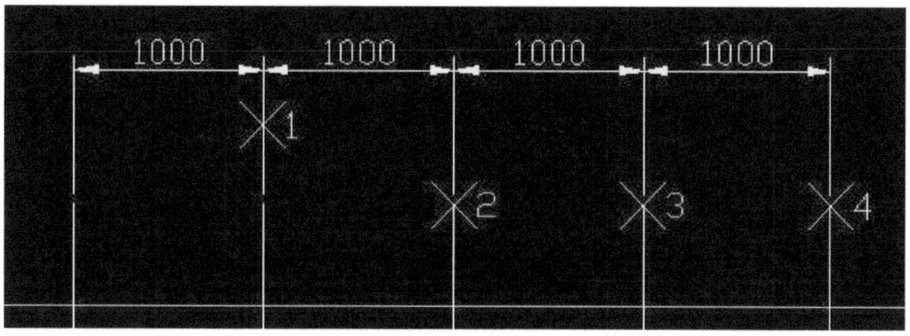

Figure 6.39

BASELINE
Baseline dimensions will measure the distance from a starting point to multiple other points.
Baseline only works on an existing Linear or Aligned dimension. It takes the settings and height from the existing dimension chosen.

Note: The direction is based on the extension line you click. If you select the right extension line it will measure from the right side, if you select the left extension line, it will measure from the left side.
You can create a Baseline dimension by clicking the Baseline tool from the Ribbon **Annotate** tab > **Dimensions** panel.

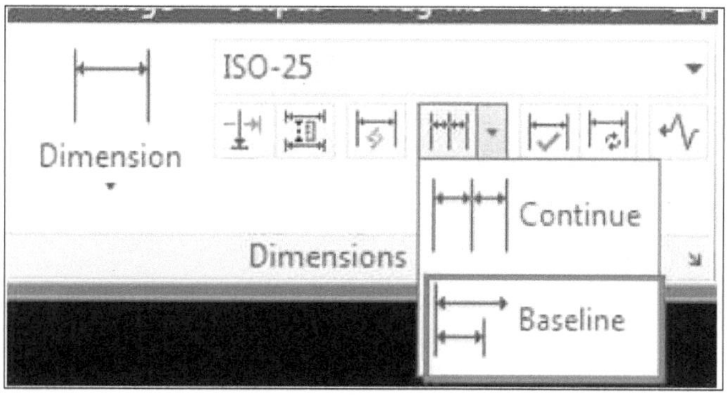

Figure 6.40

COMMAND LINE: To start the Baseline Dimension tool from the command line, type "DIMBASE" and press [Enter].
A Baseline dimension is created by selecting the already created Linear or Aligned dimension (1) then by clicking the subsequent points (2, 3 and 4).

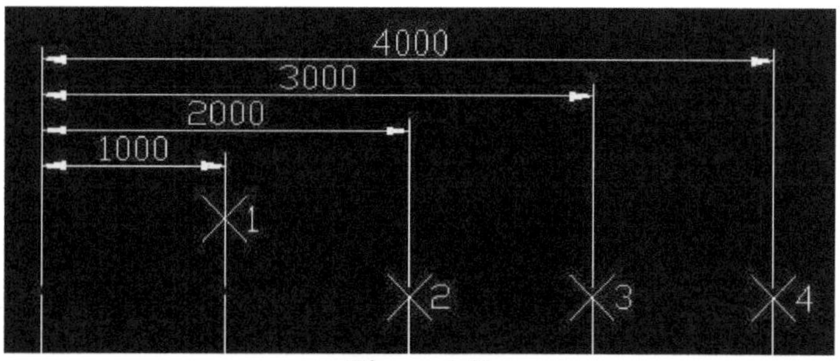

Figure 6.41

MULTILEADER
The Multileader dimension tool creates arrows and text pointing to objects.

To create Multileader click the Ribbon **Annotate** tab > **Leaders** panel > **Multileader** button.

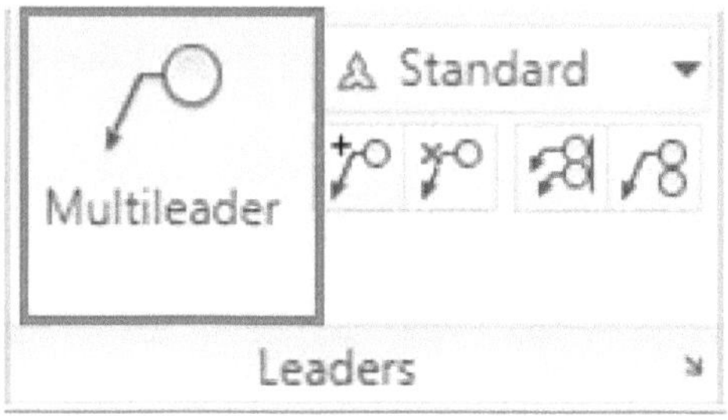

Figure 6.42

COMMAND LINE: To start the Multi Leader tool from the command line, type "MLD" and press [Enter].
Click point 1 to locate the start of your Multileader. Click point 2 to locate the end of your Multileader. 3 enter the text you wish to display.

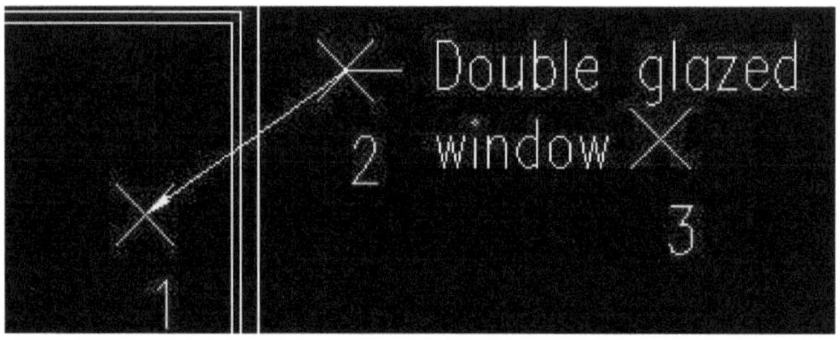

Figure 6.43

To add additional Leaders to your Multileader click the Add Leader button from the Ribbon **Annotate** tab > **Leaders** panel.

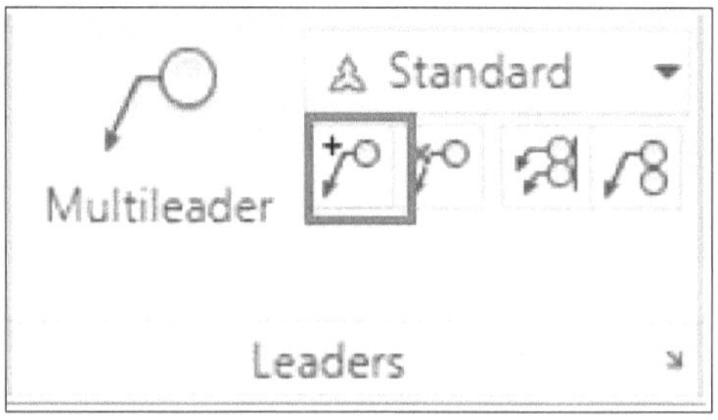

Figure 6.44

Click an existing Multileader (1) to identify what you want to add a Leader to. Click point 2 to identify the location of the arrow of the additional Leader line.

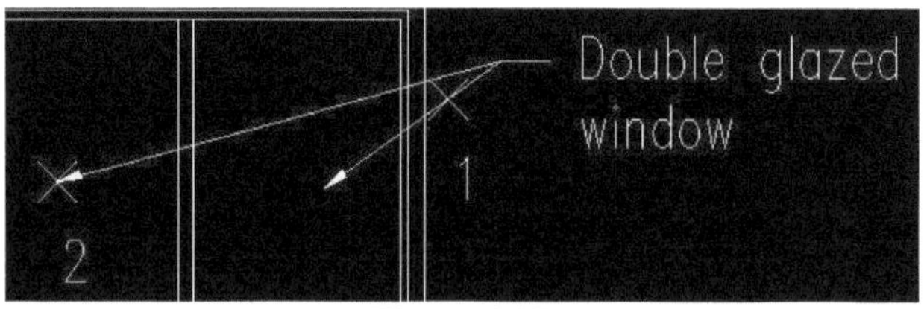

Figure 6.45

ASSOCIATIVE DIMENSIONS

When placing dimensions, by default, they are created Associative. An Associative Dimension will automatically adjust if the geometry it dimensions is modified.

Dimension associativity is controlled from the AutoCAD Options. To show the AutoCAD options, right click the command line and choose options.

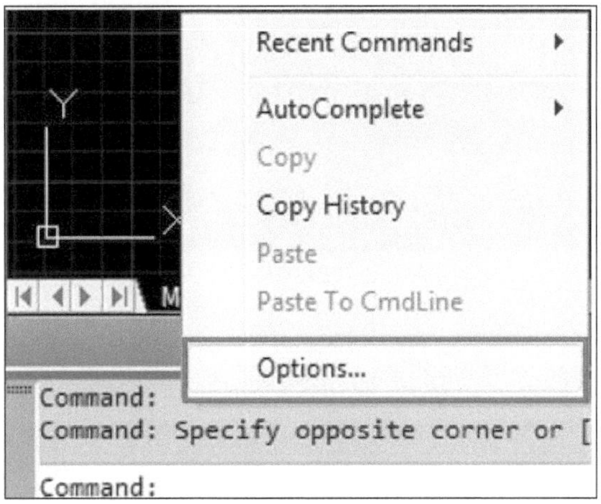

Figure 6.46

The options dialog box will show. The dimension associativity feature can be found on the User Preferences tab by ticking and un-ticking the Make new dimensions associative tick box.

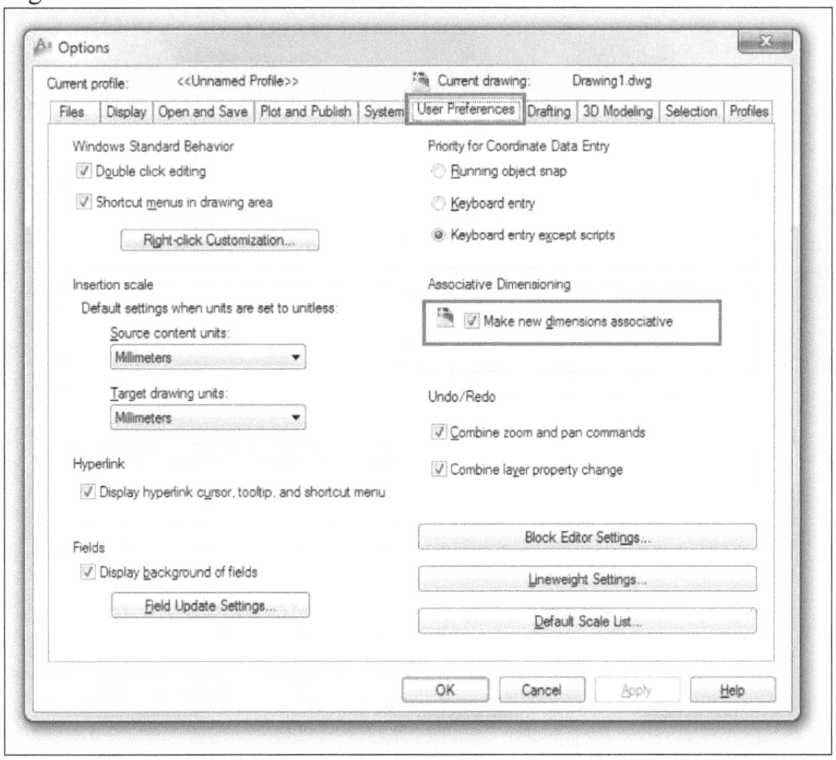

Figure 6.47

When this option is ticked (the default setting), Dimensions will be associated to objects. If the object is modified, the dimension will adjust to match. This is the preferred method of dimensioning because it keeps the correct dimension value when geometry is edited.

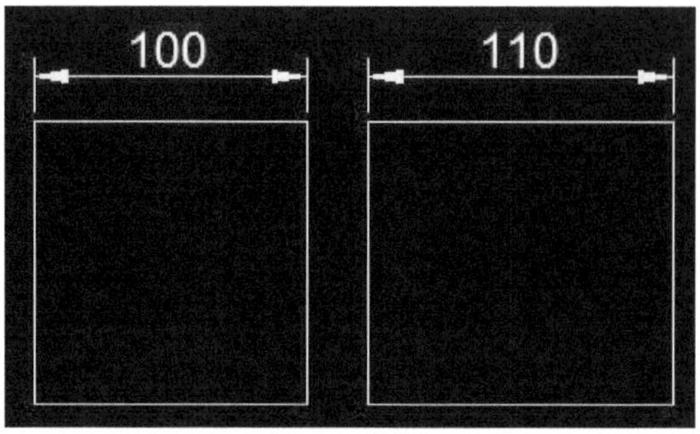

Figure 6.48

If unticked, the dimension is not associated to any objects. If the geometry is later adjusted, the dimension will remain in its original location and will need to be modified manually.

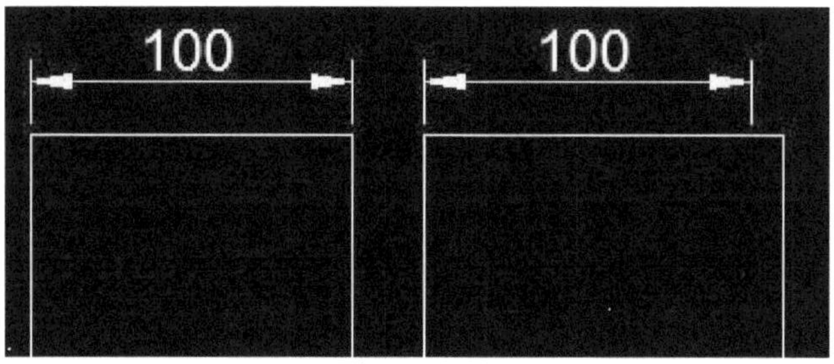

Figure 6.49

DIMENSION STYLES

The appearance of individual dimensions can be altered using the properties palette but using styles will ensure consistency.

To view the Dimension Style Manager dialog box, click the Dimension Style button from the **Ribbon Annotate** tab > **Dimensions** panel.

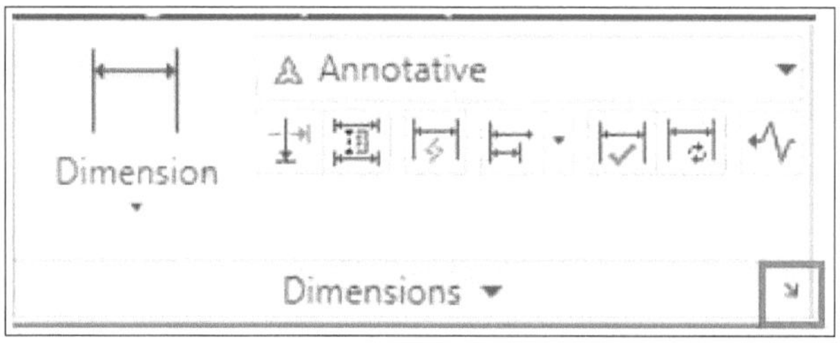

Figure 6.50

COMMAND LINE: To open the Dimension Style Manager from the command line, type "DDIM" and press [Enter].

The available styles will be listed in the Styles section of the Dimension Style Manager dialog box.

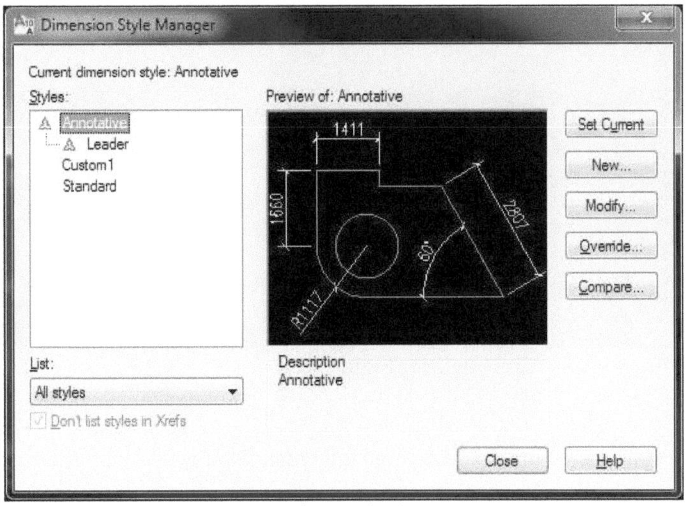

Figure 6.51

Sub-styles will show linked to a Style. Sub-styles are used where different properties for a specific dimension tool are required e.g., a different arrow head for Leaders than the other dimensions in the parent Style.

CREATING DIMENSION STYLES

New dimension styles are created by clicking the New button. The Create New Dimension Style dialog box will show allowing you to enter a New Style Name for your dimension style. You have to choose an existing dimension style as a starting point.

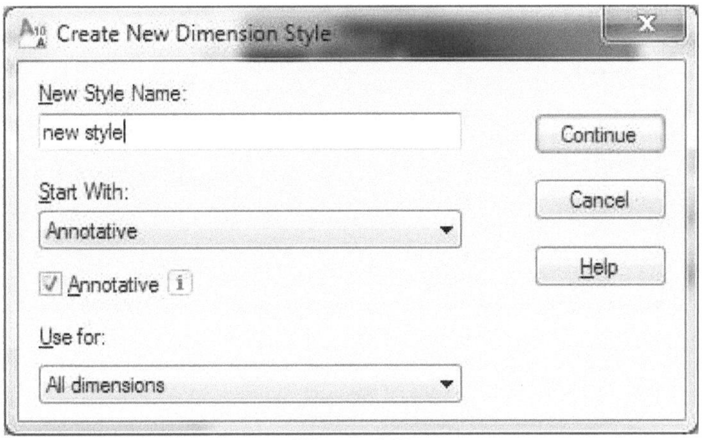

Figure 6.52

Once the new style has been created, the New Dimension Style dialog will show with all the properties of the starting dimension style.
The tabs across the top of the dialog box control the different aspects of the dimensions.

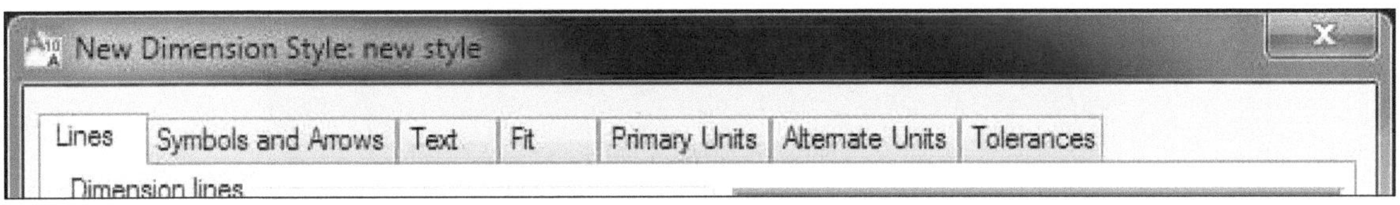

Figure 6.53

The dimension below shows which tab is used to modify each part of the dimension.

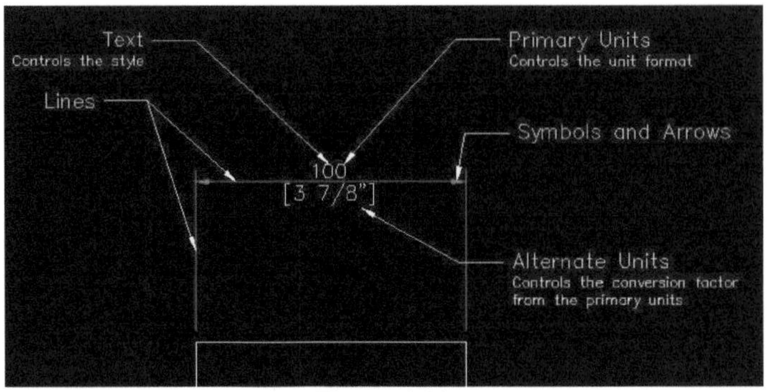

Figure 6.54

LINES

The Lines tab controls the appearance of the dimension and extension lines.

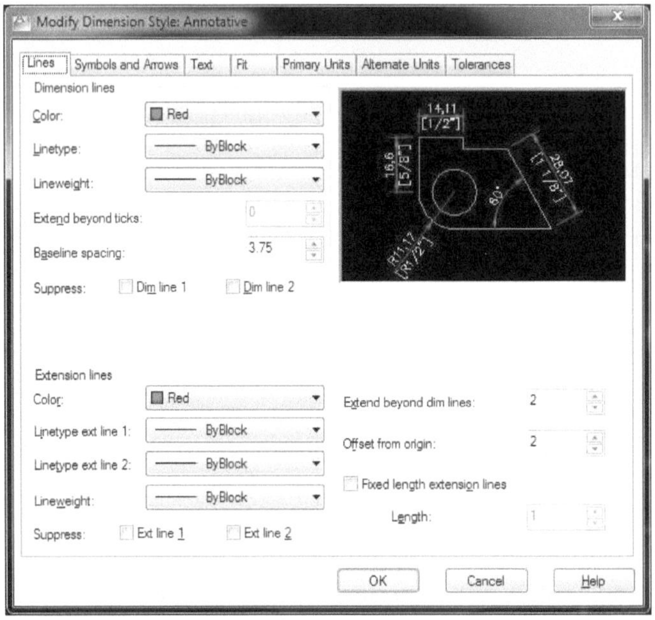

Figure 6.55

The image shows the extension lines and dimension lines

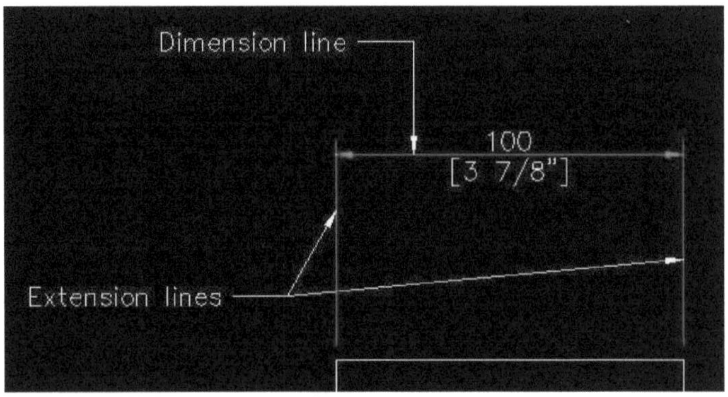

Figure 6.56

Colour controls the colour of the Dimension lines or Extension lines

Linetype controls the line type of the Dimension lines. There are 2 line type options in the Extension lines section: Line type ext line 1 and Line type ext line 2, where the 1 is the first extension line placed.
Lineweight controls the line weight of the Dimension lines or Extension lines.
Baseline Spacing is the space between the text and the dimension line.
Supress Dim Line / Ext Line allows you to remove the dimension or extension lines so they are not shown. This can be useful when placing dimension or extension lines over other geometry such as grid lines so the dimension doesn't obscure the dimension line.
Extend beyond dim lines is the distance the Extension lines extend above and past the Dimension lines.
Offset from origin is the distance from the object being dimensioned to the start of the Extension lines.
Fixed length extension lines is the length of dimension lines can be fixed to help keep the appearance of dimensions tidy. To change the dimension line length, on the Lines tab, tick the Fixed length extension lines option and enter a length for your dimension lines in the Length box. Each new dimension will now have extension lines fixed to the length you specified in the Length box.

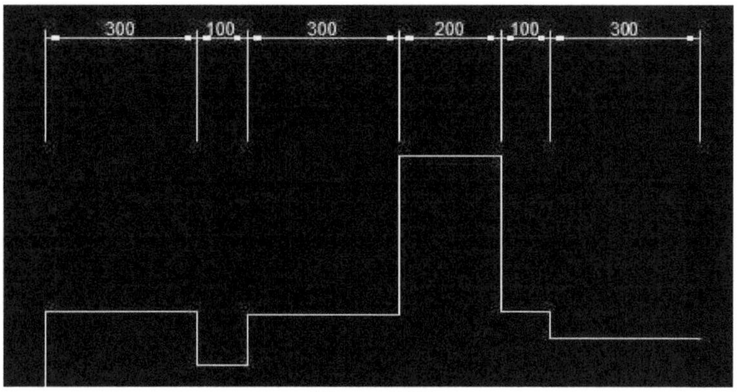

Figure 6.57

SYMBOLS AND ARROWS

The Symbols and Arrows tab controls the arrow heads at the end of the Dimension lines and Centre marks.

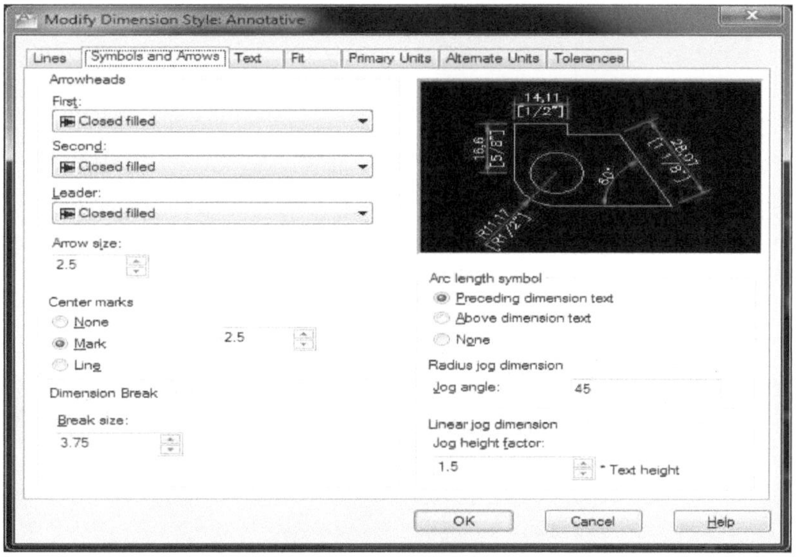

Figure 6.58

Arrowheads controls the arrow head type and size of the arrowheads on the end of dimension lines.
Centre marks controls the appearance when marking centre points of circles and arcs. It is quite often important to use centre marks to set out a curved element.
Dimension Break is the size of the break mark when using Jogged dimensions.

Arc Length Symbol controls the location of the arc length dimension symbol. In the example, the symbol is shown Proceeding dimension text.

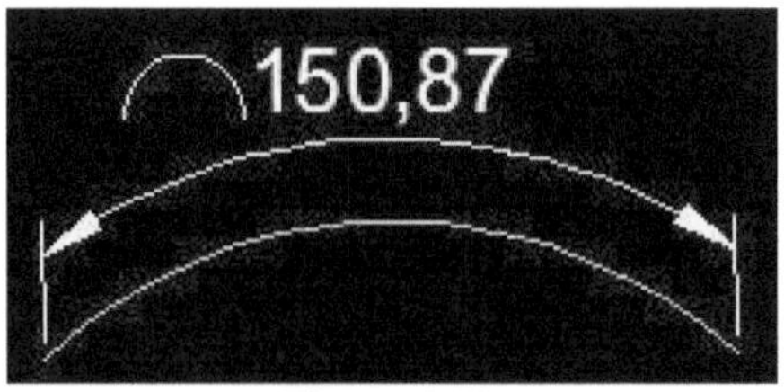

Figure 6.59

Radius jog dimension controls the jog angle on Jogged dimensions.
Linear jog dimension controls the jog height factor as a ratio of the text size for jogged dimensions.

TEXT

The text tab controls the appearance of text. This shouldn't be confused with the Primary and Secondary Units tabs which control the value of the text.

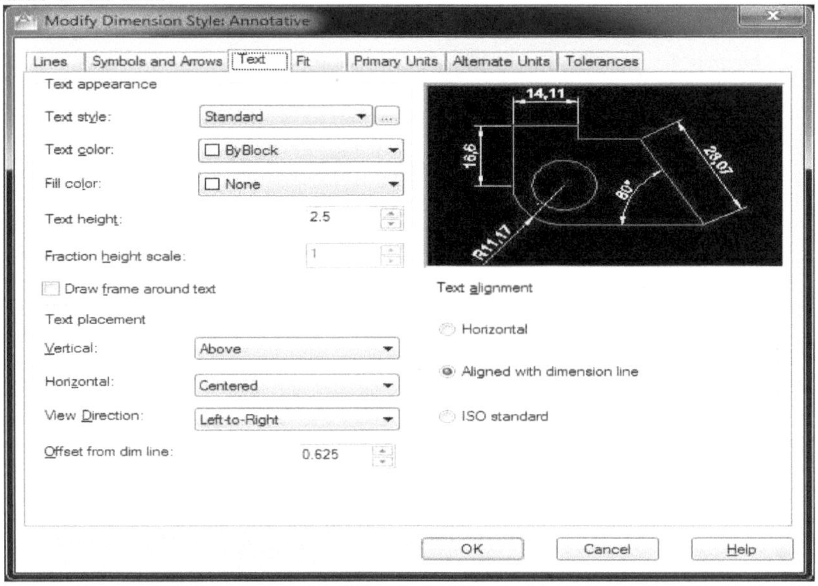

Figure 6.60

Text appearance controls the style of text used, the colour and the height. (For more details on Text and Text Styles, see module 06-Text)
Text Placement controls the location of the text in relation to the Dimension line.
Vertical

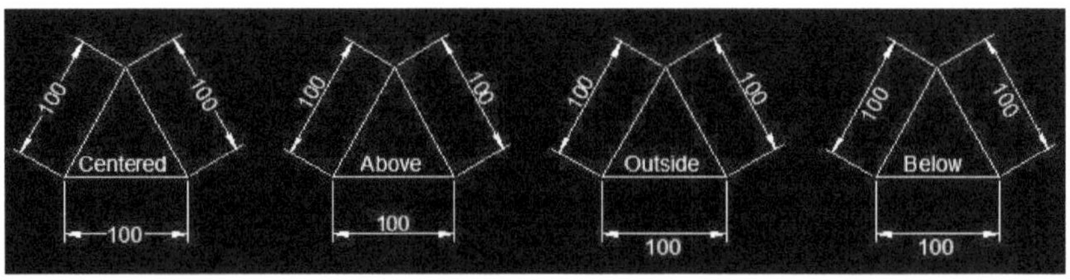

Figure 6.61

Note: The JIS (Japanese Industry Standard) placement isn't shown but is the same as the Above option.
Horizontal

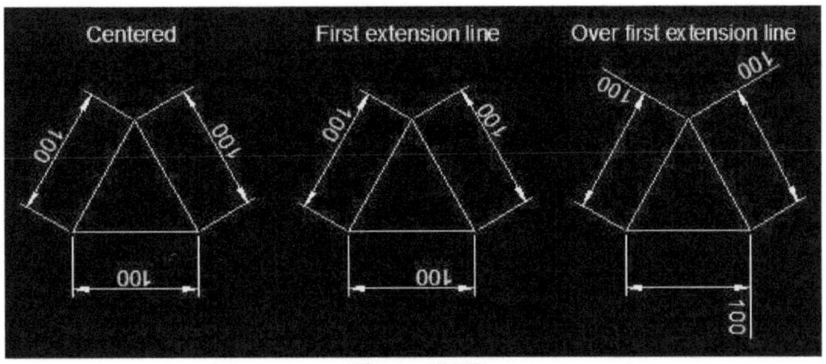

Figure 6.62

View direction Controls whether the text is read from left to right or right to left.
Text Alignment controls the rotation of the text: whether it is rotated to the view orientation, whether it aligns with the dimension being created, or whether it follows ISO standard (which maintains all text horizontally).

FIT

The fit options control how the text in a dimension reacts when space is limited and the overall scale of the text.

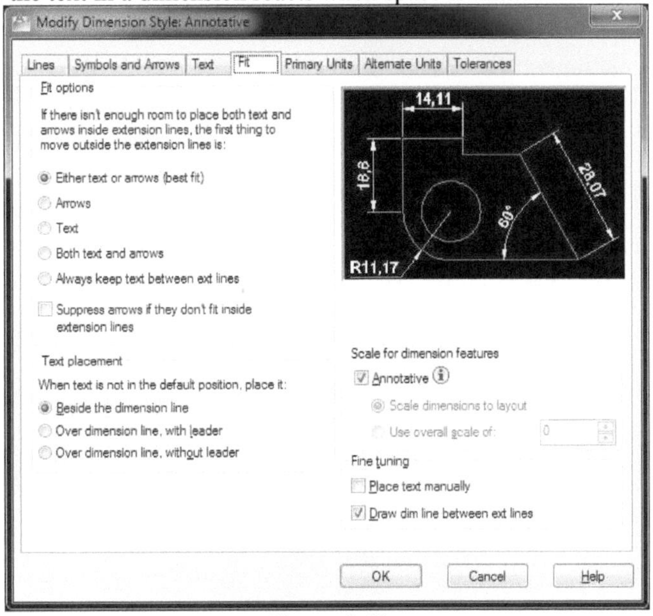

Figure 6.63

Fit options controls how the text and arrows relocate around the extension line if there isn't enough space. When dimensioning smaller objects, AutoCAD will relocate the text, arrows or both depending on the selected option.
Text Placement controls where the text is located when it is reposition by the Fit options.
Scale for dimension features controls dimension scale in one of two ways: the manual method and the annotative method.

MANUAL METHOD

On the **Fit** tab of the **Modify Dimension** dialog box, enter the scale of your drawing e.g. if your drawing will be scaled 1:100, enter a value of "100" in the Use overall scale of box. This will scale the dimension attributes such as text and arrows by 100.

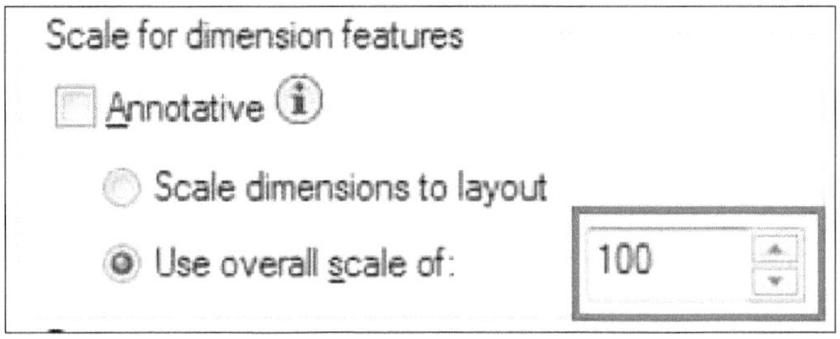

Figure 6.64

ANNOTATIVE METHOD

On the Fit tab of the Modify Dimension dialog box, tick the Annotative option to use the file's Annotation Scale to scale your dimensions.

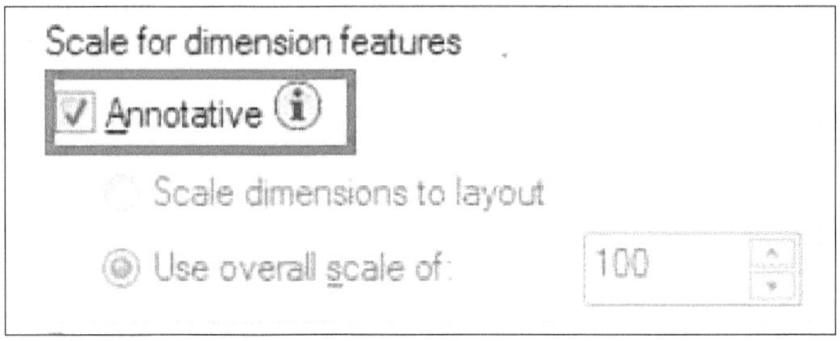

Figure 6.65

ANNOTATION SCALE

Annotative Dimension Styles are based on the Annotation Scale. Select your Annotation Scale as required from the Drafting Settings at the bottom of your application window.

Figure 6.66

If the Add Scales button is active, annotative objects will change size when the Annotation Scale drop-down value is changed. The selected Annotation Scale will be assigned to the annotation.

Figure 6.67

When Annotation Visibility is activated, the annotation will re-size to display at the current Annotation Scale, if that scale is assigned to the object. If the select Annotation Scale has not been assigned, the annotation will remain at its current size. For example, if an annotative dimension is only assigned to 1:50 and the Annotation Scale is changed to 1:100, the dimension will remain at 1:50.

Figure 6.68

When an annotative object is selected, the different scales the annotative object uses are displayed. In the diagram below, an annotative dimension is selected. Its text appears at **1:20** and **1:100** scales.

Figure 6.69

PRIMARY UNITS
These are the units used by the dimensions.

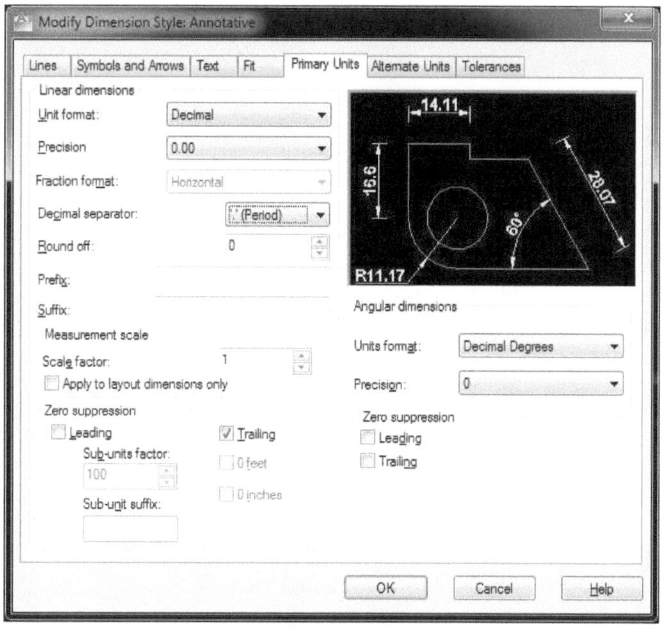

Figure 6.70

Linear Dimensions is the unit format of distance dimensions. You can apply prefixes and suffixes to indicate the unit format being used. For example, if you are dimensioning in millimetres, you would enter a Suffix of "**mm**".
Measurement scale is the scale that the dimension length will be multiplied by to give the final dimension value. When dimensioning an object drawn in metres in millimetres, entering a Scale factor of "1000" will give the correct result. i.e., if the object if 5m long, using a Dimension Style with a Scale factor of 1000 will result in a "5000" (5 x 1000) dimension value.
Zero Suppression removes leading and trailing zeros. If a measured value was 100.00, ticking the Trailing option would give a dimension of 100.
Angular Dimensions controls the unit format of angles.

ALTERNATE UNITS
You can display two different unit values on your dimensions, such as metres and feet, by ticking the Display alternate units option from the Alternate Units tab.

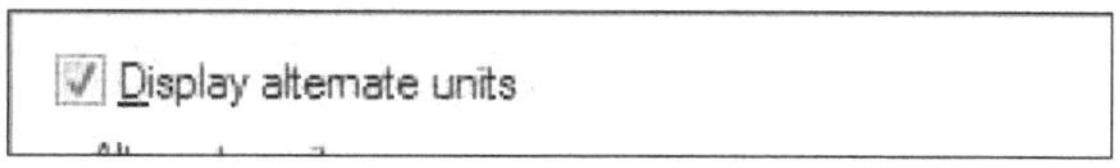

Figure 6.71

Set a scale factor from the original units. E.g., If you drew in mm and want to convert to inches, enter a scale factor of "0.03937" (this is 1 / 25.4, the scale factor to convert from mm to inches) in the Multiplier for alt units box.

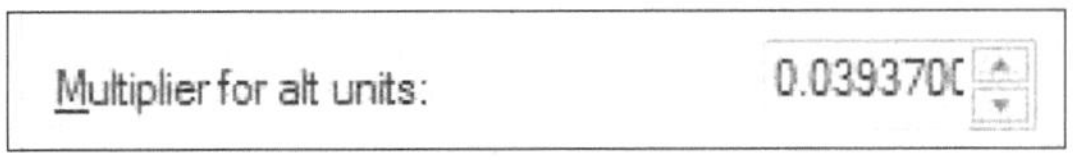

Figure 6.72

Note: When using Alternate Units, it is recommended to add a suffix (e.g., in mm) to help clarify which units are being used.

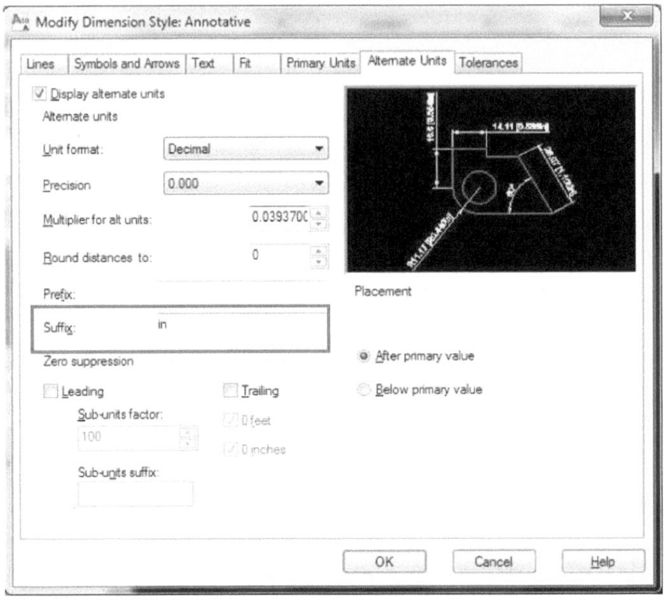

Figure 6.73

The Placement section controls the location of the Alternate Unit in comparison to the original units.

MOVING DIMENSION TEXT

To move dimension text select the dimension to be modified to show the dimension's Grips. Click the centre Grip over the text. Move the text to the required location.
The dimension line will also move to maintain the alignment between the text and the dimension line.

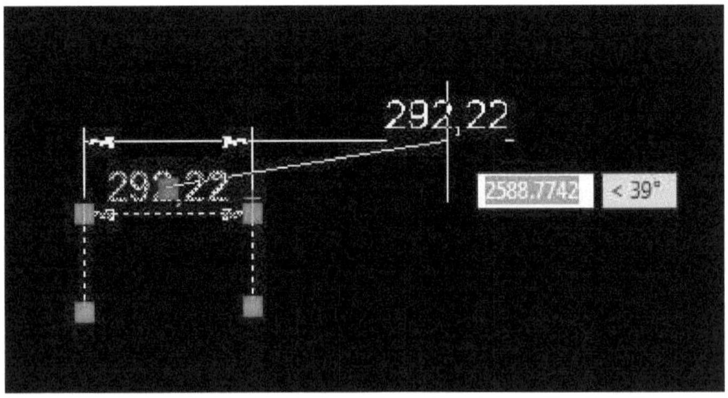

Figure 6.74

EDITING DIMENSION TEXT

The value of the dimension text can be modified as required.
Note: It is not recommended to change the value. If you require a different value, the geometry should be amended. Editing dimension text should only be used to add suffixes, prefixes and if dimensioning "Not to scale" drawings.
To change the text, select the dimension to be modified. Right-click and choose Properties to show the Properties palette. From the Text Override field enter a new value to be displayed.

In the image below, note the difference between the Measurement (100), the actual dimension value, and the Text override (120 NTS). The dimension is actually 100 long, but has been edited to read "120 NTS".

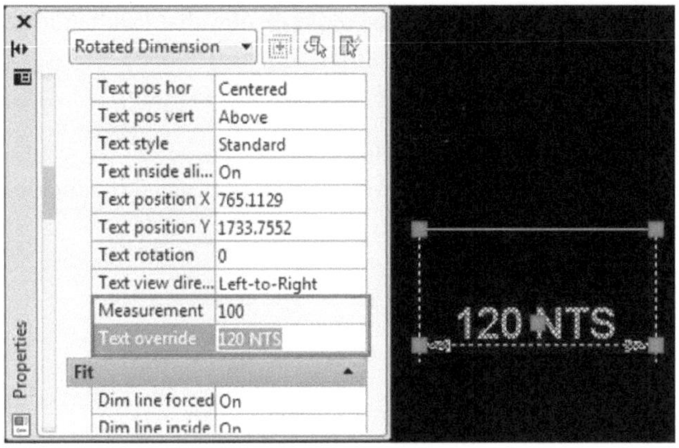

Figure 6.75

Using the <> symbols will display the dimension value.
For example, to change a dimension from "500" to "Width = 500", enter the value:
"**Width = <>**"

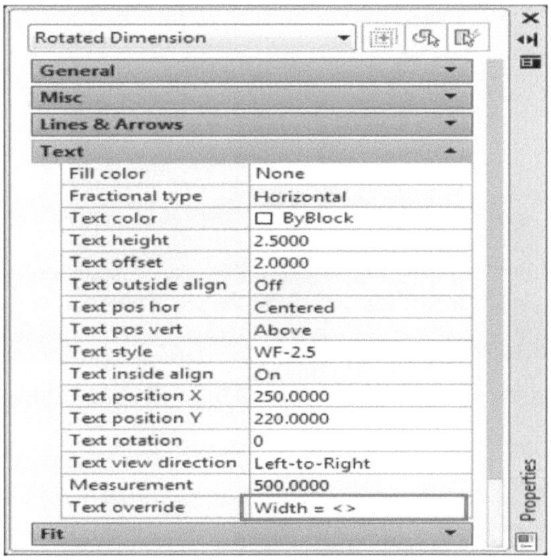

Figure 6.76

Figure 6.77

To change the dimension value back to the default of just the value, delete any Text override value.

6.5 TYPES OF TEXT

There are two kinds of text that can be used in AutoCAD: **Multiline** and **Single Line**.
You have more control over the appearance of **Multiline** Text and if you move the text, multiple lines stay grouped together. **Single Line** text is useful for writing out a collection of notes; each return creates a separate entity. This is useful for writing out multiple single line notes, such as room names, that can be moved around the screen independently from each other.

MULTILINE TEXT
The most flexible form of text for formatting is **Multiline Text.**
Multiline Text can be accessed from the Ribbon **Annotate** tab > **Text** panel by clicking the **Multiline Text** button.

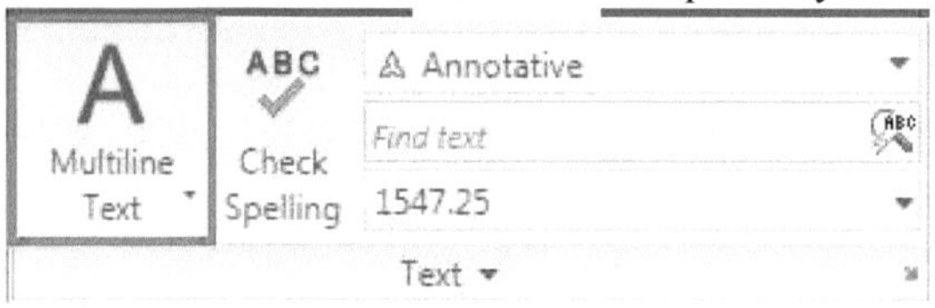

Figure 6.78

COMMAND LINE: To start the Multiline Text tool from the command line, type “T” and press [Enter].
When creating Multiline Text, a rectangle shape is drawn to identify the margins. When text reaches the edge of the margin, the text will wrap onto the next line without the need to enter a carriage return.
After clicking the Multiline Text button, AutoCAD will prompt:
Specify first corner:
A rectangle to identify the text box size can be drawn. The Abc notation in the top-left hand corner is an indication of the text size.

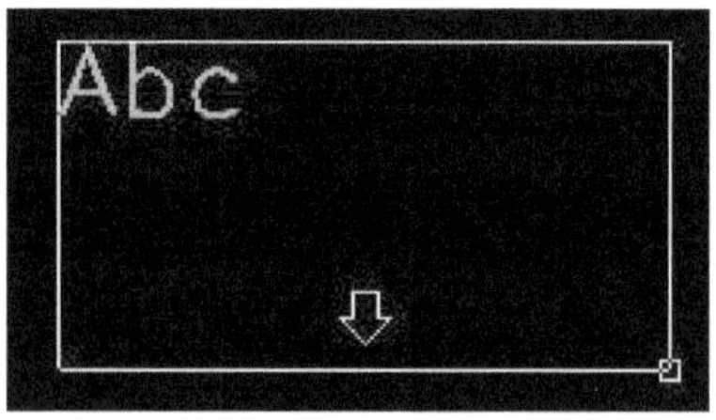

Figure 6.79

Once the text box has been placed, the Text Editor contextual tab will show all the available options.

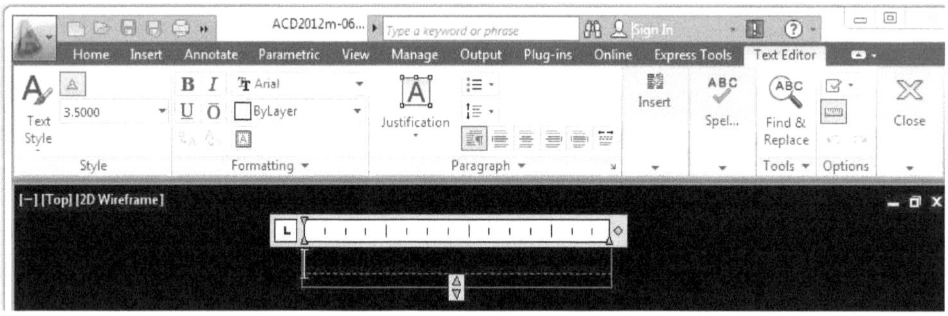

Figure 6.80

Note: A contextual Ribbon panel is one that will appear only when you have a specific tool selected. In this case, the Text Editor is active so the Text Editor tab is added to the end of all the Ribbon tabs.

Type the required text. Notice the text will wrap to stay inside the drawn text box.

Figure 6.81

STYLE

The style options for **Multiline** Text are shown under the Style panel of the Text Editor contextual tab.

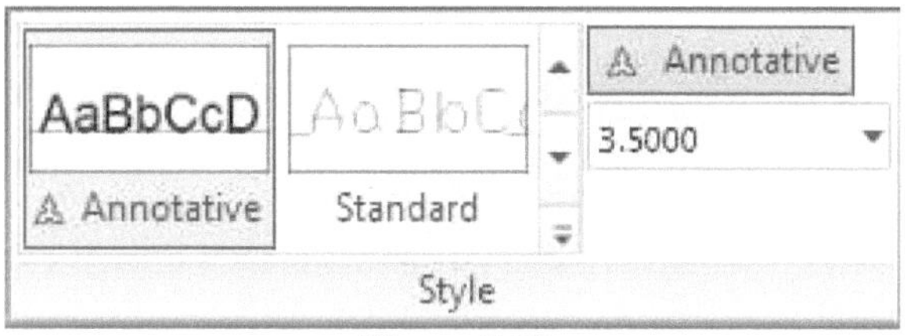

Figure 6.82

In the **Style** section of the Style panel, the defined text styles can be selected. AutoCAD by default has two styles, Annotative and Standard.

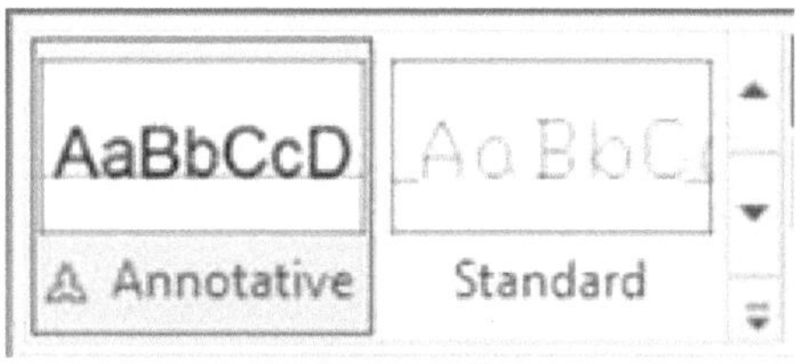

Figure 6.83

The **Annotative button** turns Annotation Scale on or off for the selected text but doesn't change the text style.

.

The **Text Height** field allows changing the size of the text without modifying the text style. Clicking the drop-down arrow will allow recently used text heights to be selected.

ANNOTATION SCALE

Annotation scale is used to control the size and scale of annotative objects such as text and dimensions.
From the Drafting Settings, the Annotation Scale drop-down allows you to choose the required annotation scale.

If the **Add Scales** button is active, annotative objects will change size when the Annotation Scale drop-down value is changed.

When **Annotation Visibility** is activated, the annotation will re-size to display at the current Annotation Scale, if that scale is assigned to the object. If the select Annotation Scale has not been assigned, the annotation will remain at its current size. For example, if an annotative dimension is only assigned to 1:50 and the Annotation Scale is changed to 1:100, the dimension will remain at 1:50.

When an annotative object is selected, the different scales the annotative object uses are displayed. In the diagram below, annotative text is selected showing text will appear at 1:20, 1:50 and 1:100.

Figure 6.84

FORMATTING
The formatting options for Multiline Text are shown under the Formatting panel of the Text Editor contextual tab.

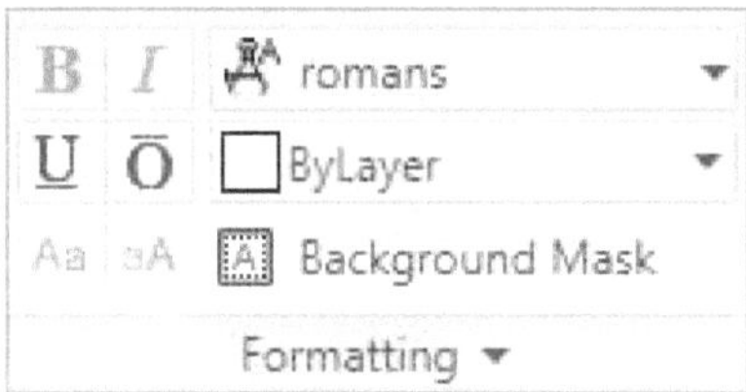

Figure 6.85

 Creates **BOLD** text

 Creates *ITALIC* text

 Creates UNDERLINED text

 Creates overlined text

 Changes selected text from lowercase to UPPERCASE

 Changes selected text from UPPERCASE to lowercase

 Changes the text font

 Changes the text colour

 Adds a BACKGROUND MASK to text. Background masks can be used to obscure information behind a text object.

BACKGROUND MASK
Clicking the **Background Mask** button will show the Background Mask dialog box.

Figure 6.86

Tick the Use background mask tick box to turn on the mask.
The Border offset factor is the width around the text that the background mask will cover. A border offset is a multiplication of the text height. If the value is set to 1, the border will cover the extents of the text. If the border offset is set to 5, the actual offset will be 5x the height of the text.

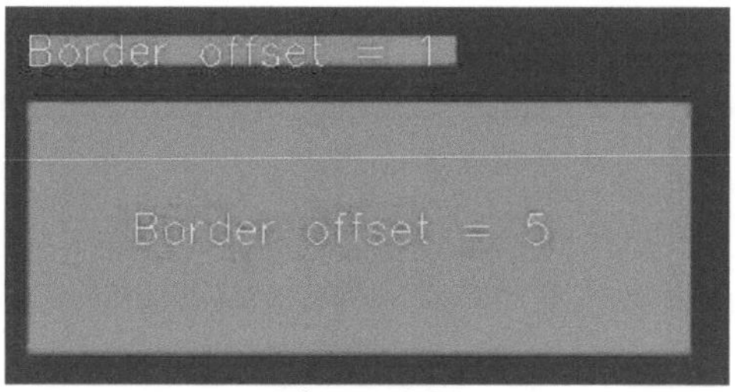

Figure 6.87

The **Fill Colour** is the colour of the background mask. If Use drawing background colour is ticked, the colour will be ignored and simply hide any geometry behind it.

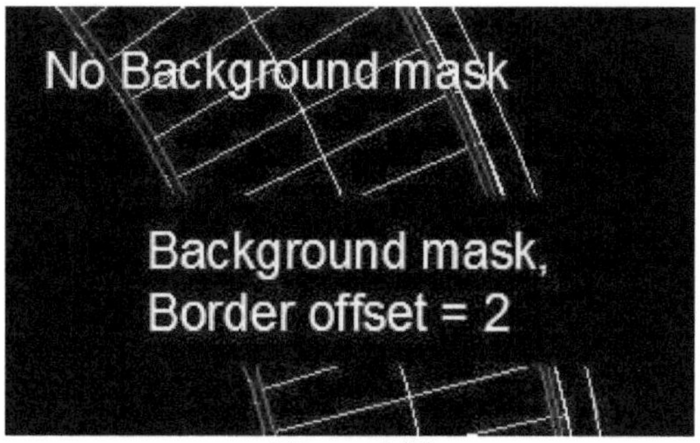

Figure 6.88

PARAGRAPH
The **Paragraph panel** has options for Multiline Text to allow custom alignment, spacing and bullet points.

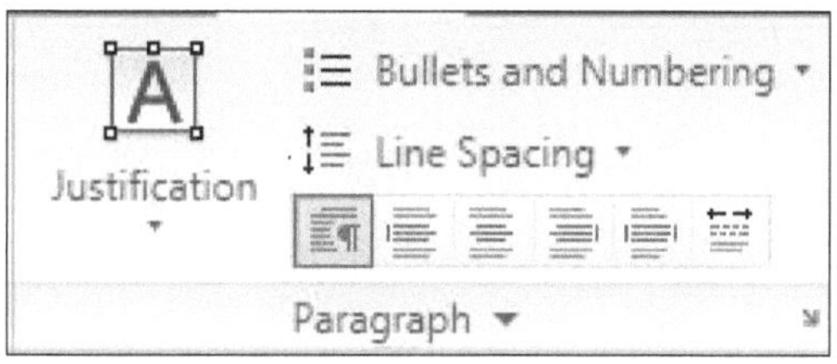

Figure 6.89

INSERT
The **Insert panel** has options to allow for the addition of Columns, Symbols and Fields.

Figure 6.90

INSERTING COLUMNS
Once text has been created, columns can be added to give a newspaper column format.
From the Ribbon **Text Editor** contextual tab > **Insert** panel, click the Columns drop-down, select Static Columns and choose the number of columns you require.
To adjust the amount of text in each column, the grip below the annotation can be adjusted as required.

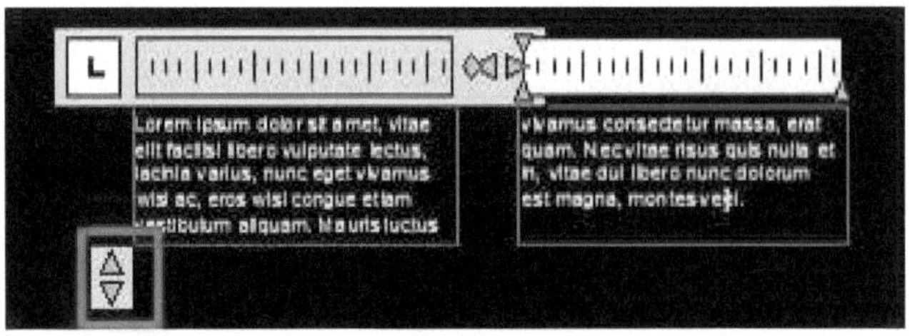

Figure 6.91

INSERTING SYMBOLS

From the Ribbon **Text Editor** contextual tab > **Insert** panel, click the **Symbol** drop-down and choose the symbol you wish to use.

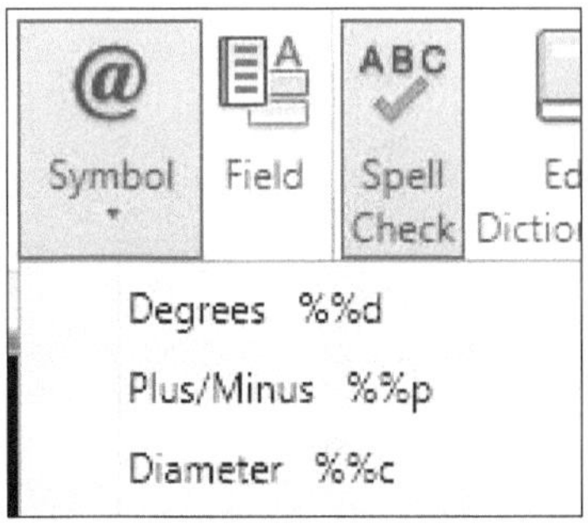

Figure 6.92

If the symbol you need isn't shown, such as the ³ or ç symbol, click the **Other** button at the bottom of the list.

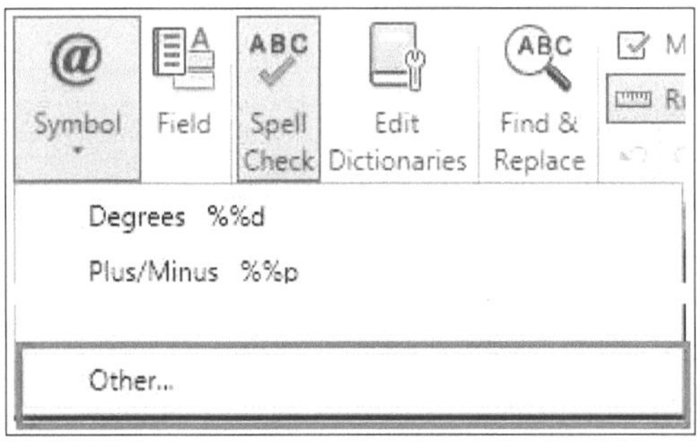

Figure 6.93

The **Character Map** dialog box will show.

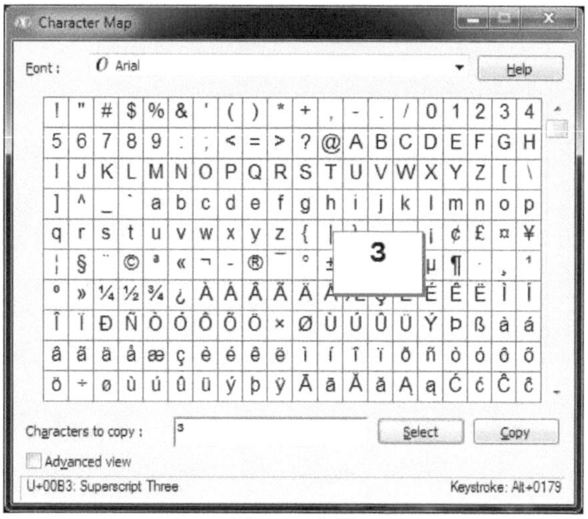

Figure 6.94

Double-click the symbol and it will appear in the Characters to copy box at the bottom of the dialog box.
Click the Copy button to copy the symbol to the clipboard.
Return to the AutoCAD text window, right-click and choose Paste to add the symbol.

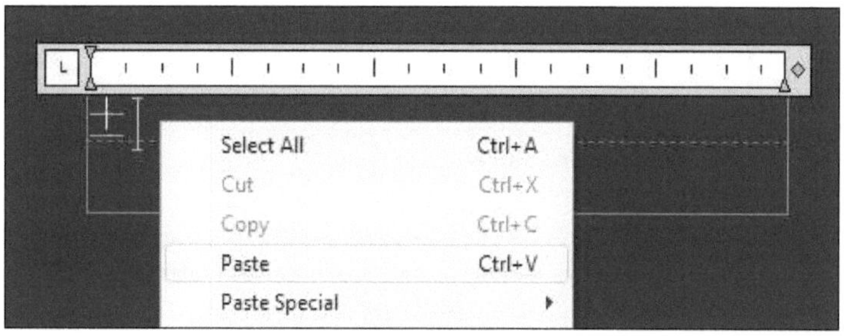

Figure 6.95

Note: At the bottom of the Character Map dialog box, the keystroke required to insert the symbol is shown. In order to reuse this symbol without having to find it in the Character Map, type the keystroke. In the example of the ± symbol, hold the [**ALT**] key on your keyboard and type "**0177**". When you release [**ALT**] the symbol will show.

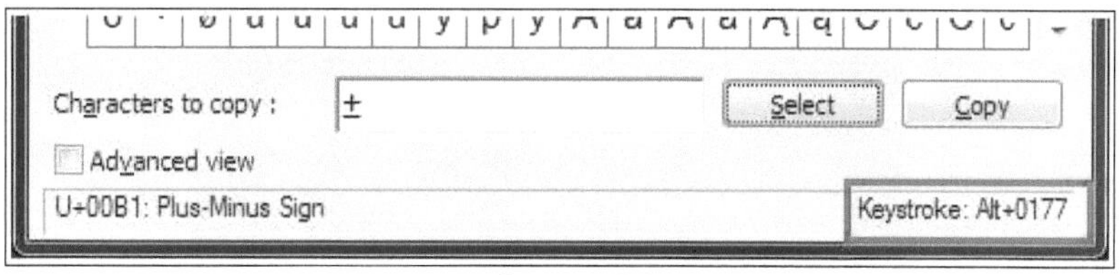

Figure 6.96

INSERTING FIELDS

Fields are text that display information about an AutoCAD object such as a circle or Polyline and can provide information including area and diameter. If the circle changes size, the field automatically updates the text to match.

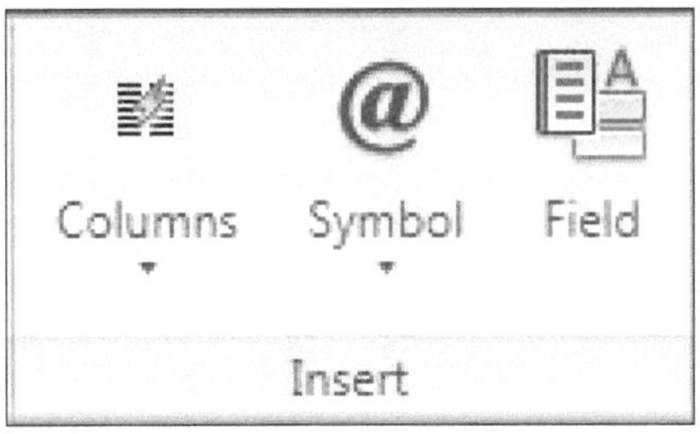

Figure 6.97

To place a field, click the Field button to show the Field dialog box and select the required field.

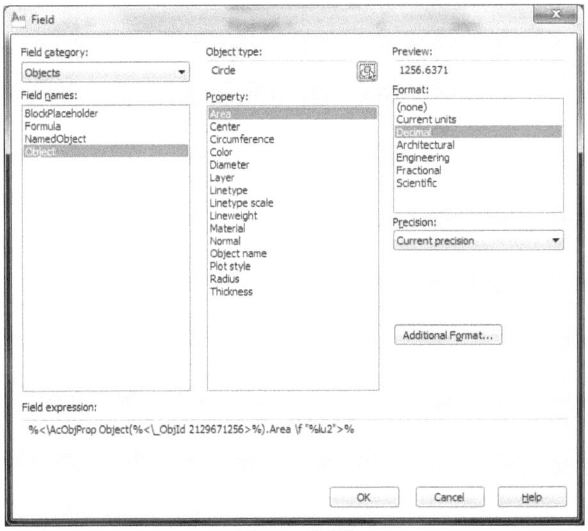

Figure 6.98

Once a field has been created, the value will be shown in the Multiline Text editor.

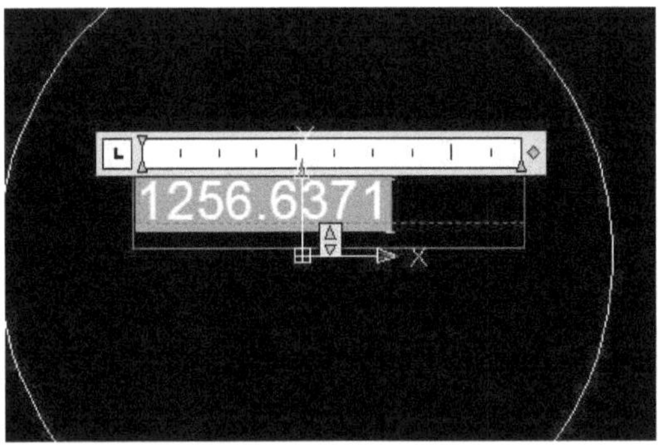

Figure 6.99

The grey background indicates the field will update automatically each time the area of the shape changes. Field backgrounds don't print.

SPELL CHECK

The **Spell Check** button is used to check the spelling of the text currently being inserted. If a word is spelled incorrectly, the word will be underlined in red.

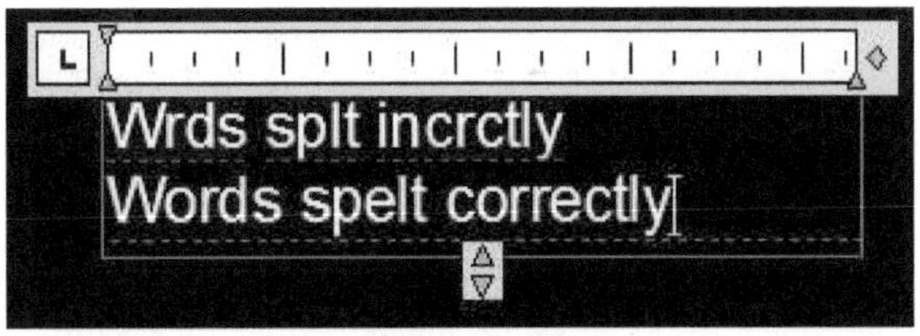

Figure 6.100

To view suggestions for the correct spelling, right click the underlined word. At the top of the right-click menu will be a list of suggestions. Select the required spelling.

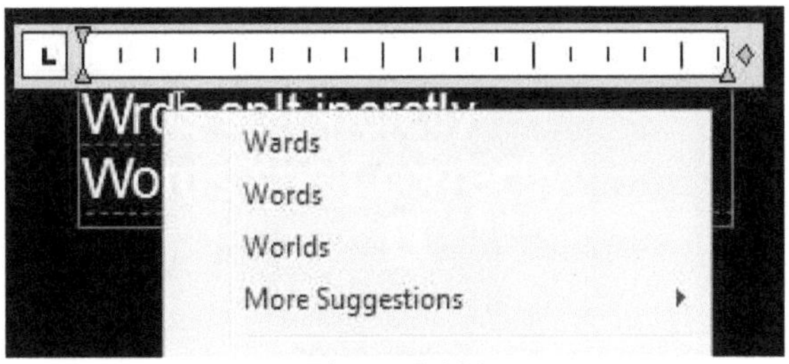

Figure 6.101

If you want to check the spelling on your entire drawing without starting a new Multiline Text object, the Ribbon **Annotate** tab > **Text** panel has a **Check Spelling** tool.

Figure 6.102

FIND AND REPLACE

The **Find & Replace** tool is used to find and replace text in your drawing. 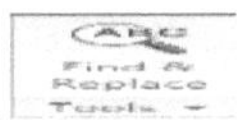
Click the Find & Replace button to show the Find and Replace dialog box.

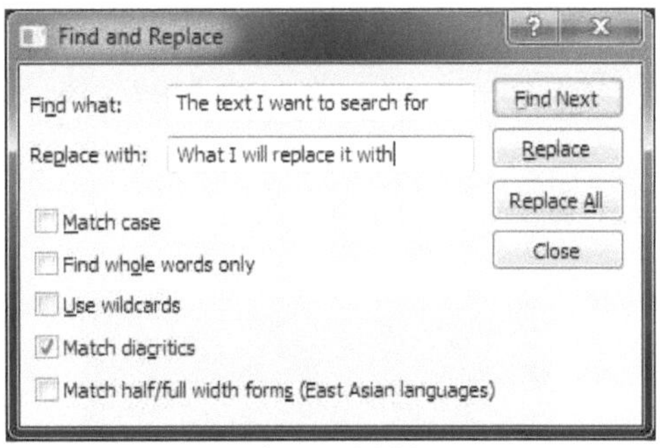

Figure 6.103

Enter the words to find. If required, enter the words to replace with.
There are a number of options when using Find and Replace that can ensure only the words you want to be replaced, are.
Match case only finds the search string if it matches the same upper or lower case values.
Find whole words only will ignore words with the find string contained within. For example, if you entered in the Find what: box "LED" find whole words would only find LED and not compiled

Use wildcards allows the use of the * character within the string to make your search criteria more specific. An example would be, if you entered "*LOCK" in the Find what box, use wildcards would not show the text "BLOCKWORK" unless you changed the search criteria to "*LOCK*"
Match diacritics uses diacritics, the accents added to letters such as the C in fascade.

SINGLE LINE

The **Single Line** tool is a much simpler form of text with fewer options than Multiline Text but, is useful for adding many short notes to a drawing.
Single Line text can be accessed from the Ribbon **Annotate** tab > **Text** panel, by clicking the drop-down below the Multiline Text button and choosing Single Line.

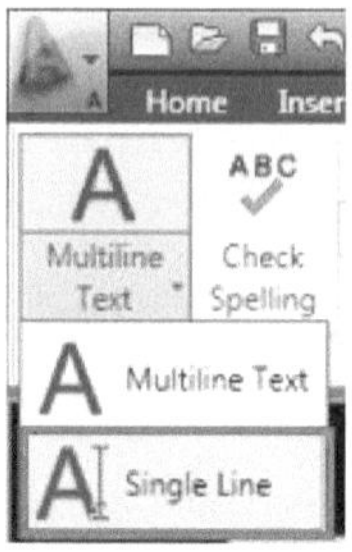

Figure 6.104

COMMAND LINE: To start the Single Line tool from the command line, type "DT" and press [Enter].
When you click on the **Single Line** tool, you will be prompted with:
Specify start point of text or [Justify/Style]:
You can either click a point on the screen as the insertion point for your text or, choose one of the options in the "[ ]" brackets.
Justify is the alignment of the text, the command line prompts with the available justification methods:
Enter an option [Align/Fit/Centre/Middle/Right/TL/TC/TR/ML/MC/MR/BL/BC/BR]
Each of the two letter codes are acronyms for common alignment options such as TL = Top Left, MC = Middle Centre, BR = Bottom Right.
Style is the text style. The available text styles won't be listed but if the name of the style is known, it can be typed.
Once the initial point has been clicked for the insertion point, if the text style doesn't have a pre-defined height, AutoCAD will prompt for the size of the text. If the current text style is annotative, the prompt will ask for the paper height. If the style is not annotative, the prompt will ask for the height for the new text.
You will then be prompted for the rotation of the text.
Start typing your text. It will be displayed on the screen as it is entered.

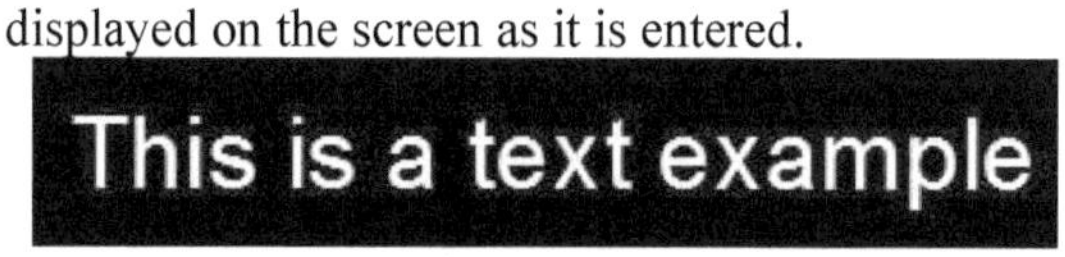

Figure 6.105

Pressing [**Enter**] will take you to a new line but, will also create a separate text object which can be selected individually.

Figure 6.106

To relocate multiple Single line text elements, select all the text elements to show the grips and, using the grips, move them individually to their required location.

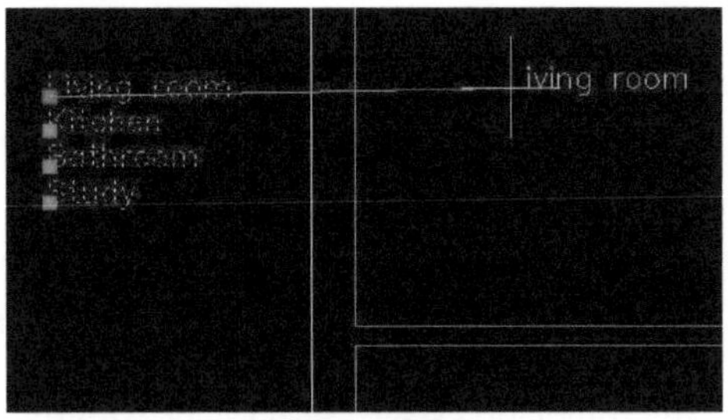

Figure 6.107

SELECTING TEXT STYLES

Text Styles are pre-defined properties for text. Styles can be used to help standardise the appearance of text on drawings. All text placed will use a text style, you can change the text style and the appearance of the text will change to match the style.

From the Annotate **Ribbon** tab > **Text** panel, click the **Text Style** drop-down and choose the required text style.

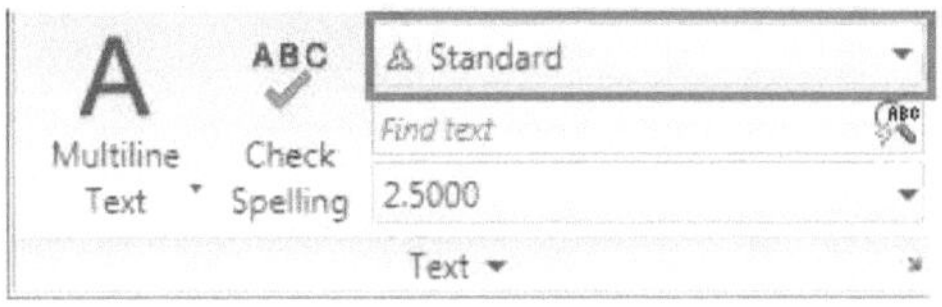

Figure 6.108

COMMAND LINE: To start the Test Style tool from the command line, type "ST" and press [Enter].

DEFINING TEXT STYLES

Text Styles can be edited from the Ribbon **Annotate** tab > **Text** Panel by clicking the **Text Styles** button.

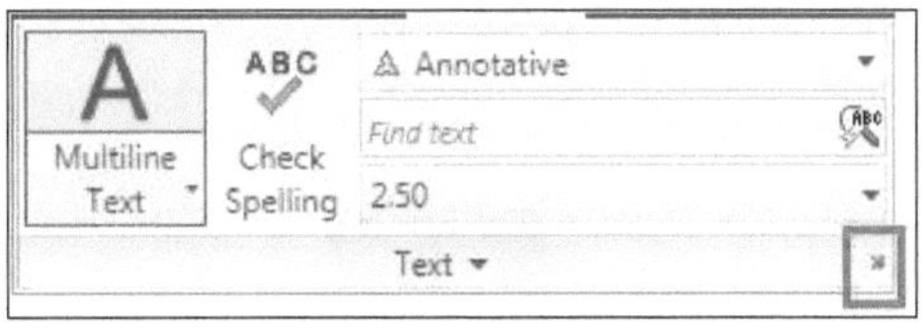

Figure 6.109

The text styles available will show in the **Text Style** dialog box.

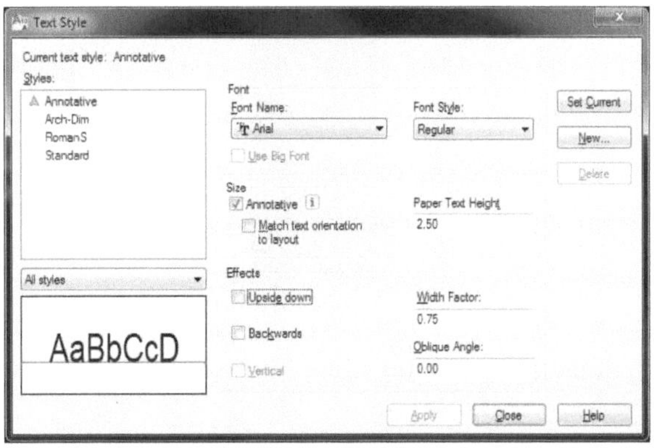

Figure 6.110

Each of the properties in the text style can be amended unless the font doesn't support that feature. In the dialog, Vertical is greyed out because the selected, "**Arial**" font doesn't support it.

FONT
This controls the font used by the style. AutoCAD will use the fonts installed on your PC.
Certain fonts such as the "**txt.shx**" font don't have **Bold** and **Italic** style options, in these cases, the Font Style drop-down will be greyed out.

SIZE
Controls the size of the text. If the annotative option is ticked the font will multiply the **Paper Text Height** by the **Annotation scale** to insert text at the correct size for the selected scale.
Annotative Text Styles are based on the Annotation scale. Select your **Annotation scale** from the **Drafting Settings** below the application window.

EFFECTS
Provides special formatting to create non-standard appearances:
Upside Down places text characters upside down.

Figure 6.111

Backwards places text backwards.

Figure 6.112

Vertical displays the characters aligned vertically. Vertical is only available if the selected font supports dual orientation.

Figure 6.113

Width Factor adjusts the width of the font as a ratio of the Text Height. 1 is full size.

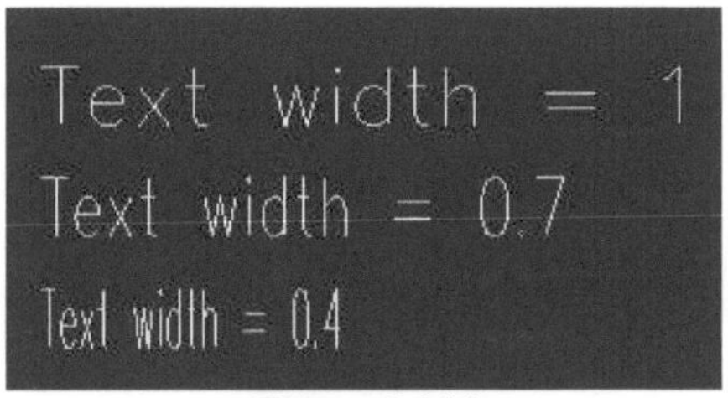

Figure 6.114

Oblique Angle allows text to be italicised by a specified angle.

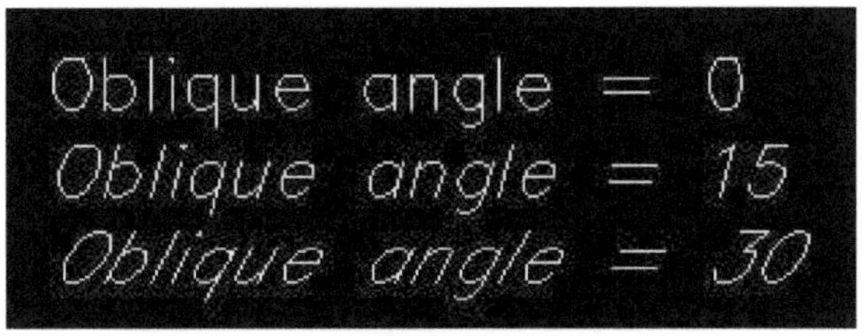

Figure 6.115

MODIFYING TEXT
Inserted text can be modified using the **Properties palette**.

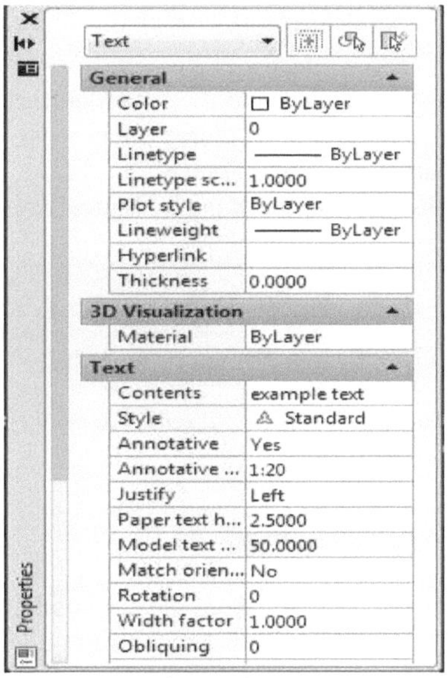

Figure 6.116

Select the text you wish to make a change to, right-click and choose Properties to show the Properties palette.
Adjust the properties as required, such as the Font or Obliquing angle.
If you wish to change individual parts of the text, double click the text to open it in the text editor. The contextual Text Editor tab will show.
Highlight the text you wish to change.

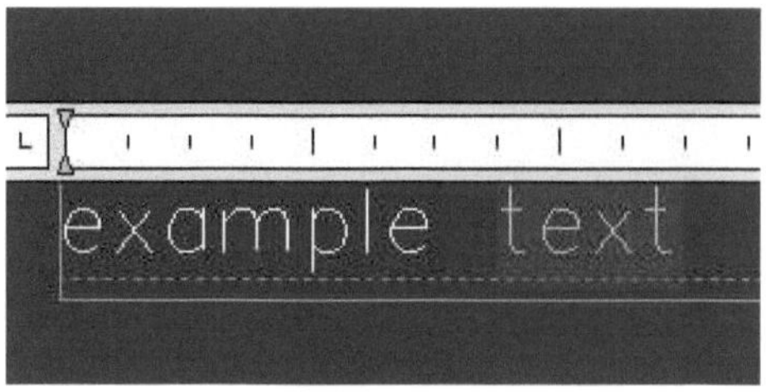

Figure 6.117

From the **Text Editor** contextual tab, change the values in the Formatting panel to change just the highlighted words.
In the example, the word "text" has had the font, colour and underline value changed.

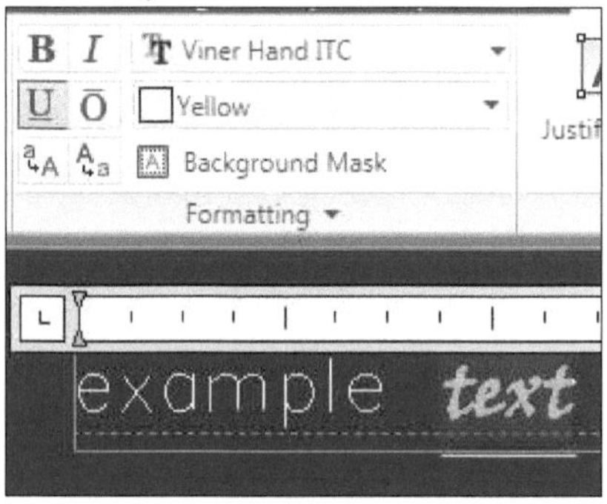

Figure 6.118

6.6 STANDARD SIZES OF SHEET

AutoCAD uses page setups to control layout settings and plotter properties. Page setups can be saved and applied to multiple layouts in the same drawing file. This guide explores the process of creating setups for printing that can be used over and over again each time the same drawing is issued.
To open the Page Setup Manager click the Page Setup Manager button from the Ribbon **Output** tab > **Plot** panel.

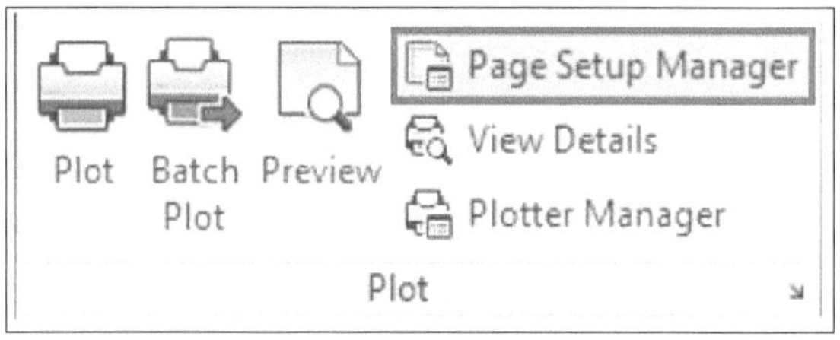

Figure 6.119

COMMAND LINE: To start the Page Setup Manager from the command line, type "PAGE" and press [Enter].
Note: If you have the Model and Layout tabs shown, you can also right-click on the layout tab and choose Page Setup Manager
Different page setups are depending on whether you are currently in Model space or Paper space. Before creating any page setups, ensure you are in the correct 'space'.

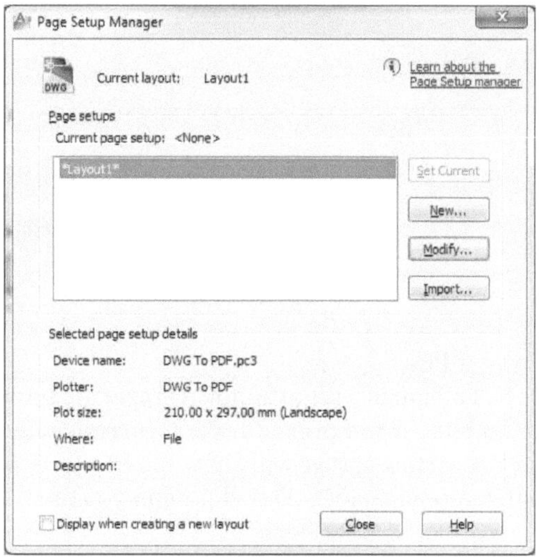

Figure 6.120

The default page setup will have the same name as the layout.

CREATING PAGE SETUPS

To create a new page setup, click the **New** button. The **New Page Setup** dialogue will display. Enter a name for your page setup. Logical names help find your page setup later.

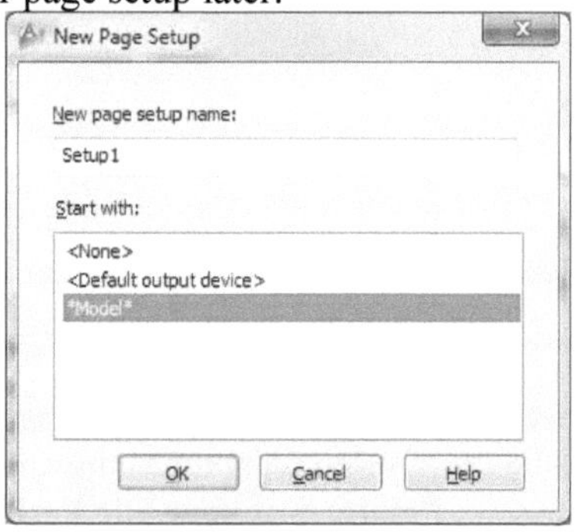

Figure 6.121

Click **OK** to accept. The **Page Setup** dialogue box will show:

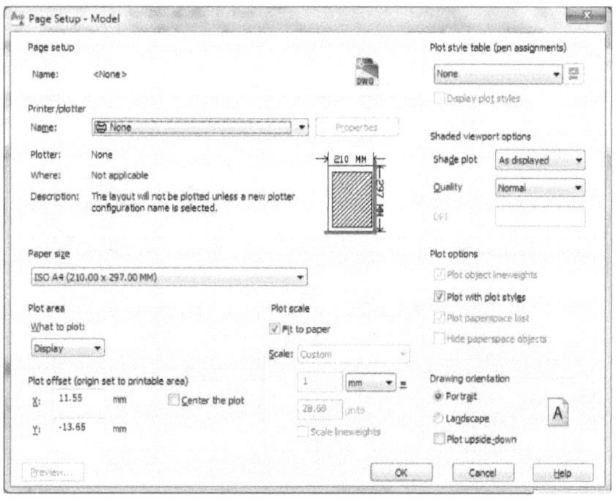

Figure 6.122

The **Printer/plotter** section will allow you to choose from installed printers and AutoCAD PC3 files.

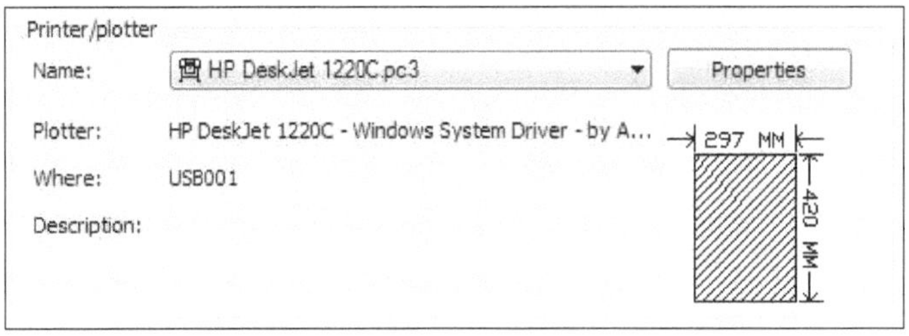

Figure 6.123

An installed printer is installed on your PC.
PC3 files are printer configurations that maintain a set of AutoCAD specific settings for a printer. PC3 files will either be configured to point to an installed printer or to a printer driver that creates file-based prints such as PDFs or DWFs Selecting a printer and clicking the **Properties** button will show the **Plotter Configuration Editor** dialogue box. If changes are made to the printer configuration, AutoCAD will prompt you to save the changes to a PC3 file.

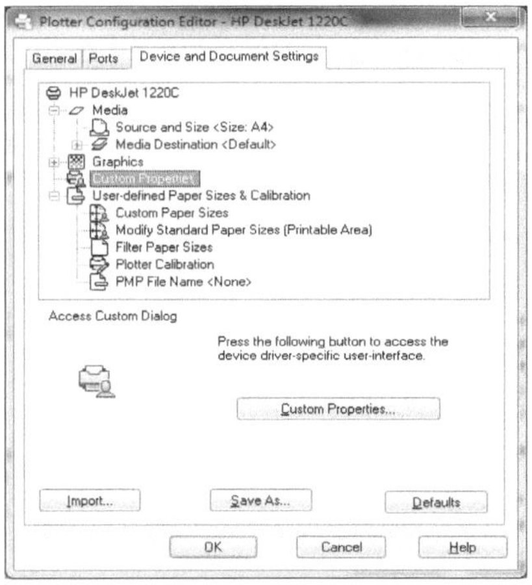

Figure 6.124

Note: PC3 files must be stored in a specific folder for AutoCAD to be able to pick them up. This folder can be configured by closing any open dialogue boxes right-clicking the command line and choosing ***Options***. In the options dialogue box, click the **Files** tab and browse to **Printer Support File Path** > **Printer Configuration Search Path.**

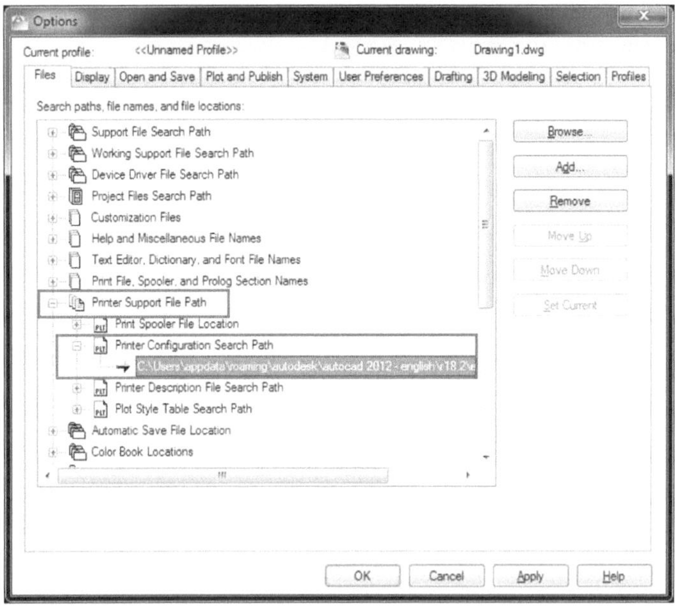

Figure 6.125

The **Paper size** section allows the selection of the paper size. Only paper sizes valid for the selected printer will be shown, although it is possible to add custom paper sizes by modifying a PC3 file for a selected printer.

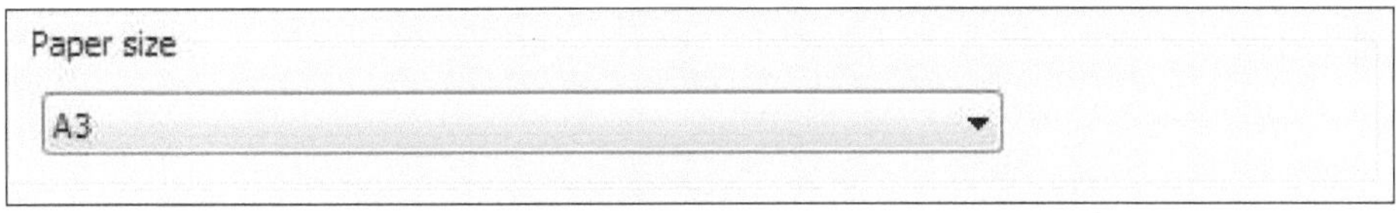

Figure 6.126

The **Plot area** section allows a choice of area to print.

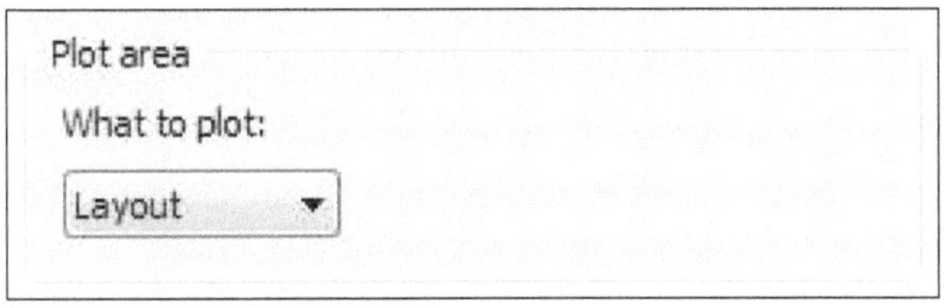

Figure 6.127

Display, Print what is visible on the screen.
Extents, Prints everything in Paper space even if it sits outside of the shown printable area.
Layout, Prints the contents inside the printable area (recommended method).
The window, Allows an area of the screen to be specified.

The **Plot offset (origin set to the printable area)** can be used to adjust the margins. This can be useful to reposition a drawing to fit on a piece of paper. Most printers have a margin around the page that cannot be printed (to allow for the paper to be moved through the printer) so, should the edges of the drawing not print, these offsets can be used to shift the drawing across and up the paper.

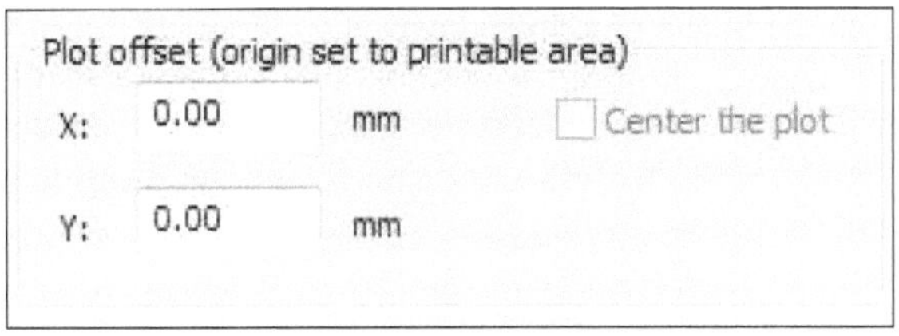

Figure 6.128

The **Plot style table (pen assignments)** section allows you to select your CTB/STB file. These are used to control which elements print colour and which print black and white. Plot styles can also be used to control line weights, line styles and whether the end of lines is rounded, square, etc.

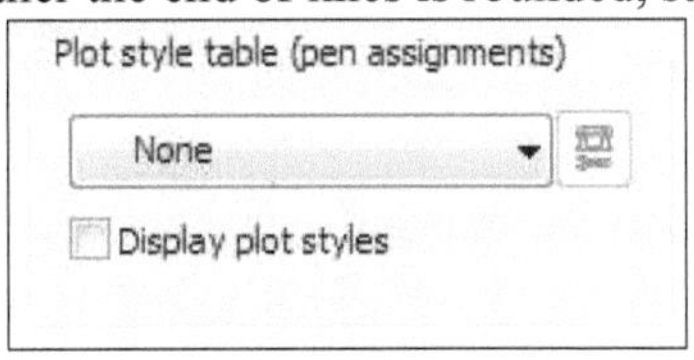

Figure 6.129

STB FILES

STB or **Object Dependent Style Tables** define settings that control how objects on a particular layer print out. An example STB Plot Style would use the layer's properties except for the colour which would change to black when printing. The STB method is the most flexible, and works best for exchanging data with other CAD/BIM systems – you are not limited to using colour to define thickness.
Each layer can be assigned a **Plot Style** in the **Layer Manager**.

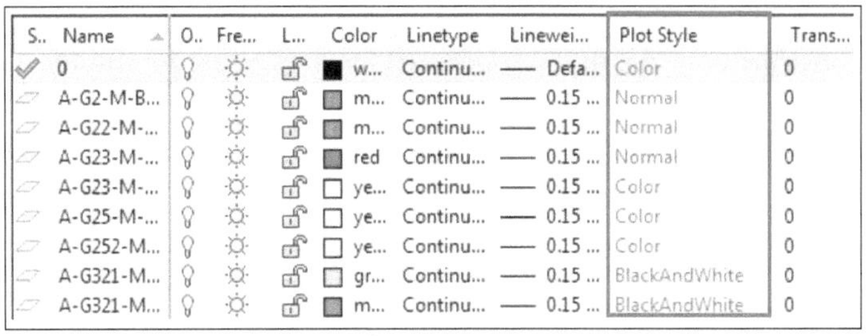

Figure 6.130

CTB FILES
CTB or Colour Dependant Styles Tables define settings that are applied to a specific colour. An example would be anything drawn using colour **6** (Magenta) would print out black.

SWITCHING THE TYPE OF PLOT STYLE
To change a drawing from using **CTB** files to **STB**, or vice versa, type "**CONVERTPSTYLES**" in the command line and press [**Enter**]. You will be prompted that the styles are being converted.

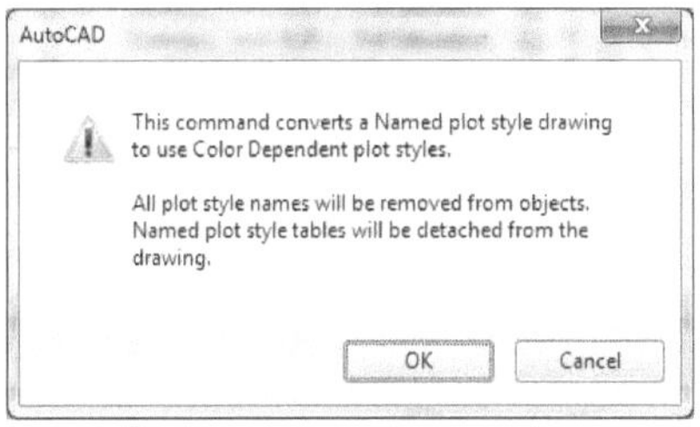

Figure 6.131

When printing 3D views, the **Shaded viewport options** will allow you to select the shade mode such as wireframe or shaded from the Shade plot drop-down. If the drawing doesn't show a 3D view, the drop-down will be greyed out. The Quality drop-down controls the resolution of the 3D shaded views. If "DPI" is selected, the DPI box will allow a custom Dots Per Inch resolution to be entered.

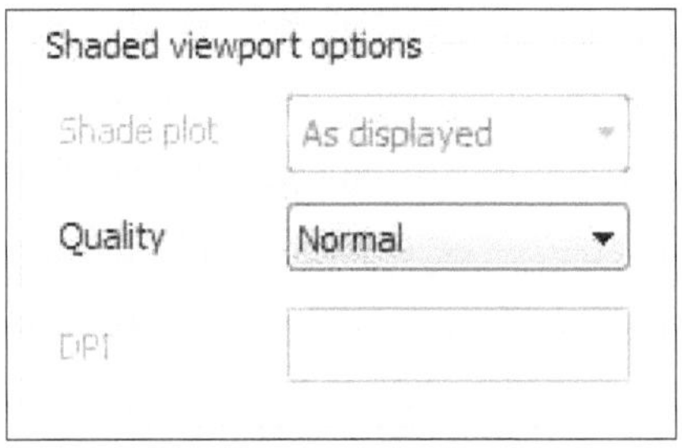

Figure 6.132

The **Plot options** further adjust how the drawing will appear.
To make sure your drawing is printed using the Plot Styles you have defined, you need to ensure that Plot styles are ticked "ON" in the Plot Options section.

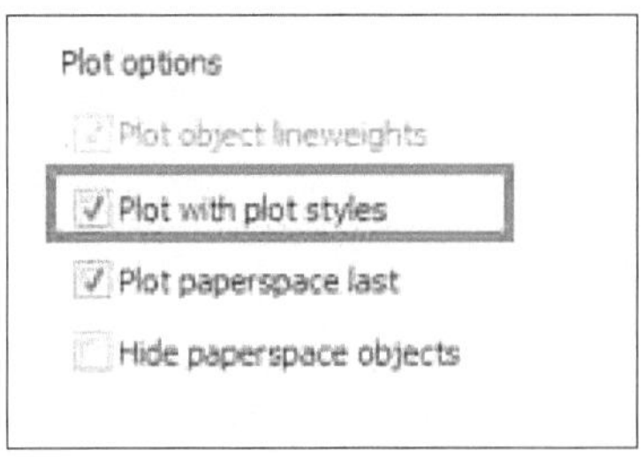

Figure 6.133

Plot with plot styles OFF: *Plot with plot styles ON:*

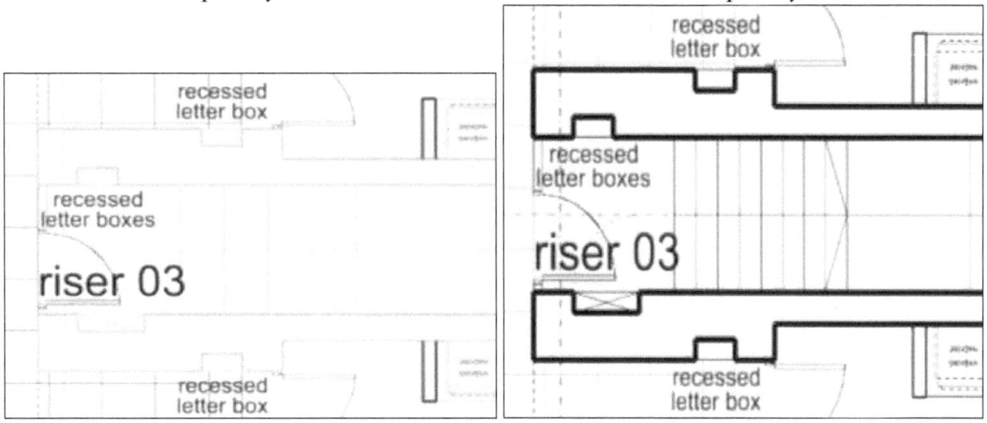

Figure 6.134

The **Plot Scale** section allows the specific scale for the drawing to be set. Usually, in Paper space, a scale of 1:1 would be used but, if printing an A1 drawing at A3 size, the scale can be set to 1:2. The **Fit to paper** checkbox can be useful to quickly print out a portion of a drawing or Model space without having to worry about scale.

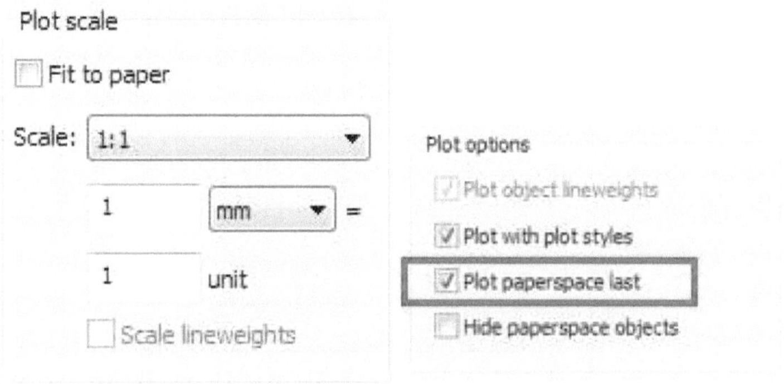

Figure 6.135

If you experience problems with objects printing out above annotation or other objects in Paper space, use Plot paper space last. This prints the information on Model space first then overlays the items in Paper space.

Plot paper space last OFF: *Plot paper space last ON:*

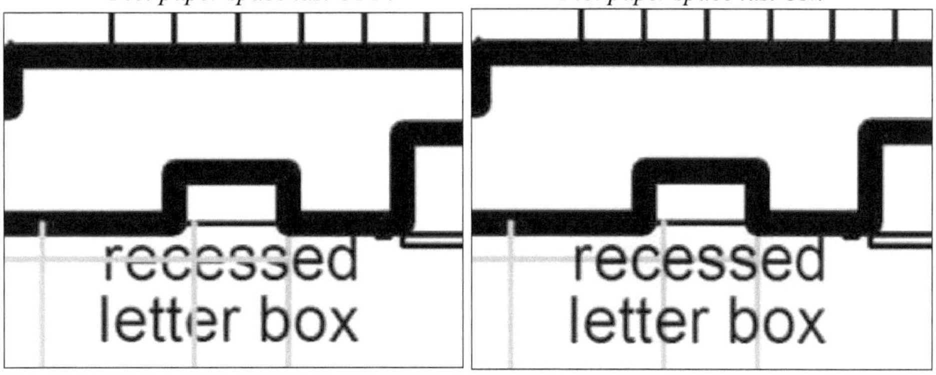

Figure 6.136

This requires annotation to be added in the Paper space layout.

If you are asked to make a quick "informal" print of a drawing for a meeting without any sheet annotation, or perhaps without any title information, you can use the Hide paper space objects tick box.

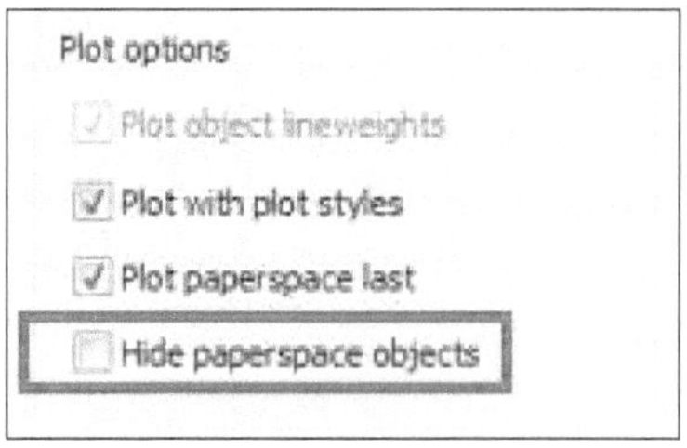

Figure 6.137

This will simply ignore any objects in Paper space, plotting only the information contained in Model space.
Note: You may wonder why you wouldn't just plot from Model space, but remember that your Layout will be already set up for the correct printer, Plot Styles, and laid out correctly on the paper.
Drawing orientation controls the page and whether it is printed in a portrait or landscape manner.

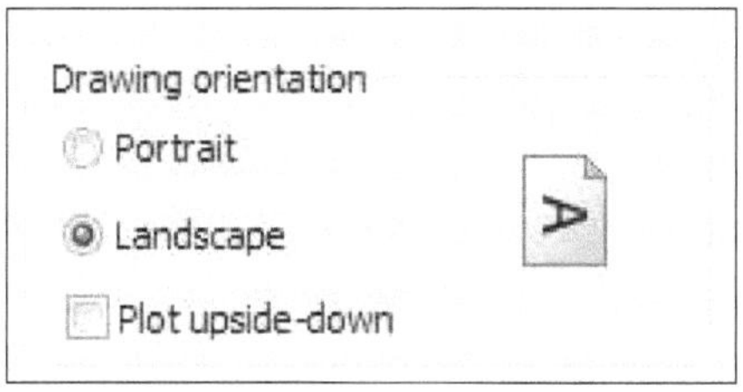

Figure 6.138

The **Plot upside-down** option can be used when printing to large format plotters. If printing an A3 drawing using an A1 or A0 printer, this option will not only rotate the drawing being printed but will also print it to the other side of the page saving paper.

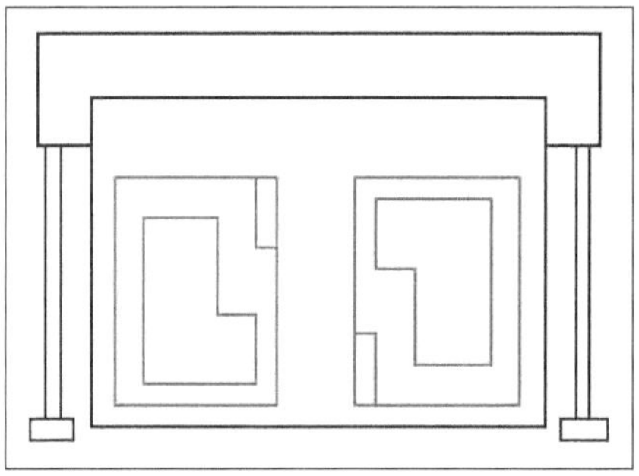

Figure 6.139

Once you have set all your settings, click **OK** to accept and return to the **Page Setup Manager**.
Select your newly created page setup and click **Set Current** to apply the settings to the current layout.

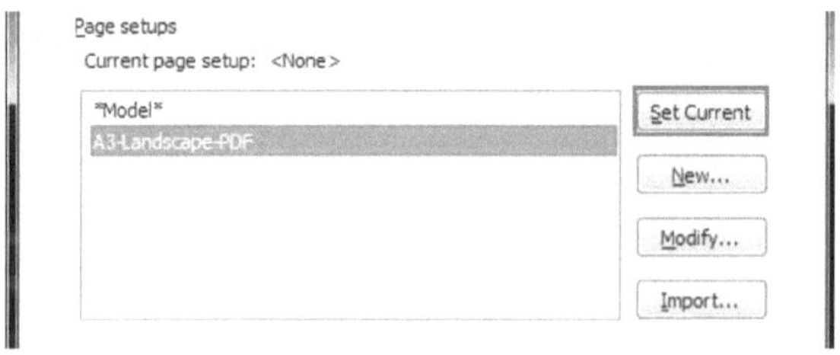

Figure 6.140

LAYOUT APPEARANCE

Once the **Page Setup** has been configured, Paper space will show a preview of the paper size (the white area) and the printable area (Inside the dotted black line). The margins between the printable area and the paper size are used for feeding the paper through the printer.

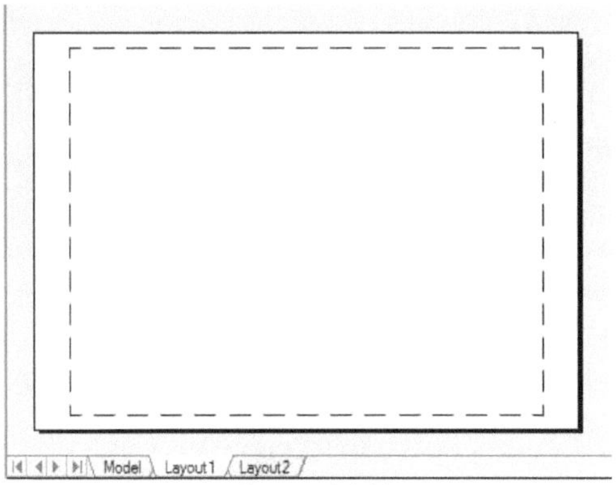

Figure 6.141

IMPORTING PAGE SETUPS

Page setups can be imported from other files to save having to continually define them.
From the Page Setup Manager, click the Import button. The Select Page Setup From File dialogue box opens.
Browse to an AutoCAD .dwg file that contains the page setup you want to import and click **Open**.
The available page setups will show in the Import Page Setups dialogue box. Select the required page setup or setups and click **OK**.

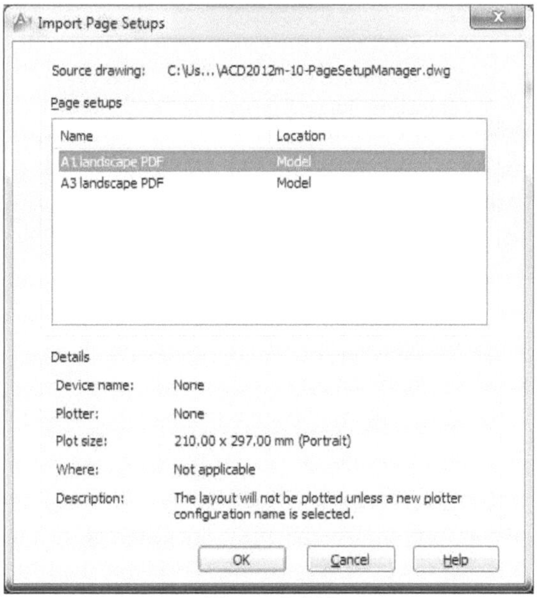

Figure 6.142

The page setup will now be available in your file.

WHAT IS THE STANDARD BLUEPRINT PAPER SIZE?

Blueprints and house plans will come in several standard sizes. Two of the most common architectural drawing sizes are 18" x 24" and 24" x 36", but you can also find them in 30" x 42" and 36" x "48" sizes. Large sizes are necessary on bigger and more expensive properties. But regardless of the blueprint paper size being used, its purpose is to show a trained person how to build that particular home. Feel free to look at all of the drawing paper sizes we have at Engineering Supply. We're sure you'll be able to find something that will meet your specific needs.

WHAT ARE THE STANDARD DRAWING SIZES?
Here are the standard US architectural drawing sizes:
ARCH A - 9 x 12 inches (228.6 x 304.8 millimeters)
ARCH B - 12 x 18 inches (304.8 x 457.2 millimeters)
ARCH C - 18 x 24 inches (457.2 x 609.6 millimeters)
ARCH D - 24 x 36 inches (609.6 x 914.4 millimeters)
ARCH E - 36 x 48 inches (914.4 x 1219.2 millimeters)
And here are the standard US engineering drawing sizes:
ANSI A - 8.5 X 11 inches (215.9 x 279.4 millimeters)
ANSI B - 11 x 17 inches (279.4 x 431.8 millimeters
ANSI C - 17 x 22 inches (431.8 x 558.8 millimeters)
ANSI D - 22 x 34 inches (558.8 x 863.6 millimeters)
ANSI E - 34 x 44 inches (863.6 x 1117.6 millimeters)
Feel free to look at what we have at Engineering Supply, so you can find the right type of blueprint paper size for your specific needs.

WHAT IS THE STANDARD SIZE FOR CONSTRUCTION PLANS?
The drawing sizes for construction plans will come in several different formats — with 18" x 24" and 24" x 36" being the most common. But you can also find larger plans at 30" x 42" and 36" x 48". Larger sizes are necessary for bigger properties. But regardless of the blueprint paper size, the plan will have the following information:
Exterior elevations
Floor plans
Building sections and details
Roof plan
General notes

WHAT IS A FULL-SIZE DRAWING?
Architectural drawing sizes come in several different formats, which consist of the following:
ARCH A - 9 x 12 inches (228.6 x 304.8 millimeters)
ARCH B - 12 x 18 inches (304.8 x 457.2 millimeters)
ARCH C - 18 x 24 inches (457.2 x 609.6 millimeters)
ARCH D - 24 x 36 inches (609.6 x 914.4 millimeters)
ARCH E - 36 x 48 inches (914.4 x 1219.2 millimeters)
Engineering drawing sizes use a different format, which consists of the following:
ANSI A - 8.5 X 11 inches (215.9 x 279.4 millimeters)
ANSI B - 11 x 17 inches (279.4 x 431.8 millimeters)
ANSI C - 17 x 22 inches (431.8 x 558.8 millimeters)
ANSI D - 22 x 34 inches (558.8 x 863.6 millimeters)
ANSI E - 34 x 44 inches (863.6 x 1117.6 millimeters)
Feel free to look at what we have at Engineering Supply. We're sure you'll be able to find the right type of blueprint paper size for your specific needs.

WHAT ARE THE DIMENSIONS FOR D-SIZE PAPER?
The standard sizes for architectural plans and engineering plans are different, so it will depend on what type of drawing you're looking at. Architectural drawing sizes are based on the ARCH scale, while engineering drawing paper sizes are based on the ANSI scale. D-size paper for architectural plans is 24" x "36" inches (or 609.6 x 914.4 millimeters), while D-size paper for engineering plans is 22" x 34" (or 558.8 x 863.6 millimeters). There are, of course, other drawing paper sizes. So, if you're looking for D-size paper or any other blueprint paper size, feel free to look at what we have at Engineering Supply.

Assignment

1. In order to measure the dimensions of a land, the surveyor uses which of the following system
a). No one method
b). Related cartesian coordinates
c). Absolute polar coordinates
d). None of the above
2. The command that is used to convert a block into a drawing object
a). EXPLODE
b). OVERKILL
c). REDEFINE
d). UNGROUP
3. The function of a polar array is to create object
a). In a grid pattern
b). In a circular pattern
c). In a straight line
d). All the above
4. Fillet in various corners can be created simultaneously using which command
a). Loop
b). Chain
c). Multiple
d). Edge
5. The command used to convert separate entities into a polyline
a). Union
b). Subtract
c). Join
d). Polyline
6. Command used to create a rectangle 8 units in length and 6 units in width, starting not from the origin
a). 8,6
b). 6,8
c). @8,6
d). @6,8
7. AutoCAD commands can be accessed in
a). Drawing area
b). Status bar
c). Tool bars
d). Command window
8. Various methods to access PAN command
a). Right click and select PAN from contextual menu
b). PAN command
c). Press and hold middle mouse button
d). All of the above
9. The file extension that cannot be accessed by AutoCAD
a). dwg
b). dxf
c). doc
d). dws
10. Plagiostomi angle can be created by which angle
a). chamfer
b). fillet
c). offset
d). mirror
11. In AutoCAD, polar coordinate is used for creating
a). arc
b). ellipse
c). angular lines

d). none of the above
12. To print the entire project, you will choose to regulate what to plot
a). Display
b). Extends
c). Limits
d). Window
13. The layer that can be deleted from the layer property manager
a). Layer 0
b). Defpoints
c). Layer containing objects
d). Frozen layer not containing objects
14. The maximum number of layers that a drawing can have?
a). 1
b). 2
c). As many as depending on the complexity
d). None of the above
15. The hatch pattern used to fil a closed area,
a). ANSI31
b). ANGLE
c). SOLID
d). Filled area can't be made with hatch command
16. Which state grid is used to design perspective?
a). Parametric
b). Isometric
c). Pro-optic
d). Rectangular
17. The position of the camera can be tuned accurately using
a). Pan
b). Zoom
c). Orbit
d). Navigation Wheel
18. What is the highest value of layer transparency?
a). 100
b). 50
c). 30
d). 90
19. An inclined angle can be produced using which command.
a). <
b). >
c). @
d). #
20. The command that is not suitable to start print command
a). PRINT
b). PR
c). PLOT
d). ctrl + P
21. If I want to draw a line in the direction 6 pm (local time) will give an angle
a). -135 degrees
b). 270 degrees
c). -225 degrees
d). None of the above
22. The number of ways a circle can be created in AutoCAD
a). 4
b). 5
c). 6
d). 8

23. An object has how many snaps in AutoCAD?
a). 1
b). 4
c). 5
d). Depend on object
24. Which factor of hatch command you will change to correct the spacing between hatch lines or pattern?
a). Pattern
b). Angle
c). Scale
d). Layer
25. The extension for AutoCAD drawing file is
a). DWG
b). DWT
c). DWF
d). DXF
26. What is the usefulness of viewports?
a). Allows us to see the screen or on paper different views of the same project
b). Give us the ability to see projects have become a newer version of AutoCAD from our
c). We can make a change in one part of the plan, without affecting the rest
d). None of the above
27. The shortcut key to enable snap n AutoCAD
a). F2
b). F3
c). F5
d). F8
28. In AutoCAD, scaling function is used for making an object
a). Smaller
b). Bigger
c). Either smaller or bigger
d). None of the above
29. Block Control should be used
a). To write text block
b). To fix outer block
c). To fix dynamic block
d). To store it in another version of AutoCAD
30. Which of the following AutoCAD command does not have a thickness property?
a). TEXT
b). MTEXT
c). LINE
d). CIRCLE
31. Identify the command that does not belong to the Boolean group in AutoCAD
a). Union
b). Subtract
c). Intersect
d). Divide
32. Rectangle command can be defined using how many commands
a). One
b). Two
c). Three
d). Four
33. Which of the following tool cannot be used to make 2D views from 3D drawing?
a). Flatten
b). Flatshot
c). Layout Views
d). Section Plane

34. When in absolute Cartesian coordinates have points A (10.8) and B (6.5), then to make a line from A -> B with relative polar coordinates will write
a). @ -5 <36.88
b). @ 4 <30
c). @ 5 <216,88
d). @ 3 <60
35. An animation video in AutoCAD cannot be created in which of the following format?
a). 3GP
b). WMV
c). AVI
d). MPG
36. The shortcut used to save a drawing in AutoCAD
a). CTRL + S
b). SHIFT + S
c). ALT + S
d). CTRL + ALT + S
37. Print scale 1:100 means that
a). The draft is 100 times less expensive than the original
b). A 3 cm corresponds to a meter
c). A measure corresponds to 100 cm
d). None of the above
38. AutoCAD file is saved in which format?
a). DWG
b). DWT
c). DWF
d). DXF
39. The shortcut command for a rectangle
a). R
b). RE
c). REC
d). RET
40. To bring a particular area of the workspace in focus, what command should be used?
a) Zoom Extents
b) Zoom Window
c) Zoom Dynamic
d) Zoom All

Answers

1. Answer: b
2. Answer: a
3. Answer: b
4. Answer: c
5. Answer: c
6. Answer: c
7. Answer: d
8. Answer: d
9. Answer: c
10. Answer: a
11. Answer: c
12. Answer: c
13. Answer: b
14. Answer: c
15. Answer: c
16. Answer: b
17. Answer: c
18. Answer: a

19. Answer: d
20. Answer: b
21. Answer: c
22. Answer: a
23. Answer: d
24. Answer: c
25. Answer: a
26. Answer: a
27. Answer: b
28. Answer: c
29. Answer: c
30. Answer: d
31. Answer: d
32. Answer: a
33. Answer: d
34. Answer: b
35. Answer: a
36. Answer: a
37. Answer: b
38. Answer: a
39. Answer: a
40. Answer: b